# AQUEOUS ZINC BATTERIES

# AQUEOUS ZINC BATTERIES

Editor

## Hong Jin Fan

Nanyang Technological University, Singapore

**W₩ World Scientific**

NEW JERSEY · LONDON · SINGAPORE · BEIJING · SHANGHAI · HONG KONG · TAIPEI · CHENNAI · TOKYO

*Published by*

World Scientific Publishing Co. Pte. Ltd.

5 Toh Tuck Link, Singapore 596224

*USA office:* 27 Warren Street, Suite 401-402, Hackensack, NJ 07601

*UK office:* 57 Shelton Street, Covent Garden, London WC2H 9HE

Library of Congress Control Number: 2023046442

**British Library Cataloguing-in-Publication Data**
A catalogue record for this book is available from the British Library.

**AQUEOUS ZINC BATTERIES**

ISBN 978-981-127-831-0 (hardcover)
ISBN 978-981-127-832-7 (ebook for institutions)
ISBN 978-981-127-833-4 (ebook for individuals)

For any available supplementary material, please visit
https://www.worldscientific.com/worldscibooks/10.1142/13470#t=suppl

Desk Editor: Shaun Tan Yi Jie

Typeset by Stallion Press
Email: enquiries@stallionpress.com

# Preface

Driven by the increasing demand for high-safety and low-cost energy storage technologies, researchers are putting huge efforts in developing aqueous-based batteries as a replacement for or supplementary to commercial lithium-ion batteries. Aqueous electrolytes are regarded as safer than the organic ones used in commercial Li-ion and Na-ion batteries. In particular, aqueous zinc batteries (AZB, sometimes also called Zn-ion batteries, Zn metal batteries) are under the spotlight of current research. In addition to the high theoretical capacity (820 mAh g$^{-1}$) of zinc metal, the low-cost and high-safety electrode materials and aqueous electrolyte system make AZB more favorable for large-scale energy storage compared to currently used alkali metal-ion batteries. In the past ten years, thanks to the deep research into electrode design, electrolyte engineering, redox chemistry, and flexible devices, the progress of AZB has been escalated. While there are already a large number of high-quality review articles published in various journals, I had this idea of publishing a comprehensive and inclusive collection of opinions and trends in the field of AZBs.

Despite the increasing number of papers and tremendous progress, critical questions are being asked. For example, what is the most promising application of AZBs? What are the main hurdles for commercialization? Can they replace Li-ion battery? Are AZBs really safe and fire hazard-free?

The key challenges of AZB can be summarized as two: (i) Insufficient energy density (low capacity of cathode materials, limited voltage windows); (ii) Instability of Zn metal anode (dendrites and corrosion, low utilization rate and depth of discharge). For fundamental research, the complicated reaction and charge storage mechanisms associated with oxide cathodes have been a nuisance. Conventional intercalation cathode materials such as manganese- and vanadium-based oxides are generally dissolvable and have low capacities and poor rate performance. Zn-Mn batteries have been regarded as most promising for commercialization, and particularly the two-electron process associated with $Mn^{4+}/Mn^{2+}$ conversion reaction paves a way for aqueous redox batteries. Issues associated with charge storage mechanism for Zn-Mn batteries, and new protocols for standardizing the assessment of device performance, are elaborated in this book.

Conversion-type cathodes are being explored for various advantages, including mitigation of the pH sensitivity, multiple electron reaction, and widening the operation voltage. Hence, they are good candidates for achieving high-energy-density aqueous batteries. Typical examples are $Zn-I_2$, Zn-S, and Zn-Se batteries. Take the zinc-iodine battery as example; iodine displays a conversion-type storage behavior between $I_2$ and $ZnI_2$, which substantially enhances energy density. However, the shortfalls are also obvious. The notorious shuttle effect of polyiodides during the charge/discharge process accounts primarily for the rapid capacity decay. The insulating nature of $I_2$ and $ZnI_2$ hinders redox kinetics, resulting in insufficient iodine utilization. In addition, most conversion-type aqueous batteries require a membrane separator which adds complexity to the battery structure and cost.

The safety and cycle stability of AZB is closely related to the Zn anode. Zn is a very active metal. It reacts with water (either protons or hydroxide ions) in a wide range of pH values, generating insoluble oxides or hydrogen gas. The Zn striping/plating reactions at the anode tend to be arbitrary due to non-uniform nucleation, leading to dendrite growth and battery failure. Various detrimental side reactions may occur in AZB, which makes the fabrication of long-term durable pouch cells a big challenge. In recent years, diverse innovative strategies for Zn surface protection are being published. However, while nearly all these protective methods are allegedly effective, the cost, scalability, and environmental impact should be seriously considered. Additionally, attention should be taken to improve the calendar life, Coulombic efficiency, and Zn utilization (currently around 10%). Anode-free AZBs are interesting and worthy of comprehensive exploration.

Similar to conventional alkali metal-ion batteries, the electrolyte is of vital importance to AZBs. The electrolyte condition (type of ions, pH value, additive, etc.) basically determines the ion insertion chemistry of the cathode and impacts the reversibility of the zinc anode. Electrolyte additive has been a popular and apparently effective approach to manipulate cation solvation structure and ion transport. It is very likely that any engineered electrolyte will have certain impact on both cathode and anode interfaces. However, it seems this issue has been rarely considered. In most reports on electrolyte design and engineering, the effect on only cathode or only anode surface is considered, but not both. Some researchers believe that aqueous electrolytes make it impossible for pouch cells or column cells. I am not sure about that. Driven by applications, hydrogel electrolytes can be potent materials to enable smart wearable devices, implantable medical devices, flexible displays, etc., which is articulated in this book.

Coming back to the question at the beginning, the demand for energy density, power density, service life, safety and environmental friendliness may not be all fulfilled for next-generation energy storage. So far, most of the advanced batteries are achieved only at the laboratory scale and <1 Ah level. It is time to think big and go beyond. Finally, an important criterion for all new energy technologies is carbon neutrality.

I hope the opinions and trends collected in this book are helpful to the ever-growing aqueous battery community.

Hong Jin Fan
Nanyang Technological University, Singapore
3 Oct 2023

# About the Editor

**Professor Hong Jin Fan** received his PhD degree from the National University of Singapore in 2003. After that he conducted postdoctoral research at the Max-Planck-Institute of Microstructure Physics, Germany, and the University of Cambridge, UK. In 2008, he joined the School of Physical and Mathematical Sciences, Nanyang Technological University, Singapore as a Nanyang Assistant Professor and was promoted to full Professor in 2019. He is exploring new nanomaterials and understanding their functions in energy processes, including electrocatalysis for hydrogen generation and electrochemistry in new batteries. He has co-authored more than 300 journal papers with a h-index of 106, and has been recognized as a Highly Cited Researcher consecutively since 2016. He is currently the Editor-in-Chief of *Materials Today Energy*, and editorial/advisory board member of a number of prestigious journals. He is an elected Fellow of the Royal Society of Chemistry.

# Contents

Chapter 1

# Roadmap for Advanced Aqueous Batteries: From Design of Materials to Applications[1]

Dongliang Chao,[a,*] Wanhai Zhou,[a] Fangxi Xie,[a] Chao Ye,[a] Huan Li,[a] Mietek Jaroniec,[b] Shi-Zhang Qiao[b]

[a]*School of Chemical Engineering, The University of Adelaide, Adelaide, SA 5005, Australia*
[b]*Department of Chemistry and Biochemistry, Kent State University, Kent, OH 44242, USA*

Safety concerns of organic media-based batteries are the key public arguments against their widespread usage. Aqueous batteries (ABs), based on the environmentally benign water, provide a promising alternative for safe, cost-effective, and scalable energy storage with high power density and tolerance against mishandling. Research interests and achievements in ABs have surged throughout the world in the past five years. However, their large-scale application is plagued by the limited output voltage and inadequate energy density. Herein, we present the challenges in the fundamental research of ABs focusing on the design of advanced materials and practical applications of the whole devices. Potential interactions of the challenges in different AB systems are established. A critical appraisal of recent advances in ABs is presented for addressing the key issues, with special emphasis on the connection between advanced materials and emerging electrochemistry. Finally, we provide a roadmap starting with the design of materials and ending on the commercialization of next-generation reliable ABs.

---

* Corresponding author: chaod@fudan.edu.cn
[1] Adapted with permission from D. Chao, W Zhou, F. Xie, C. Ye, H. Li, M. Jaroniec, S.-Z. Qiao, *Sci. Adv.* **2020**, 6, eaba4098.

## 1.1 Introduction

Because of the dwindling supplies and pollution caused through burning of fossil fuels, the search for alternative clean energies is becoming the spotlight of worldwide research. This has led to an upswell in demand for storage of electrical energy, particularly in advanced batteries that have practical potential for grid-scale applications. Of particular research interest are the rechargeable lithium ion batteries (LIBs).[1] However, despite a high energy density, safety remains a ubiquitous issue that has impeded LIBs in security-critical applications. Incidents in recent years include Boeing 787 battery fires in 2013, Samsung Note 7 explosions in 2016, and the Tesla Model S combustions in 2019. These have caused serious threats to human health or life, which remind us continuously that safety is a prerequisite for batteries.[2] Additionally, the scarce abundance and increasing cost of Li (and Co) also pose challenges for large-scale applications.[3] On the other hand, the development of resourceful sodium ion batteries (SIBs) and potassium ion batteries (PIBs) in the past decade is somewhat hindered by safety risks and environmental challenges due to the use of volatile, flammable and toxic organic electrolytes.[4] The aforementioned drawbacks of the organic media-based systems have stimulated the pursuit for alternative advanced batteries with possibility for their grid-scale applications.

Aqueous batteries (ABs) are safer alternatives compared with current LIBs, SIBs, and PIBs. The use of aqueous electrolytes also offers tremendous competitiveness in terms of i) low cost, the electrolyte and manufacturing costs are reduced by excluding oxygen-free and drying assembly lines; ii) environmental benignity, due to the non-volatility, non-toxicity and non-flammability of water; iii) aqueous systems are capable of fast charging and high power densities due to the high ionic conductivity of aqueous media; and iv) ABs also exhibit high tolerance against electrical and mechanical mishandlings, i.e., survival after fast discharging, bending, cutting and washing, which will not cause any disastrous consequences. Until now, various types of ABs (e.g., Ni-Fe, Ni-Cd, Pb-acid ABs) have been successfully fabricated. Recently, owing to the advances in the development of materials and their better electrochemical performance, ABs as one of the ideal candidates are being revitalized, especially for large-scale energy storage.

We have witnessed an astounding increase in publications regarding ABs, especially in the last five years (Figure 1.1a and 1.1b). However, the grid-scale applications of ABs

(a)

(b)

(c)

non-metal-ion type:
H+, NH4+, OH-, F-, Cl-

metal-ion type:
Li+, Na+, K+
Zn2+, Mg2+, Ca2+, Al3+

Aqueous Battery

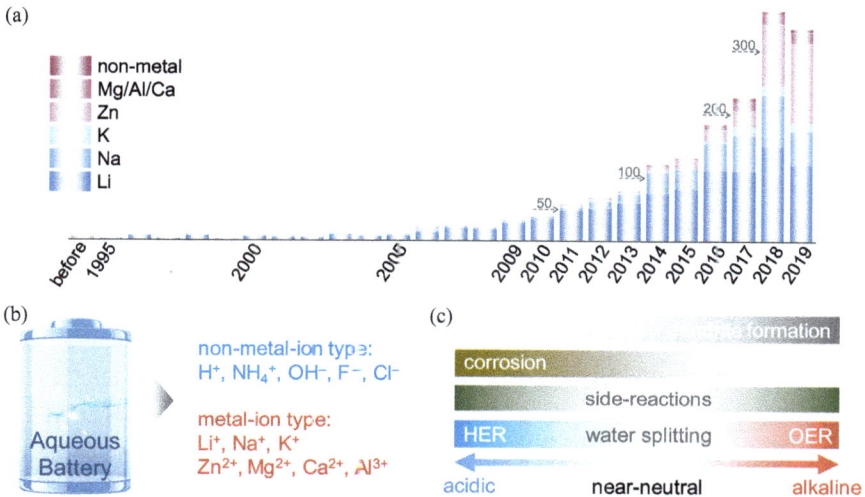

**Figure 1.1.** Status and challenges of current ABs. (a) Number of publications devoted to different ABs. Data are collected from Web of Science in October 2019. Inset numbers represent typical scale labels. (b) Classification of ABs. (c) Summary of key challenges that are limiting energy/power densities and lifespan of current acidic, near-neutral and alkaline ABs.

have been impeded by two ubiquitous issues, i.e., limited energy density and unsatisfactory lifespan. Fundamentally, water has an inherent thermodynamic oxidation potential (OER) and reduction potential (HER), which differ by a narrow voltage window of 1.23 V. The narrow electrochemical stability window (ESW) of water would suppress the operating voltage (Figure 1.1c), leading to insufficient energy density of ABs.[5] Practically, aqueous electrolytes with different pH ranges have been adopted in various ABs, which may trigger the water-based side reactions and greatly restrict the lifespan of ABs. Early applications of alkaline rechargeable batteries such as Ni-based Ni-Cd, Ni-Fe, and Ni-metal hydride and Zn-based Zn-Ni/Co and Zn-MnO$_2$ batteries, undergo either low deposition/dissolution Coulombic efficiency (CE) by dendrite formation, corrosion and irreversible by-products, or volume contraction/expansion during discharge/charge cycles at the anode side. While at the cathode part, the risk of OER further decreases CE, resulting in severe capacity fading of ABs (Figure 1.1c). A decrease in pH of acidic electrolytes (Pb-acid, Zn-Ce and all-V ABs) would introduce HER at the anode and induce low CE by side reactions, such as corrosion and by-product formation (Figure 1.1c), leading to sustained consumption of electrolyte and capacity decay. For the near-neutral electrolytes ($Li^+$, $Na^+$, $K^+$, $Zn^{2+}$, $Mg^{2+}$, $Ca^{2+}$, $Al^{3+}$ metal-ion ABs), limitations in the capacity and redox potential at the anode side further lower the output energy density of ABs. The high capacity and low-voltage redox reactions of metals (except for $Zn/Zn^{2+}$) are out of the scope of the ESW of water, which cannot be directly used as anodes in aqueous electrolytes. Moreover, the dendrite formation and decrease of CE by side reactions are inherently unavoidable even in near-neutral electrolytes (Figure 1.1c), especially for achieving a long-term lifespan. Obviously, different AB systems may pose different challenges, while their potential interactions could be exploited, and more importantly, the successful strategies established for a specific system could benchmark the success of others. To the best of our knowledge, the existing reviews of ABs either specialize in acidic Pb-acid batteries,[6] alkaline batteries,[7] or in neutral monovalent ion,[5] multivalent ion[8] and hybrid ion batteries.[9] It is desirable to provide an overview with integrated strategy in the broader context of different ABs.

Despite the recent research efforts in understanding the electrochemistry of novel ABs and achieving high electrochemical performance in terms of various materials design, the gaps between expectations and reality have plagued their grid-scale application. In this review, instead of compiling recent achievements, the key issues that limit electrode operation in different AB systems are critically analyzed, from the perspective of both fundamental research and their practical application. This review also draws a timely generalized understanding, with potential relationships and integrated strategies, to the rapid advances in the development of different AB systems. Furthermore, considering that large-scale application of ABs is still in the incipient stage, it is timely to present a perspective on the design principles and roadmap to the practical use of next-generation reliable ABs.

## 1.2 Challenges of Aqueous Batteries

The promising combination of safety, low cost of raw materials and manufacturing, and environmental benignity should allow ABs to become leading candidates for energy storage solutions. To date, considerable progress on ABs has been achieved. We have witnessed an explosive growth of publications regarding advanced ABs especially in the recent five years, but there are still some limitations that should be overcome. The design and application of electrode materials that are compatible with aqueous electrolytes are of vital importance to achieve high performance and producible energy storage systems.

### 1.2.1 Nature of Different Charge Carriers

Depending on the nature of migration ions, ABs can be classified into two types: metal-ion ABs and non-metal-ion ABs. So far, a variety of metal-ion ABs, such as $Li^+$, $Na^+$, $K^+$, $Zn^{2+}$, $Mg^{2+}$, $Ca^{2+}$, and $Al^{3+}$, have been demonstrated based on metal-ion intercalation chemistry. $Li^+$-based ABs (LiABs) are first extensively developed due to the solid research basis in conventional non-aqueous Li-ion batteries, and benefits in terms of cost, safety, power capability, etc. As sodium and potassium are more abundant than lithium, $Na^+$-based ABs (NaABs) and $K^+$-based ABs (KABs) are considered as more attractive power sources than LiABs for large-scale energy storage. However, the radii of $Na^+$ (0.95 Å) and $K^+$ (1.33 Å) are much larger than that of $Li^+$ (0.60 Å) (Figure 1.2a), so the selection criterion was specialized to only a few compounds showing capability of $Na^+$ or $K^+$ deintercalation/intercalation in aqueous media. Due to the smaller hydrated radius of solvated $K^+$ (3.31 Å), KAB-based electrolyte exhibits much higher ionic conductivity, which enables higher rate capability in $K^+$ storage than $Li^+$ and $Na^+$ (Figure 1.2a).[10]

New opportunities are emerging in other multivalent metal-ion charge carrier-based ABs such as $Zn^{2+}$, $Mg^{2+}$, $Ca^{2+}$, and $Al^{3+}$, not only because they use earth-abundant metals but also due to their improved safety and high volumetric energy density. Nevertheless, the development of Mg, Ca, and Al-ABs has been stagnant because of even tougher criteria for hosting their large-size solvated cations (Figure 1.2a) and poor reversibility in plating/stripping of Mg, Ca and Al.[11] In contrast, zinc is particularly advantageous in superior $Zn/Zn^{2+}$ reversibility and its proper redox potential of –0.763 V vs. standard hydrogen electrode (SHE) compared with other metals, e.g., Li, Na, K, Mg, Ca, Al, etc. in aqueous media.[12] These virtues have fueled astounding development of Zn-based ABs in the past five years and provide a potential candidate for large-scale electrical energy storage.

Non-metal-ion charge carriers include anions such as hydroxyl ($OH^-$) and halides ($F^-$, $Cl^-$), and cations such as proton ($H^+$) and ammonium ($NH_4^+$).[13] The attractiveness of rechargeable non-metal-ion ABs resides in the employment of sustainable and unlimited

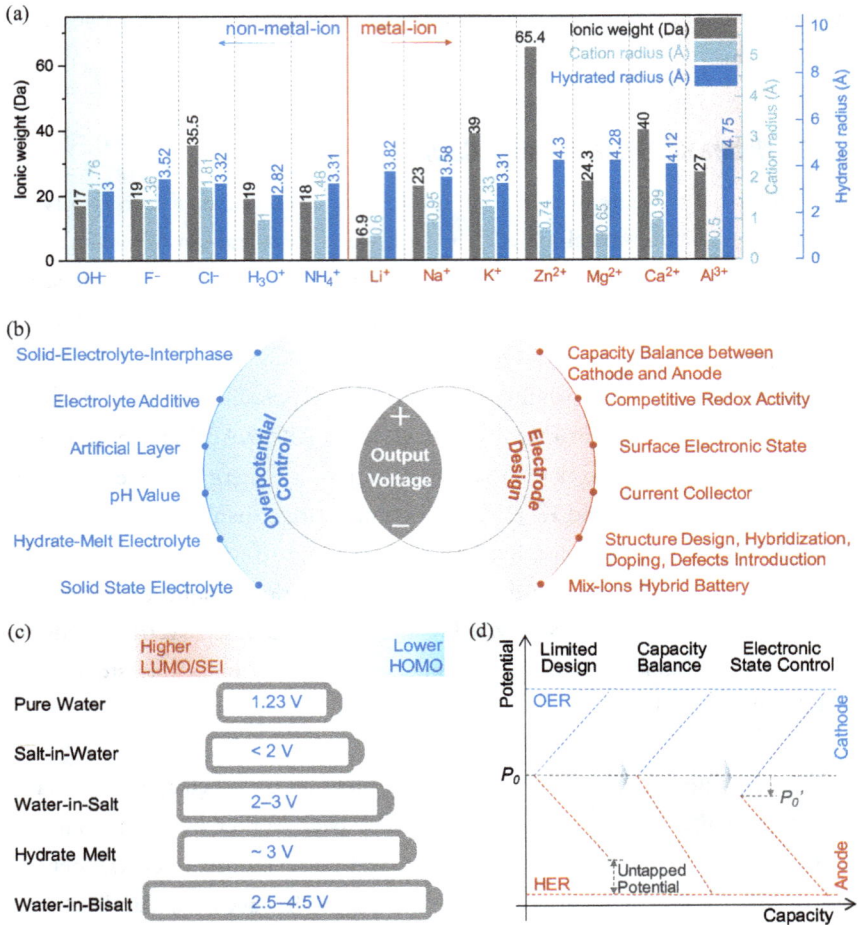

**Figure 1.2.** Summary of electrolyte and electrode engineering strategies for the design of high-performance ABs. (a) Comparison of ionic weight, cation radius, and hydrated radius of the typical metal-ion and non-metal-ion charge carriers. (b) Summary of strategies to improve the output voltage of ABs, which can be categorized as either overpotential control or electrode design. (c) Illustration of output voltage extension in a perspective of overpotential control with 'Water-in-Salt' or hydrate-melt electrolytes. The 'Salt-in-Water' electrolyte refers to traditional aqueous electrolyte; the 'Water-in-Salt' electrolyte corresponds to $Mo_6S_8$/$LiMn_2O_4$ cell in $LiTFSI \cdot H_2O$ electrolyte;[23a] the hydrate-melt electrolyte is $Li_4Ti_5O_{12}$/$LiNi_{0.5}Mn_{1.5}O_4$ cell in $Li(TFSI)_{0.7}(BETI)_{0.3} \cdot 2H_2O$ electrolyte;[25] the 'Water-in-Bisalt' electrolyte can be C-$TiO_2$/$LiMn_2O_4$ cell in LiTFSI+LiOTf electrolyte,[24] graphite/$LiVPO_4F$ cell in LiTFSI+LiOTf gel electrolyte,[55] or graphite/lithium halide salts cell in LiTFSI+LiOTf gel electrolyte.[23b] (d) Illustration of output voltage extension in a perspective of electrode design with capacity balance between cathode and anode, and surface electronic state control.

charge carriers on Earth. Compared with metal-ion charge carriers (Figure 1.2a), non-metal ions not only deliver lighter molar mass (only 17 g mol$^{-1}$ for OH$^-$, 18 g mol$^{-1}$ for NH$_4^+$, and 19 g mol$^{-1}$ for F$^-$ and H$_3$O$^+$) but also exhibit smaller hydrated ionic size (2.82 Å for H$_3$O$^+$, 3.00 Å for OH$^-$, 3.31 Å for NH$_4^+$, 3.32 Å for Cl$^-$, and 3.52 Å for F$^-$), resulting in fast diffusion in aqueous electrolytes.[14] Different from high corrosion damage and pollution hazards of the strong alkaline electrolytes, the halide ABs take advantage of the mild salt electrolyte such as NaCl solution.[15] Compared with proton or hydronium, NH$_4^+$ is less corrosive and less prone to HER, which may deliver superior cycling performance.[16]

## 1.2.2 Risk of Water Splitting and Limited Output Voltage

Water splitting involves two half-reactions of the oxygen evolution reaction (OER) and hydrogen evolution reaction (HER), which require four- and two-electron transfer, respectively. The reaction pathways are sensitive to 1) the pH of electrolyte, for example, the overall reaction pathways of both OER and HER vary at different pH (OER in acidic solution is $2H_2O \rightarrow O_2 + 4H^+ + 4e^-$ while in alkaline solution is $4OH^- \rightarrow O_2 + 2H_2O + 4e^-$; HER in acidic solution is $2H^+ + 2e^- \rightarrow H_2$ while in alkaline solution is $2H_2O + 2e^- \rightarrow H_2 + 2OH^-$); 2) the structure of electrode surface, for example, the materials with different geometric structures (e.g., facets and crystallinity) and electronic structures (e.g., d-band center) exhibit different reaction mechanisms. Consequent generation of gases (H$_2$ and O$_2$) may destroy the structure, isolate the electrolyte, and induce large polarization, instead. The evolution of H$_2$/O$_2$ from aqueous electrolytes is a crucial issue that should be considered when designing materials for advanced ABs with long lifespan and high energy density. Moreover, HER/OER occurring on the surface of electrodes consume a portion of electrons, which should be provided to the active materials, resulting in inferior CE, battery swelling and continuous consumption of the electrolyte. It is important to extend the operating window and suppress the water splitting process of both cathodes and anodes in ABs.

Generally, efficient ways to address this problem involve either the overpotential control or electrode design (Figure 1.2b). From the perspective of overpotential control, the passivation of water electrolysis is evident. Materials with evident catalytic effect on water splitting should be avoided in ABs to maximize the voltage window. For example, Li$^+$-intercalated vertically aligned MoS$_2$,[17] S-vacancy and edge-activated 1T-phase

$MoS_2$,[18] and Li conversion reactions-induced ultra-small transition metal oxides[19] are demonstrated as (bifunctional) catalysts showing high activity for water splitting. However, some philosophies for the design of catalysts could involve the passivation of catalytic processes, which in turn optimizes output voltage for ABs. It was found that the rate of HER is closely associated with exchange current density, overpotential, and electrolyte concentration.[20] Strategies to decrease the exchange current density and increase the overpotential have been developed to improve the CE and suppress self-discharge in ABs. The pH of the electrolyte, as an indicator of the acidity or alkalinity, plays a pivotal role in affecting the dynamics of redox reactions involving $H^+$ and $OH^-$. It is speculated that the concentrated, high pH, $O_2$-free electrolytes are better for low-voltage insertion electrodes.[21] Generally, neutral electrolytes show evident higher operating voltages than acidic or alkaline electrolytes, attributed to the larger HER and OER overpotentials.[22]

Different from the decomposition of organic electrolytes in which the protective solid electrolyte interphase (SEI) on the surface of the active material is generated for the electrode stabilization after solvent decomposition, the dissociation of aqueous electrolyte cannot form traditional SEI and stabilize the electrode. Recent efforts have successfully transplanted the concept of SEI into aqueous media, leading to the significant expansion of ESW in ABs from 1.23 V to beyond 4.0 V.[23] The tailored design of ion-conductive and electron-insulating SEI has enabled high-voltage ABs (Figure 1.2c), such as $Mo_6S_8/LiMn_2O_4$ cell in lithium bis(trifluoromethylsulphonyl) imide (LiTFSI) 'Water-in-Salt' electrolyte (WISE),[23a] and $C-TiO_2/LiMn_2O_4$, graphite/ $LiVPO_4F$ and graphite/lithium halide salts cells in LiTFSI+LiOTf (lithium trifluoromethane sulfolate) 'Water-in-Bisalt' electrolyte (WIBSE),[24] that deliver energy densities and cycling stability approaching the state-of-the-art organic LIBs. Similarly, the thermodynamic and kinetic extension of the potential window and the upward shift of the electrode potentials can be achieved in $Li_4Ti_5O_{12}/LiNi_{0.5}Mn_{1.5}O_4$ cell in $Li(TFSI)_{0.7}(BETI)_{0.3}\cdot 2H_2O$ hydrate-melt electrolyte,[25] in which all water molecules are coordinated with hydrate melt of Li salts and uncoordinated water molecules therefore are eliminated.

Apart from the overpotential control strategies of pH adjustment and introduction of SEI and hydrate melt, electrolyte additives such as cathode additive tris(trimethylsilyl) borate (TMSB)[26] and anionic surfactant sodium dodecyl sulfate,[27] protective artificial layers of graphene on $LiMn_2O_4$,[28] polymer-coated Mg,[29] $AlCl_3$-ionic liquid-treated Al,[30] hydroxylated interphase-protected $Mn_5O_8$,[31] and solid-state electrolytes[20a,32] are also effective in suppressing HER and/or OER and further stabilizing and extending the output voltage of ABs.

From the perspective of electrode design, first of all, the capacity between cathode and anode should be well balanced to take full usage of ESW. As can be found in Figure 1.2d, the untapped potential could be readily exploited by capacity balance, which has been proven to be the easiest way to expand the operating voltages of electronic double-layer capacitor and pseudocapacitors.[22,33] It is known that pseudocapacitive behavior has frequently been demonstrated in metal-ion intercalation of ABs to boost the high-rate and long-cycling performances.[14,34] The capacity balance strategy would be efficient in improving the voltage profiles of ABs after careful speculation of the capacity contribution (mass times specific capacity) from anode and cathode within the voltage window. Similarly, control of the equilibrium potential $P_0$ by surface charge control method could be used to enlarge both the operating voltage and the specific capacity of the aqueous devices.[35] It is concluded that the surface charge control can be achieved by charge injection (pre-charge), thermo-electrochemical effect (heating), and functional group introduction.

Introduction of extra redox activity that is competitive with HER or OER has been considered to be an effective approach to enhance both the capacity and output voltage of ABs. Hopefully, the selected redox couples should present redox potentials approaching the potentials of HER or OER, so that aqueous electrolyte decomposition could be suppressed and the output voltages are then boosted, attributed to the fast kinetics of the extra redox activity relative to those of HER or OER. Inspired by this, our group introduced a competitive high-voltage redox couple of $Mn^{2+}/Mn^{4+}$ in conventional Zn-ion battery (ZIB) by facile proton activity control method to significantly extend the output voltage of aqueous ZIBs.[36] Other competitive redox couples have been investigated contributing to the high voltage and energy density of ABs, like negatively charged couples of $[Fe(CN)_6]^{4-}/[Fe(CN)_6]^{3-}$, $I^3-/I^-$, $Br^{3-}/Br^-$ to suppress OER on the cathode and positively charged couples of $MV^{2+}/MV^+$ and $HV^{2+}/HV^+$ (MV and HV refer to methyl viologen and heptyl viologen, respectively) to replace the lower limit of HER.[37] More trivially, the use of current collectors is important to prevent electrolysis of water and avoid corrosion. Currently, Ti, stainless steel, and bare current collectors with unreactive coating like epoxy resin are popular choices. Aqueous mixed ions or hybrid batteries are proposed to take full advantage of the cathode and anode voltage limitations with different ionic electrochemistry, such as InHCF/Na$^+$ + K$^+$/NaTi$_2$(PO$_4$)$_3$ mixed-ion battery with a voltage of ~1.6 V[38] and Zn/Zn$^{2+}$ + Li$^+$/ LiMn$_2$O$_4$ hybrid battery with voltage plateau ~1.8 V.[39] Furthermore, other electrode design techniques have also been adopted, such as 3D structure design of Zn foam,[32] atomic layer deposition of amorphous TiO$_2$ on the surface of Zn anode,[40] doping Al into VO$_{1.52}$(OH)$_{0.77}$,[41] and introduction of cation deficiencies into ZnMn$_2$O$_4$ cathode.[42]

### 1.2.3 Dendrite Growth

Dendrite has shown potential hazards of piercing separators and detaching from the electrode, which would deteriorate the CE and cycling stability of ABs. Apart from Zn, most metals cannot be directly used as anodes in aqueous media, so dendrites/ protrusions are not an obvious problem in other metal-ion ABs. A uniform current and potential distribution have been believed traditionally in promoting homogeneous and compact zinc electrodeposits. The amphoteric zinc shows high solubility, electrochemical activity and thermodynamical instability in alkaline electrolytes. As a result, the formation of zinc dendrites/protrusions is especially serious in alkaline media compared with acidic and near-neutral electrolytes (Figure 1.1c), such as alkaline Zn/ Ni, Zn/Mn, and Zn/air batteries, which significantly limited their lifespans and commercial development prospect. The redox couple of Zn$^{2+}$ ↔ Zn in mild aqueous electrolyte may reduce the Zn-dendrite formation, but zinc dendrite after repeated plating/stripping is inherently unavoidable, especially for achieving long-term cycle life. According to the general principle that the deposits should have good adhesion to the substrate but be easily dissolved during stripping, the factors that affect the metal deposits mainly include a) electrolyte composition or additives; b) electrode design involving 3D nanoarchitectures or surface engineering; and c) design of membrane separators.

Usually, zinc dendrite growth involves repeated processes of nucleation and plating/ stripping. The formation of Zn with a high curvature due to the inhomogeneous nucleation should enhance local electrical fields, leading to strong adsorption of $Zn^{2+}$ and further exacerbate dendrite/protrusion growth. The addition of inert components, such as $Mn(CF_3SO_3)_2$[43] and $LiTFSI$[44], into aqueous electrolyte has been proved to be effective in suppressing the formation and growth of detrimental dendrites. The adsorbed inert cations on the surface of dendrites/protrusions can repel extra incoming $Zn^{2+}$ cations. However, the high cost of these additives blocked their practical application. Exploring more efficient and low-cost electrolyte additives like $MnSO_4$[45] and $Na_2SO_4$[46] presents more tangible benefits. Other affordable attempts of using solid-state electrolytes were shown to minimize the formation of dendrites/protrusions. Our recent studies showed that the growth of Zn dendrite can be thoroughly eliminated by adopting low-cost fumed silica-based solid-state electrolyte,[32] allowing dendrite-free Zn deposition due to its excellent compatibility with Zn anode and partially wetted interface between Zn and solid-state electrolyte. Due to the sufficient surface area and active sites for Zn plating/stripping, the derived Zn electrode showed a depth of discharge of 66%, which is much higher than those reported for Zn foil-based ZIBs. Hence, the battery can deliver impressive long-term durability up to 2,000 cycles at 20 C.

Another effective strategy is to coat a protective surface layer on the Zn foil to facilitate homogeneous current distribution and Zn accommodation, such as carbonaceous materials like reduced graphene oxide and active carbon, metal oxides/hydroxides like $Bi_2O_3$, $CaCO_3$, $Ca(OH)_2$, and alloying with other metals, e.g., Sn, In, Bi, or Ni. In addition, the topography of the Zn anode is crucial in regulating dendrite formation. By increasing the surface area or designing 3D porous architecture one can increase the concentration of electrochemical active sites, minimize local current or overpotential to suppress the formation of zinc dendrites/protrusions. Moreover, the void space of 3D Zn architecture would restrain the fresh Zn deposits and diminish the potential for shape change after repeated plating/stripping. To date, Zn metal anodes with various nanoarchitectures have been investigated, including fine powders, spheres, 1D wires, 2D flakes, and 3D foams.[32,47] Similar to restraining Li dendrite growth, the design of separators is considered effective for regulating uniform Zn deposition. It was reported by Yuan et al. that the low-cost polybenzimidazole membrane could effectively prevent the zinc dendrite growth and deliver lower resistance than Nafion 115 membrane in the Zn-Fe battery.[48] The selective ion conductive membrane with heterocyclic rings guarantees fast transportation of hydroxyl ions and mitigates ion concentration gradients, which are believed as the primary source of dendrites/protrusions.

### 1.2.4 *Corrosion, Passivation and Other Side Reactions*

The corrosion, passivation and other side reactions during charge/discharge process are closely related to the CE and cycling stability. Corrosion (self-discharge) is an influential side reaction in strong acidic or alkaline aqueous electrolytes, which not only decreases CE, but also leads to the irreversible consumption of water and concentration change in electrolyte. The corrosion of electrodes commonly exists in ABs and usually takes place at the interface between the electrode surface and the electrolyte environment. Thus, chemistry at the electrode-electrolyte interface plays a critical role in protecting the electrode against corrosion. The interfacial chemistry and corrosion mechanisms are closely related to the composition and physicochemical properties of the electrolyte such as the pH value. Therefore, depending on these factors, different strategies have been used for certain systems, which resulted in reducing corrosion of zinc anode and improving battery performance. The mainstream strategies reported so far to reduce the corrosion rate of electrode materials include the addition of inorganic (like Zn-Bi alloy and silica coating) or organic (e.g., polyethylene glycol and polyaniline (PANI) coating) corrosion inhibitors to the electrode or electrolyte.[49] On the other hand, inhibiting the corrosion of the current collectors is also a very important issue. Considering the charge effect, which can accelerate electrochemical corrosion, the stable potential of metal current collectors should also be considered based on the Pourbaix (pH vs. potential) diagram. For example, conventional Al and Ni cathodic current collectors are prone to the corrosion in weak acidic media at a relatively positive potential, and therefore should be avoided for cathodic current collectors in Zn/Al-based ABs.

Passivation, which is related to the formation of secondary insulating species, such as ZnO and Zn(OH)$_2$ in alkaline Zn-based ABs, would increase internal resistance and reduce the reversibility and energy efficiency of the electrodes. Precipitation of the insulating layer on the surface of active electrodes would block the migration of charge carriers and formation of reversible products. It is important to reduce the passivation effect in Zn anodes to increase zinc utilization efficiency.

The thermodynamic stability of the host materials after charge injection is detrimental to the long-term cycling stability in ABs. The thermodynamic relationship to evaluate the stability of the electrode materials in a Li-based aqueous medium was evaluated by Dahn and coworkers[50] with the help of the following reactions:

$$Li(s) + H_2O \leftrightarrow Li^+ + OH^- + 0.5H_2 \tag{1.1}$$
$$V = 3.885 - 0.118 \times pH \tag{1.2}$$
$$Li(s) + 0.25\ O_2 + 0.5\ H_2O \leftrightarrow Li^+ + OH^- \tag{1.3}$$
$$V = 4.268 - 0.059 \times pH \tag{1.4}$$

where $V$ (vs. $Li^+/Li$) is the delithiation potential of the electrode material, pH refers to the electrolyte, and Li(s) refers to the intercalated Li. From Eqs. (1.1) and (1.2), one electrode material is expected to be stable in the aqueous solution when its redox voltage is higher than $V$. For example, the lithiated $Li_2Mn_2O_4$ with redox voltage of ~2.97 V vs. $Li^+/Li$ is unstable in neutral or weak acidic electrolyte, while it would be stable in electrolytes with pH higher than 8. Unwanted side reactions of elemental dissolution may occur in $LiFePO_4$ electrode at concentrated alkaline media,[5] or $Zn^{2+}$-intercalated $MnO_2$ even at near-neutral electrolyte.[36] In addition, side reactions might be worse at similar voltages in the presence of $O_2$ as can be concluded based on Eqs. (1.3) and (1.4). Hence, strategies such as elimination of residual $O_2$ by $N_2$ ejection or assembling batteries in inert atmosphere and fine optimization of the cutoff voltage are effective ways for stabilizing electrode in ABs Surface coatings like carbon and introduction of electrode additives are also helpful in preventing side reactions.[5]

It is worth noting that the challenges in ABs, such as water splitting, corrosion, dendrite growth, passivation and other side reactions are closely interconnected. As can be concluded based on earlier discussion (Eqs. (1.2) and (1.3)), the corrosion or other side reactions may occur concurrently with hydrogen evolution. Formation of the passivation layer should suppress the charge transport kinetics on the electrode surface, which accelerates dendrite growth with shape change. An integrated strategy should be encountered to overcome the difficulties in fabricating advanced ABs with high output voltage and long-term cyclability.

## 1.3 Materials Design and Improvement of Advanced Aqueous Batteries

The search for materials that are electrochemically reversible with large ESW and chemically robust towards acidic/alkaline/near-neutral aqueous media is essential for improving the comprehensive performance of ABs. Based on the knowledge of the aforementioned challenges and strategies in ABs, many advanced materials used in ABs have been investigated, which is summarized in the following sections.

### 1.3.1 *Advances in Li/Na/K-based Aqueous Batteries*

#### 1.3.1.1 *Li-based Aqueous Batteries*

Li-ion aqueous batteries (LiABs) were first proposed in 1994 by Dahn *et al.*,[51] using $LiMn_2O_4$ cathode and $VO_2$ anode in 5 M $LiNO_3$ aqueous electrolyte (see Figure 1.3a). The first reported LiABs provided an average operating voltage of 1.5 V, with energy density (~55 Wh kg$^{-1}$) larger than the Pb-acid batteries (~30 Wh kg$^{-1}$), although the cycling was very poor. Since then, extensive efforts have been made toward developing various electrode materials for LiABs. Unlike the electrode materials used in organic systems, the redox potentials of electrode materials in aqueous systems should be within or near the electrolysis potentials of water. As shown in Figure 1.3b, electrodes with redox potential out of range of the red box (ESW of aqueous electrolytes at neutral pH conditions) cannot function properly due to the continuous participation of water splitting. The electrode materials can be categorized into three classes based on their reaction mechanism: desertion/insertion, conversion and alloying. Initially, all the electrode materials were based on deinsertion/insertion mechanism.[5,52] Cathode candidates, including oxides ($LiMn_2O_4$, $LiCoO_2$ and $LiNi_{1/3}Co_{1/3}Mn_{1/3}O_2$ (NCM)), polyanionic compounds ($LiFePO_4$, $FePO_4$, $LiMnPO_4$, $Li(Fe,Mn)PO_4$), and Prussian blue analogs (FeHCF, NiHCF, CuHCF, MnHCF) have been developed.[8,52] Meanwhile, various anode candidates have been considered, including oxides ($VO_2$, spinel $Li_2Mn_2O_4$, layered $\gamma$-$LiV_3O_8$, $H_2V_3O_8$, $Na_{1+x}V_3O_8$, $V_2O_5$ and $TiO_2$), polyanionic compounds (pyrophosphate $TiP_2O_7$ and NASICON-type $LiTi_2(PO_4)_3$) and organic polymeric compounds (polypyrrole and polyimides).[8,52]

Although great achievements have been made in increasing discharge capacity and rate performance via materials design (e.g., surface coating, architecture and composite) and electrolyte optimization (e.g., concentration, pH, oxygen elimination and additives), LiABs are still limited because of their low voltage (<1.5 V), low energy density (<70 Wh kg$^{-1}$), and poor cycling.[52] For instance, advanced LiABs proposed by Sun *et al.*[53] based on $LiTi_2(PO_4)_3$/C anode and $LiMn_2O_4$ cathode only presented an energy density of 68 Wh kg$^{-1}$ (based on the total mass of active materials) with 90% of capacity retention after 300 cycles at 0.2 C.

WISE, first proposed in 2015 by Wang and coworkers,[23a] represents a revolutionary step in the development of LiABs, in which concentrated LiTFSI aqueous solution

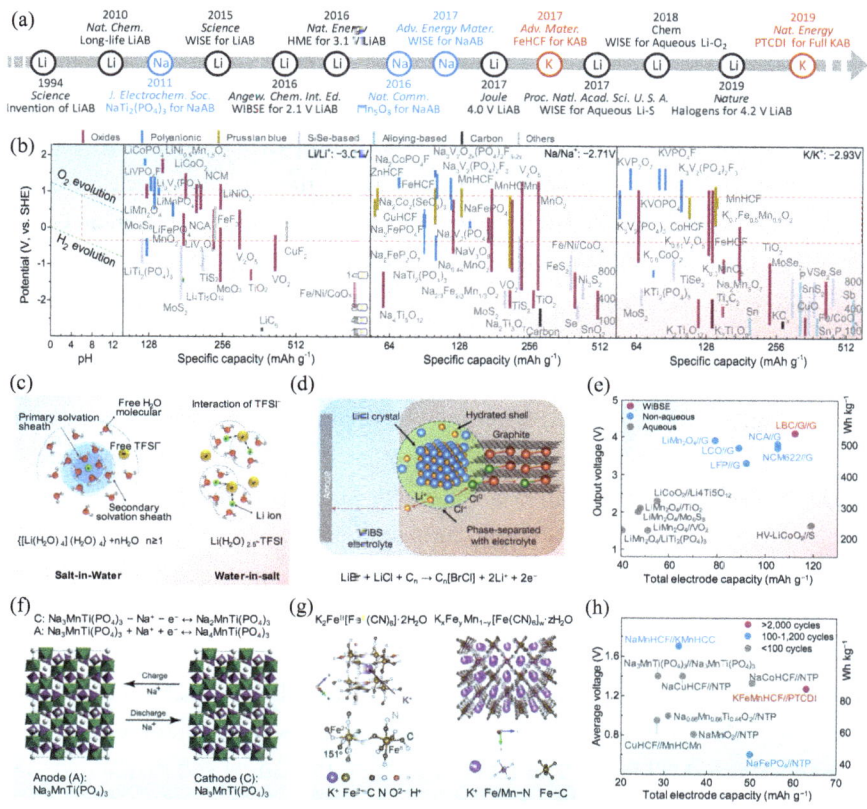

**Figure 1.3.** Summary of advanced materials for Li/Na/K-ion based ABs. (a) Main progress in LiABs, NaABs and KABs. (b) Summary of ESW of water and redox potentials of various electrode materials in organic LIBs, NIBs and KIBs. (c) Illustration of the difference between Li⁺ solvation sheath in diluted and 'Water-in-Salt' solutions. Reproduced with permission from the American Association for the Advancement of Science.[23a] (d) Schematic illustration of the conversion-intercalation mechanism occurring in the LBC-G composite in WIBS electrolyte. (e) Actual (red star) energy density of the LBC-G//G full cells compared with various state-of-the-art commercial and experimental Li-ion chemistries. (d) and (e) are reproduced with permission from the Nature Publishing Group.[23b] (f) Schematic illustration of the symmetric NaAB with the NASICON-structured Na₃MnTi(PO₄)₃ as the anode and the cathode. Reproduced with permission from John Wiley and Sons.[62] (g) Refined crystal structure of K₂Fe^II[Fe^II(CN)₆]·2H₂O (left). Reproduced with permission from John Wiley and Sons.[64] (h) Comparison of average voltage, total electrode capacity, lifespan and energy density for the KAB full battery with reported NaABs. (g) at right side and (h) are reproduced with permission from the Nature Publishing Group.[10]

(20 mol of LiTFSI in 1 kg of water) was chosen as the electrolyte. In such concentrated electrolyte, the $H_2O/Li^+$ ratio drops from 11 (5 m LiTFSI, m refers to molality) to 2.67 (21 m LiTFSI), decreasing the amount of free water molecules in the electrolyte, which contributes to restraining the electrochemical activity of water (Figure 1.3c). Interestingly, the SEI on the anode surface, just like in the case of organic electrolytes (see Figure 1.3d), can be effectively formed with the reduction of $TFSI^-$ anion before water decomposition, which further expands ESW to around 3.0 V (1.9–4.9 V vs. Li/Li$^+$). As a result, the assembled $LiMn_2O_4/Mo_6S_8$ AB shows a high average discharge voltage of 1.8 V, an impressive energy density of 100 Wh kg$^{-1}$, together with high capacity retention (68% with 1,000 cycles at 4.5 C). WISE provides wide opportunities for developing high-energy LiABs, not only by widening the ESW value of aqueous electrolytes but also by broadening the scope of materials selection. For instance, 21 m LiTFSI electrolyte affords LiAB coupled with $TiS_2$ and $LiMn_2O_4$ with an energy density of 78 Wh kg$^{-1}$.[54] In addition, formation of cathode electrolyte interphase (CEI) was proposed to stabilize the cathode.[26] With the formed CEI through electrochemical oxidation of TMSB in 21 m LiTFSI, the capacity of $LiCoO_2$ was extended to 170 mAh g$^{-1}$, thus a 2.0 V $LiCoO_2/Mo_6S_8$ battery with 120 Wh kg$^{-1}$ was achieved. Apparently, the formation of SEI highly depends on the salt concentration, that is, more concentrated electrolytes are needed to suppress water electrolysis. WIBSE containing 21 m LiTFSI and 7 m LiOTf, i.e., 28 m Li$^+$ with $H_2O/Li^+$ ratio of 2, was also proposed to obtain a wider ESW (1.83–4.9 V vs. Li/Li$^+$).[24] As a result, the system allows the use of $TiO_2$ as the anode material and endows $LiMn_2O_4/TiO_2$ LiAB with a high discharge voltage of 2.1 V and an energy density of 100 Wh kg$^{-1}$. In the same year, a $Li(TFSI)_{0.7}(BETI)_{0.3} \cdot 2H_2O$ hydrate-melt electrolyte was proposed by dissolving LiTFSI and $LiN(SO_2C_2F_5)_2$ (LiBETI) salts in water.[25] Attributing to the expanded ESW (3.8V, 1.25–5.05 V vs. Li/Li$^+$), various high-voltage LiABs such as 3.1 V $LiNi_{0.5}Mn_{1.5}O_4/Li_4Ti_5O_{12}$ with 90.6 Wh kg$^{-1}$ and 2.4 V $LiCoO_2/Li_4Ti_5O_{12}$ with 130 Wh kg$^{-1}$ were obtained.[25]

Although the above WISE, WIBSE and hydrate-melt electrolytes can effectively expand ESW to around 3.0 V, its enlargement is still limited, especially for the cathodic stability (only >1.2 V vs. Li/Li$^+$). Consequently, most anode materials with low redox potential and high capacity, such as metallic Li (0 V vs. Li/Li$^+$, 3861 mAh g$^{-1}$), graphite (0.1 V vs. Li/Li$^+$, 372 mAh g$^{-1}$), silicon (0.30 V vs. Li/Li$^+$, 4200 mAh g$^{-1}$), etc., are excluded.

Simply increasing salt concentrations cannot resolve this severe 'cathodic challenge'. As electrode potential is polarized to 0.50 V vs. Li/Li$^+$ or below, these fluorinated salt anions experience increasing expulsion from the anode surface, and water molecules start to adsorb with hydrogen pointing toward the surface, leading to energetically favorable HER.[55] Minimization of the number of water molecules at the anode surface before the SEI forms, that is, using a hydrophobic 'inhomogeneous SEI additive' (LiTFSI-HFE gel) as a thin coating gel on the anode surface, is confirmed to be an effective strategy.[55] Through minimizing the competing water decomposition, a conformal and dense SEI rich in inorganic LiF/organic C–F species will form upon lithiation of the anode. Combining the proposed LiTFSI-HFE gel with 'solidified' WISE (gel-WISE, hydrogel of WISE using either polyvinyl alcohol or polyethylene oxide), a series of 4.0 V-class LiABs were achieved, including 4.1 V LiVPO$_4$F/Li, 4.0 V LiVPO$_4$F/graphite, and 4.0 V LiMn$_2$O$_4$/Li, whose energy densities approached those of the state-of-the-art LIBs but with significantly enhanced safety.[55] These works represent a fundamental breakthrough across the gap separating ABs and non-ABs, although the cycling stability of these 4.0 V class LiABs needs to be further improved. More recently, Wang and coworkers in *Nature* [23b] introduced a halogen conversion-intercalation mechanism in graphite to improve the cathode capacity. The cathode consisting of a mixture of solid LiBr and LiCl with graphite (LBC-G) undergoes a two-stage charge and discharge process, which can be defined as a 'conversion-intercalation' mechanism. During charging, bromide and chloride ions in LiBr and LiCl are consecutively 'converted' into their nearly neutral atomic states, i.e., Br$^{-0.05}$ and Cl$^{-0.25}$, and intercalated into a graphite interlayer (Figure 1.3e), forming a densely packed stage-I graphite intercalation compound C$_{3.5}$[Br$_{0.5}$Cl$_{0.5}$] with a highly reversible electrochemical performance. Notably, this LBC-G cathode can deliver a high capacity of 243 mAh g$^{-1}$ with an attractive potential (4.2 V vs. Li/Li$^+$), outperforming most conventional LIB cathode materials, such as LiNi$_{0.8}$Co$_{0.1}$Mn$_{0.1}$O$_2$ (3.8 V, 200 mAh g$^{-1}$) and LiNi$_{0.8}$Co$_{0.15}$Al$_{0.05}$O$_2$ (3.7 V, 200 mAh g$^{-1}$). By coupling this cathode with a passivated graphite anode, a 4 V-class LiAB with an energy density of 460 Wh kg$^{-1}$ (total mass of cathode and anode) can be achieved.

### 1.3.1.2 Na-based Aqueous Batteries

Na-ion aqueous batteries (NaABs) appear to be far more economically competitive than LiABs and have attracted intense interest for large-scale electric energy storage because of their natural abundant resources, low cost, high safety, and environmental friendliness. Unfortunately, similar to LiABs, most of the electrode materials that can serve well in organic Na-ion batteries system cannot work in aqueous media, due to the narrower ESW of water. Together with more difficult desertion/insertion of larger Na$^+$ (0.102 nm), the choice of electrode materials for NaABs is limited (see Figure 1.3f). In recent years, most of the reported works in developing NaABs focused on the filter electrode materials. Up to now, several kinds of cathode materials, including Mn-based oxides (e.g., $MnO_2$, $Na_{0.44}MnO_2$, $Na_{2/3}Ni_{1/4}Mn_{3/4}O_2$, $Na_{0.44}[Mn_{1-x}Ti_x]O_2$, $Co_xMn_{3-x}O_4$), polyanionic compounds (e.g., $Na_2FeP_2O_7$, NASICON-type $Na_3V_2(PO_4)_3$, $NaVPO_4F$, $Na_3MnTi(PO_4)_3$, $Na_3V_2O_2(PO_4)_2F$, $Na_3V_2(PO_4)_2F_3$, $Na_3Fe_2(PO_4)_3$, $Na_2FePO_4F$, $Na_2VTi(PO_4)_3$) and Prussian blue analogs (e.g., NiHCF, CuHCF, CuNiHCF, CoHCF) have been explored for NaABs. However, it must be pointed out that apart from the limited electrode potential, their delivery capacity (<150 mAh g$^{-1}$) is far from satisfaction. More challengingly, for NaAB anode, only a few inorganic materials ($NaTi_2(PO_4)_3$, $Na_3Fe_2(PO_4)_3$, $NaV_3(PO_4)_3$, $Na_3MgTi(PO_4)_3$, $NaV_3O_8$, KMn[Cr(CN)$_6$], $MoO_3$) have been developed. Recently, some works turn to organic anodes, such as PPTO (−0.07 V vs. SHE with 201 mAh g$^{-1}$),[56] and PNTCDA (−0.6 V vs. SCE with 140 mAh g$^{-1}$).[57]

From experience in developing LiABs, the WISE strategies are also feasible to broaden ESW in NaABs. Differently, only a lower salt concentration of 9.26 m $NaCF_3SO_3$ (NaOTf) is applicable to form a Na$^+$-conducting SEI,[58] because of the much more intense ion aggregation between Na$^+$ and $CF_3SO_3^-$ than that between Li$^+$ and TFSI$^-$, thus making a 2.5 V (1.7–4.2 V vs. Na/Na$^+$) ESW for water. After that, several other types of WISE including 35 m NaFSI with a ESW of 2.6 V (1.8–4.4 V vs. Na/Na$^+$),[59] 17 m NaClO$_4$ with a ESW of 2.7 V (1.7–4.4 V vs. Na/Na$^+$),[60] and hybrid electrolyte of 25 m NaFSI + 10 m NaFTFSI[61] have been developed. Unfortunately, although the cycling stability of the NaABs is enhanced, their delivered voltage and energy density are still far from satisfactory, owing to the limited capacity and output voltage of electrode materials. Recently, another interesting work by Goodenough and coworkers is attracting increasing attention in which symmetric NaABs are reported based on NASICON-type $Na_3MnTi(PO_4)_3$[62] with redox couples of Mn$^{2+}$/Mn$^{3+}$ and Ti$^{4+}$/Ti$^{3+}$. This symmetric design not only reduces the manufacturing costs, but also buffers the inner stress of battery with one electrode shrinkage accompanied by the expansion of another.

### 1.3.1.3 *K-based Aqueous Batteries*

K-ion aqueous batteries (KABs) along with LiABs and NaABs are gaining a lot of attention. Owing to the larger ionic radius of $K^+$ (0.138 nm) than $Li^+$ and $Na^+$, the development of desertion/insertion-type electrode materials for KABs is impeded. Up to now, the choice of cathode materials for KABs is mainly focused on Prussian blue analogues. In 2011 Cui and coworkers investigated the desertion/insertion behavior of $K^+$ into CuHCF and NiHCF.[16,53] Typical, nanostructured CuHCF with 59.14 mAh $g^{-1}$ exhibited an impressive performance, including high redox potential (0.946 V vs. SHE), high rate (67.8% at 83 C), and long cycling life (83% after 40,000 cycles at 17 C). Subsequently, based on high $K^+$ content and two one-electron redox processes, a high-capacity $K_2Fe^{II}[Fe^{II}(CN)_6]\cdot 2H_2O$ nanocubes cathode with 120 mAh $g^{-1}$ of discharge capacity and high redox potential (double platform at 0.9 and 0.3 V vs. SCE) was developed (Figure 1.3g).[64] For other cathode candidates, only V-based materials including $V_2O_5$ and $K_{0.22}V_{1.74}O_{4.37}\cdot 0.82H_2O$ have been reported.[65] Regrettably, sparse work has been reported on KAB anodes. Yao and coworkers[56] used an organic compound, poly(anthraquinonyl sulfide) in 10 M KOH, and reported capacity of 200 mAh $g^{-1}$ with –0.6 V vs. SHE. In virtue of the wide ESW (–1.7 to 1.5 V vs. Ag/AgCl) achieved by 30 m KAc WISE, the first inorganic anode $KTi_2(PO_4)_3$ was reported by Leonard *et al.*,[66] while its capacity was limited below 58 mAh $g^{-1}$. In 2019[10] a full KAB was achieved by Jiang *et al.* using KFeMnHCF cathode and PTCDI anode in 22 M $KCF_3SO_3$ WISE. As shown in Figure 1.3h, the KFeMnHCF-3565//PTCDI full battery features a capacity of 63 mAh $g^{-1}$ with an average voltage of 1.27 V, thereby an energy density of 80 Wh $kg^{-1}$ can be achieved, which even outperforms most NaABs.

## 1.3.2 *Advances in Zn-based Aqueous Batteries*

Zn-based aqueous batteries (ZnABs), the first electrochemical battery, dates back to the Voltaic pile invented by Alessandro Volta in the late 19th century. Since then, attributed to very good electrochemical reversibility of zinc, dozens of Zn-based batteries were developed. Nowadays, one-third of the world battery market is composed of Zn-based batteries, which highlights its importance as a power source for a wide range of applications. Based on the cathode and electrolyte, ZnABs can be divided into alkaline zinc batteries (such as Zn-Ni, alkaline Zn-MnO$_2$, Zn-Ag and Zn-air), near-neutral Zn-ion batteries (such as Zn-MnO$_2$, Zn-V$_2$O$_5$, and electrolytic Zn-Mn battery), and zinc-based redox flow batteries (such as Zn-Br, Zn-V, Zn-Ce and Zn-I). In recent

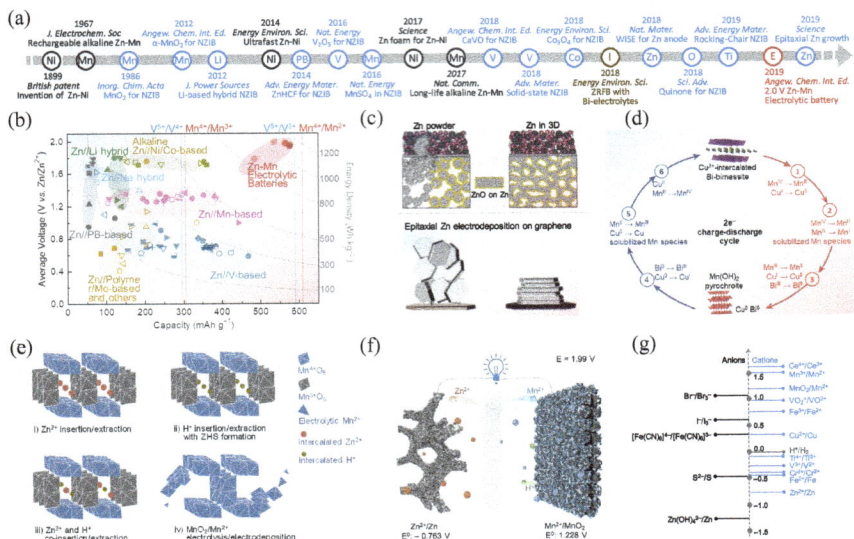

**Figure 1.4.** Summary of advanced materials for Zn-based ABs. (a) Main progress in Zn-based ABs. (b) Comparison of specific capacities and average discharge voltages of various state-of-the-art aqueous Zn-based batteries. Reproduced with permission from John Wiley and Sons.[36] (c) The above is the schematic comparison of recharging Zn/Ni battery with different zinc anodes (conventional powder zinc anode versus 3D sponge zinc anode), indicating the dendrite-free feature in 3D Zn sponge anode. The below figure shows reversible epitaxial electrodeposition of Zn on graphene. Reproduced with permission from the American Association for the Advancement of Science.[47a,75] (d) Electrochemical reactions for the regeneration cycle of $Cu^{2+}$-intercalated Bi-birnessite. Reproduced with permission from the Nature Publishing Group.[67] (e) Schematic illustration of summarized reaction mechanisms for Mn-based $Zn^{2+}$ ion batteries. (f) Schematic of the charge storage mechanism of Zn-Mn electrolytic battery with output voltage ~2 V. (g) Summary of standard redox potentials for various redox pairs that may be suitable for integrating new AB systems.

years, some breakthroughs have been achieved in aqueous Zn-based batteries (see Figure 1.4a). The specific capacity vs. operation voltage data for different types of batteries is displayed in Figure 1.4b, based on the mass of cathode.

### 1.3.2.1 *Alkaline Zn-based Aqueous Batteries*

Alkaline Zn-based batteries (AZABs), including Zn-Ni/Co, Zn-MnO$_2$, Zn-Ag$_2$O, and Zn-air batteries, which depend on the reversible redox reaction of Zn/ZnO with a redox potential of −1.35 V vs. SHE, represent an old and mature battery technology but recently they are gaining a lot of attention. This is mainly caused by weak stability arising from the unavoidable formation of Zn dendrite, shape change, corrosion, and passivation. To solve the above issues, various strategies have been proposed,[20a] including alloying with other metals (e.g., Bi, Sn and In) to suppress corrosion, hybridization or surface modification with additives (such as BaO, $Bi_2O_3$, $In(OH)_3$ and $Ca(OH)_2$) to suppress $H_2$ evolution, geometry and structure design (such as Zn fibers, rods, bars and sheets with different thickness and lengths) to mitigate shape change and zinc dendrite formation, and electrolyte additives (such as KF, $K_2HPO_4$, $K_2CO_3$, polyethylene glycol and saturated ZnO) to reduce Zn dissolution and inhibit Zn dendrite. In 2017, Rolison and coworkers[47a] investigated 3D Zn sponges as the anode materials at high depth of discharge ($DOD_{Zn}$). As shown in Figure 1.4c, thanks to the monolithic, porous, non-periodic architecture of Zn sponges, such 3D Zn anode achieves high utilization of 91% $DOD_{Zn}$ and is dendrite free with repeated 50,000 cycles at <1% $DOD_{Zn}$. Inspired by this case, various long-life Zn anodes based on 3D skeleton construction strategy (such as Ni nanowire, carbon cloth, Cu foam and graphene foam) have been developed. In addition, quasi-solid-state design not only endows the batteries with high flexibility, but also stabilizes the Zn anode via suppressing the corrosion and dissolution of Zn anode.

For cathode materials, $MnO_2$, as a non-toxic, low-cost, earth-abundant and high-capacity (617 mAh $g^{-1}$) material is a promising candidate for AZABs. In alkaline Zn-Mn batteries, by-products of spinal phase $Mn_3O_4$ (formed by $Mn^{2+}$ and mother $MnO_2$) and $ZnMn_2O_4$ (formed by MnOOH and $Zn(OH)_4^{2-}$) accumulate after repeated cycles at deep DOD, leading to the capacity fade and eventual battery failure. Generally, doping with Bi, Cu, Ni, and Co, etc., elements or integrating the corresponding oxides with the cathode and using LiOH electrolyte are effective methods to enhance the capacity and rechargeability of $MnO_2$.[20a] In 2017, Banerjee and coworkers[67] realized the two-electron utilization (617 mAh $g^{-1}$) of $MnO_2$ with 6,000 cycles of lifespan using a Cu-intercalated Bi-birnessite cathode. As shown in Figure 1.4d, the key to rechargeability relies on the redox potentials of Cu to reversibly intercalate into the Bi-birnessite-layered structure during dissolution and precipitation process for stabilizing and enhancing charge transfer characteristics. In witnessing deep utilization with high stability, such electrochemical tuning strategies unprecedentedly solved the main challenges of using $MnO_2$ cathode in alkaline batteries.

Ni/Co-based cathode materials, due to their excellent electrochemical reversibility and acceptable theoretical capacity, have been widely used for Zn-Ni batteries since the late 19th century. Among various ABs, Zn-Ni/Co batteries are particularly advantageous owing to their unique merits of higher operation voltage (around 1.7–1.8 V, see Figure 1.4b), impressive theoretical energy density (~372 Wh kg$^{-1}$), high-power ability and low cost. However, due to the poor life of zinc anode, after over 100 years, the Zn-Ni batteries commercialized by PowerGenix (now named ZincFive, Inc.) were taken off the market in 2003. Generally, the commercial Zn-Ni battery prepared from β-phase Ni(OH)$_2$ cathode delivers an energy density of 70–100 Wh kg$^{-1}$, a peak power density of 2000 W kg$^{-1}$, and a lifespan of around 500 cycles. The overall electrochemical performance is far from satisfactory for the ever-increasing demand for power storage. Aside from poor stability in zinc anodes, the researchers tend to attribute such poor electrochemical performance to the reluctant specific capacity, irreversibility, and poor electroactivity in Ni/Co-based cathode. Fortunately, some achievements in designing nano-architecture Ni/Co-based cathode materials have raised hopes in recent years. In 2014, Dai and coworkers[68] inaugurated an ultrafast high-capacity Zn-Ni battery based on ultrathin NiAlCo LDH/CNT nanoplates cathode, in which Al and Co co-doping stabilized α-Ni(OH)$_2$. Attributing to the high capacity (354 mAh g$^{-1}$), high rate (278 mAh g$^{-1}$ at 66.7 A g$^{-1}$), and good stability (94% of capacity retention after 2,000 cycles) of NiAlCo LDH/CNT cathode, the assembled Zn-Ni battery delivered an energy density of 274 Wh kg$^{-1}$ and a power density of 16.6 kW kg$^{-1}$, together with good cycling stability (85% capacity retention after 500 cycles). Inspired by this work, so far various Ni-based and Co-based materials, such as NiAlCo-LDH/CNT,[68] Ni$_3$S$_2$,[69] Co$_3$O$_4$[70] and NiCo$_2$O$_4$[71] have been extensively explored for Zn-Ni batteries.

Besides the compositional optimization, rational nano-architecture design (e.g., nanoparticles, nanowires, nanorods and nanosheets) can provide unique merits in mechanical and electrical properties, such as higher surface area and shorter pathways for transport of ions and electrons, and conquer the intrinsic challenges of bulk materials, such as poor electrical conductivity and large volume expansion. In addition, surface modification, such as surface coating of PANI and surface doping of phosphate ions, can further enhance the electrical conductivity of electrode.[69,71] However, it should be noted that the developed advanced self-standing cathodes are still far from practical application, although they have achieved remarkable gravimetric capacity, high rate and long life. Their areal capacity is generally lower than 1.0 mAh cm$^{-2}$, which is much lower than the industrial-level areal capacity of ~35 mAh cm$^{-2}$.[72] Therefore, further developments of Ni-based and Co-based materials for Ni–Zn or Co–Zn batteries, which simultaneously possess high gravimetric capacity, high rate capability and long life with high mass loading, remain difficult to achieve.

### 1.3.2.2 Neutral Zn-ion Batteries

Neutral Zn-ion batteries (NZIBs), which use neutral or weakly acidic Zn$^{2+}$-containing aqueous media as the electrolyte, are attracting increasing global attention in recent years because of their potential for large-scale electrical energy storage. As early as 1986, a rechargeable Zn-MnO$_2$ battery using MnO$_2$ cathode and Zn anode in 2 M ZnSO$_4$ electrolyte was firstly investigated by Yamamoto et al.,[73] but the reaction mechanism was unclear. It was not until 2012 that Kang and coworkers[74] found the reversible intercalation of Zn$^{2+}$ into α-MnO$_2$, and proposed the concept of NZIBs combining with a zinc anode and a mild ZnSO$_4$ or Zn(NO$_3$)$_2$ aqueous electrolyte. Intensive efforts have been devoted to NZIBs since then with the purpose of revealing the reaction mechanism and developing advanced electrode materials. Unlike AZABs, the charge storage in the anode depends on the reversible plating/stripping of Zn/Zn$^{2+}$ with a redox potential of –0.763 V vs. SHE. Although the severe corrosion and dissolution of Zn are eliminated, the biggest challenge is in suppressing the formation of zinc dendrite. Up to now, various efforts, including surface modification, structural optimization,[32] and electrolyte optimization,[42] have been explored to eliminate the formation of zinc dendrite. For instance, a high-rate flexible quasi-solid-state zinc-ion battery constructed from graphene foam-supported Zn array anode and a gel electrolyte can deliver long-term durability of 2,000 cycles with 89% of the initial capacity.[32] Moreover, it was recently pointed out by Archer and coworkers[75] that graphene, with a low lattice mismatch for Zn, is effective in driving the deposition of Zn with a locked crystallographic orientation, which prompts exceptional reversibility of Zn anode.

Another issue that has hindered the application of NZIBs is the lack of robust cathode host materials for fast and reversible Zn$^{2+}$ storage, owing to the high charge density and high hydrated ionic radius of Zn$^{2+}$. So far, although various cathode materials such as manganese oxides, V-based, Prussian blue analogs, and organic materials have been proposed,[12] the development of cathode materials for ZIBs is still in its infancy stage.

This is mainly ascribed to the following four aspects: 1) the reaction mechanism still remains controversial; 2) fast capacity decay; 3) unsatisfactory specific capacity and 4) poor rate performance. V-based compounds, especially vanadium oxides, are attractive host materials for $Zn^{2+}$ storage. Attributed to their inherent features of the multiple valence states of vanadium and the large open-framework structure, V-based materials possess merits of high capacity (even up to 400 mAh g$^{-1}$), fast dynamics, and low cost. As for the reaction mechanism, it is generally considered as the insertion/extraction of $Zn^{2+}$ in the host materials during the corresponding discharge/charge process. Recently, with the observation of zinc hydroxide sulfate in Zn-V systems, $H^+$ is also regarded as the charge carrier to participate in the electrochemical reaction.[76] Based on the simultaneous $H^+$ and $Zn^{2+}$ insertion/extraction process, the $Zn/NaV_3O_8 \cdot 1.5H_2O$ battery proposed by Chen and coworkers[46] delivers superior reversible capacity (380 mAh g$^{-1}$) and high durability (82% of capacity retention after 1,000 cycles). Apart from the study of mechanism, more work is needed on advanced materials for improving the specific capacity, rate performance and cycling life of V-based cathode. Up to now, some optimization strategies, including morphological and structural control (such as designing various nanoarchitectures, pre-insertion of Li, Na, K, Zn and Ca, etc., metal ions, and adjustment of structural water), integrating with conductive additives, designing binder-free electrode, and optimizing electrolytes have been attempted.[76] Dozens of V-based compounds, such as $V_2O_5 \cdot nH_2O$,[77] $Zn_{0.25}V_2O_5 \cdot nH_2O$,[78] and $Zn_2(OH)VO_4$[32] have been developed. Generally, the V-based oxides can present an ultrahigh discharge capacity of over 400 mAh g$^{-1}$, while their operating voltage is relatively poorer than that of Mn-based materials (Figure 1.4b). For example, the $Zn/Zn_{0.3}V_2O_5 \cdot 1.5H_2O$ battery fabricated by Wang et al.[79] delivers an average discharge voltage of 0.8 V, and a high specific capacity of 426 mAh g$^{-1}$ at 0.2 A g$^{-1}$, together with an unprecedented cycling stability (maintains 214 mAh g$^{-1}$ after 20,000 cycles at 10 A g$^{-1}$).

Prussian blue analogs (PBA), similarly to LiAB, NaAB and KAB systems, can also be used as cathode materials for NZIBs. In 2015, a PBA-based NZIB built on ZnHCF was first proposed by Liu and coworkers[80] with a relatively high operation voltage of ~1.7 V, a discharge capacity ~65.4 mAh g$^{-1}$, and an energy density of 100 Wh kg$^{-1}$. Since then, various other PBA-based NZIBs, such as CuHCF-Zn (56 mAh g$^{-1}$, 1.73 V), FeHCF-Zn (120 mAh g$^{-1}$, 1.1 V), NiHCF-Zn (56 mAh g$^{-1}$, 1.2 V) and MnHCF-Zn

(137 mAh g$^{-1}$, 1.7 V) have been developed.[12] However, note that owing to the low capacity, the energy density of PBA-based NZIBs is still not competitive. Besides the inorganic materials mentioned above, some organic ones such as PANI (191 mAh g$^{-1}$, 1.0 V)[81] and calix[4]quinone (335 mAh g$^{-1}$, 1.0 V)[82] have been developed. Up to now, the development of organic cathode materials for NZIB is still in its preliminary stage. With the abundant choices of functional group and molecular weight, there remains an immense potential to optimize the electrochemical performance of organic electrodes.

Manganese oxides, with the merits of abundant crystallographic polymorphs ($\alpha$-, $\beta$-, $\gamma$-, $\delta$-, $\lambda$-, $\varepsilon$- and todorokite-types), high theoretical capacity (308 mAh g$^{-1}$), low cost and earth abundance, have been regarded as promising cathode candidates for NZIBs. Generally, with the applicable exploration of various $MnO_2$ polymorphs for NZIBs, there are mainly four stream concepts as shown in Figure 1.4e on the energy storage mechanism: i) Zn$^{2+}$ insertion/extraction, ii) H$^+$ insertion/extraction accompanied with the deposition of zinc hydroxide sulfate ($Zn_4(SO_4)(OH)_6 \cdot nH_2O$), iii) co-insertion/ extraction of both H$^+$ and Zn$^{2+}$ in different charge/discharge steps, and iv) electrolysis/ electrodeposition of $MnO_2$/Mn$^{2+}$, which have been systematically summarized in reports.[12,36]

Although the reaction mechanism remains under dispute, the first three mechanisms assume that the severe dissolution of Mn$^{2+}$ in the discharge process is responsible for the fast capacity decay. Up to now, some effective strategies including pre-addition of Mn salt in electrolyte,[45] surface coating (such as N-doped carbon[83] and PEDOT[84]) and incorporation of closely bonded ions (such as $K_{0.8}Mn_8O_{16}$[85]) have been employed to suppress the dissolution of Mn$^{2+}$ and enhance the cycling stability of NZIBs. Specially, the cycling stability of the Zn-MnO$_2$ battery could be greatly enhanced by pre-addition of Mn$^{2+}$, achieving a 10,000 cycles lifespan without obvious capacity decay.[86] It should be noted that such excellent cycling stability is not only attributed to the stabilized $MnO_2$ by suppressing the dissolution of Mn$^{2+}$, but also from the extra capacity provided by the re-deposited $MnO_2$ in the charge process.[87] In addition, large volumetric change and structural collapse caused by repeated insertion of hydrated Zn$^{2+}$ ions also result in a rapid capacity fading. Thereby, various efforts such as morphological control of porous structure,[83] coupling with graphene and CNTs,[88] and structure stabilization by cationic doping and PANI intercalation[89] have been explored. For example, PANI-intercalated $MnO_2$ nanolayer can deliver a stable discharge capacity of around 125 mAh g$^{-1}$ over 5,000 cycles.[89]

Although great progress has been achieved as can be seen from Figure 1.4a, the current Zn-based alkaline, neutral or weakly acidic batteries have shown limited output voltages (<1.8 V) and discharge capacity below 450 mAh g$^{-1}$. In our latest research, we found a latent high-voltage $MnO_2$ electrolysis process in a conventional Zn-ion battery, and proposed a new electrolytic Zn–Mn system (see Figure 1.4f), via enabled proton and electron dynamics.[36] The four-step $MnO_2$ electrolysis process was first analyzed by density functional theory calculations. This Zn–Mn electrolytic system presents an output voltage as high as 1.95 V, an imposing gravimetric capacity of about 570 mAh g$^{-1}$, and density of ~409 Wh kg$^{-1}$ based on both anode and cathode active materials. A prototype redox flow-battery stack was also built in our Zn-Mn electrolytic battery. In summary, the output voltage (~2 V), energy efficiency (88%), and cost of the electrolyte (3–5 US$ (kW h)$^{-1}$) outperform other redox pair-integrated AB systems (Figure 1.4g), such as Zn-Fe, Zn-Br$_2$, Zn-Ce, and all vanadium flow batteries.[36] It is expected that with further judicial development, such as the use of a more selective electrolyte, Zn efficiency improvement, and efficient flow-stack battery design, this Zn–Mn electrolytic flow battery design will be applicable for practical energy storage and, particularly, for large-scale grid energy storage.

### 1.3.3 Advances in Mg/Al/Ca-based Aqueous Batteries

Other than the conventional aqueous Li/Na/K-ABs and ZnABs, there is another type of AB based on relatively abundant metal elements as charge carriers, such as magnesium, aluminum and calcium ions. Some breakthroughs and materials in Mg/Al/Ca-based ABs in recent years have been summarized in Figure 1.5a and 1.5b.

In the study of magnesium ion aqueous batteries (MgABs), Chen *et al.* recently proposed an MgAB which involves a reversible $Mg^{2+}$ intercalation-deintercalation in a nickel hexacyanoferrate cathode.[90] PBA nickel hexacyanoferrate, which is shown to be capable of reversibly intercalating various cations in its open-framework structure, was also selected as the cathode in this battery. It is worth to note that the PBA cathode goes through a replacement process of extracting the Na$^+$ ion and substituting by

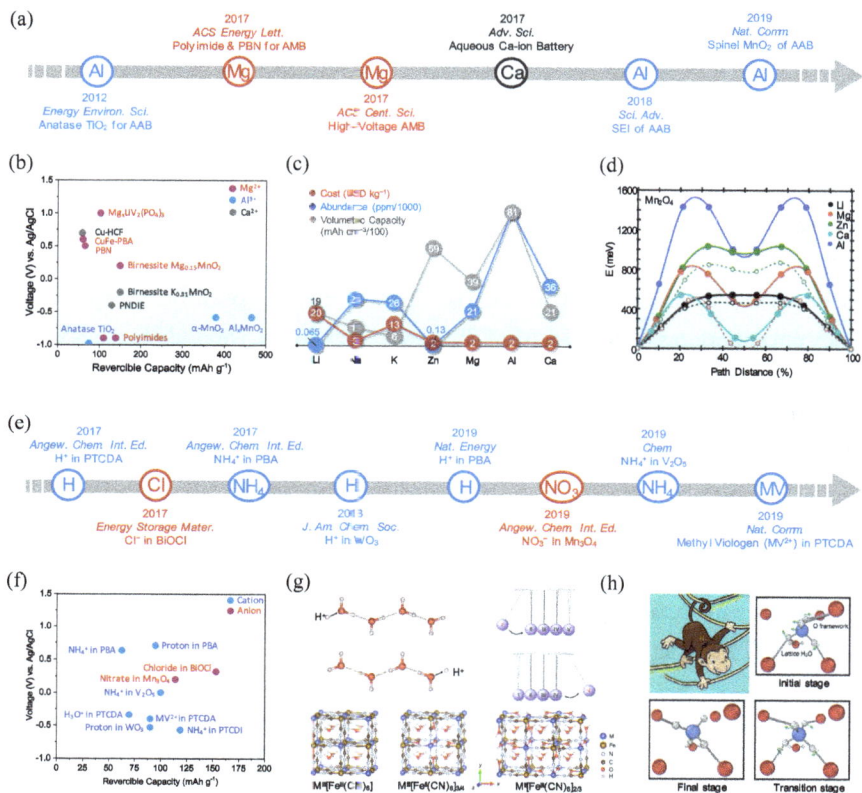

**Figure 1.5.** Summary of electrode materials for Mg/Al/Ca-ion- and non-metal-ion-based ABs. (a) Timeline of recent developments in the area of advanced aqueous Mg/Al/Ca-ion batteries. (b) Comparison of various electrode materials for aqueous Mg/Al/Ca-ion batteries. (c) Comparison of cost, abundance and volumetric capacity of metal anodes. Data sourced from Refs.[21] and [95]. (d) First-principle elastic band simulations of migration energies of metal ions in the spinel $Mn_2O_4$. Reproduced with permission from the American Chemical Society.[96] (e) Timeline of recent developments in the area of advanced aqueous non-metal-ion batteries. (f) Comparison of various electrode materials for non-metal-ion-based ABs. (g) Scheme of Grotthuss proton transportation. Reproduced with permission from the Nature Publishing Group.[98c] (h) Scheme and diffusion mechanism of $NH_4^+$ ion. Cartoon of the 'monkey swinging' process, $NH_4^+$ breaks and reforms H–O bonds one at a time during its migration within the $V_2O_5$ bilayer. Reproduced with permission from Elsevier.[14,100]

$Mg^{2+}$ ion in the open framework without structural destruction. On the other hand, an aromatic polyimide formed by condensation reaction was chosen as anode. Polyimide family has been reported as anode host materials for both $Li^+$ and $Na^+$ intercalation while effectively suppressing hydrogen evolution in aqueous electrolyte.

As a result, the corresponding battery can achieve a specific energy density of 33 Wh $kg^{-1}$, which is close to their theoretical energy density of 48 Wh $kg^{-1}$. Apart from that, Nazar and coworkers also reported the insertion of Mg ion in Birnessite in aqueous Mg-ion batteries, suggesting that aqueous Mg-ion batteries exhibit higher rate performance than the one in non-aqueous media due to the difference in anion desolvation energy.[91]

To improve the energy density of MgABs, Wang and co-authors applied the superconcentration strategy to expand ESW to 2.0 V, which is about three times higher than that of the conventional dilute $MgSO_4$ electrolyte of ~0.7 V.[23a,92] It should be noted that the wide ESW of 2.0 V enables the employment of $Mg_xLiV_2(PO_4)_3$ (LVP), which is obtained by delithiation and subsequent magnesiation process of $Li_3V_2(PO_4)_3$. Similar to most of the previous MgABs, a member of the polyimide family PNTCDA is chosen as the anode. Thanks to the accelerated Mg ion diffusion within LVP and the relatively wide ESW achieved by the superconcentration strategy, MgAB exhibited an excellent rate capability of 60 C, cycling stability within 6,000 cycles, high power density of 6.4 kW $kg^{-1}$, and high specific energy density of 68 Wh $kg^{-1}$.

Overall, in the selection of electrode materials for MgABs, two important aspects should be mentioned: i) to enable the reversible intercalation/deintercalation of magnesium, nowadays, most of the cathode materials are achieved via pre-replacement of the cations with Mg ions while some Mg-containing oxide cathodes are also reported;[90–92] ii) as can be seen in Figure 1.5b, most of the reported anodes are carbon-based polyimides and V-based oxides.[93] Although diverse polyimide and PBA anodes are available, the exploration of new types of Mg-ion host electrodes with high capacity and output voltage is essential for the future development of MgABs.

Furthermore, aluminum is earth abundant, low cost, chemically inert and, most importantly, possesses the highest volume-specific charge storage capacity (8040 mAh $cm^{-3}$) as shown in Figure 1.5c, which is approximately four times larger than that for lithium metal batteries.[30,94] The major issue for aluminum anode lies in the rapid and irreversible formation of a high bandgap-passivated oxide coating, $Al_2O_3$. Archer and coworkers employed an $AlCl_3$–1-ethyl-3-methylimidazolium chloride-treated aluminum substrate, and found that this ionic liquid-enriched film is capable of eroding the oxide film and protecting the aluminum against subsequent oxide film formation in aqueous media.[30] After coupling with $MnO_2$ cathode, the Al-based aqueous battery (AlAB) achieved an energy density of up to 500 Wh kg$^{-1}$. The major challenge in the cathode part lies in the high charge density of Al$^{3+}$ for its reversible insertion/extraction in the host materials. To address this issue, another AlAB configuration with $Al(OTf)_3$-$H_2O$ electrolyte and $Al_xMnO_2 \cdot H_2O$ cathode was reported. [94] The AlAB enabled a high discharge capacity of 467 mAh g$^{-1}$, resulting in a high energy density of 481 Wh kg$^{-1}$. Based on previous studies, it can be concluded that the main issue of AlABs remains in improving the reversibility of Al$^{3+}$ plating/striping at the anode and extraction/insertion at the cathode.[95] Breakthroughs in novel host materials and electrolytes related to the interface engineering are desired.

In contrast to Mg and Al ions, calcium ions have a larger ion radius (0.99 Å) than that of Mg$^{2+}$ (0.65 Å) and Al$^{3+}$ (0.50 Å) as shown in Figure 1.2a. Nevertheless, Ca$^{2+}$ has a low polarization strength and charge density similar to that of Li$^+$, therefore the superiority of Ca$^{2+}$ chemistry could be reflected in better kinetics in electrode materials, avoiding the kinetics issues related to Mg$^{2+}$ and Al$^{3+}$. As a result, higher electrolyte conductivity and faster ion diffusion in the electrolyte may also be observed due to the smaller Ca$^{2+}$ hydrated radius and more facile dehydration. As shown in Figure 1.5d, Ca$^{2+}$ is proven to have a smaller migration barrier than that of Mg$^{2+}$ and Al$^{3+}$ in spinel $Mn_2O_4$.[96] In spite of the kinetics superiority, the development of Ca-based aqueous batteries (CaABs) is stagnated due to the limited success in Ca-ion storage materials. Recently, Gheytani et al. firstly reported a full CaAB with organic polyimide as the anode and copper hexacyanoferrate as the cathode,[97] showing specific energy of 54 Wh kg$^{-1}$ and outstanding stability within 1,000 cycles. More interestingly, identical electrodes have also been applied in a MgAB to comprehensively understand the differences of the charge carriers The kinetics analysis showed faster kinetics in $Ca(NO_3)_2$ electrolyte than in $Mg(NO_3)_2$.

### 1.3.4 Advances in Non-Metal-Ion-Based Aqueous Batteries

Other than the metal-ion batteries, recently there is a novel kind of aqueous 'Rocking-Chair' battery that employs non-metal ions as the charge carrier. Some breakthroughs and materials in non-metal-ion-based ABs in recent years have been summarized in

Figure 1.5e and 1.5f. Compared with the metal-ion batteries, the most significant feature of non-metal-ion batteries is that the ions employed in these systems are based on abundant elements; thus, the limited reserves of the elements used is no longer the bottleneck to an energy storage system.

Specifically, proton and hydronium, its simplest hydrate, have been investigated as the charge carrier in different electrodes. Ji and coworkers have investigated the proton storage capability in hydrate oxide, highly crystalline organic electrode and PBA.[98] The H-based aqueous battery (HAB) is advantageous because of its excellent rate performance when hydrated oxide and PBA are employed.[98c] It is suggested that the diffusion-free Grotthuss topochemistry is the key to this superior rate performance of proton batteries. Akin to Newton's cradle (as shown in Figure 1.5g), the transfer of protons with correlated local displacement enables their fast long-range transport. The motion process is highly different from conduction of metal ions where solvated metal ions diffuse long distance individually. As a consequence, the battery based on the so-called Grotthuss proton conduction exhibited excellent rate performance of 78, 67, 56 and 49 mAh g$^{-1}$ at 20 C, 200 C, 2000 C and 4000 C, respectively. It is noteworthy that half of its capacity at 1 C can retain even at the extremely high current density of 4000 C. The rate capability potential of HABs remains enormous due to the present limitation in the electrical resistance of the testing cells.

Ammonium is another attractive charge carrier due to its light molar mass of only 18 g mol$^{-1}$ and the small hydrate ionic size, which facilitates its fast diffusion in the electrolyte. Compared with proton-based batteries, employment of ammonium ions would avoid the usage of corrosive acidic electrolyte and correspondingly suppress the elemental dissolution of cathode and hydrogen evolution at the anode.[99] Initially, PBA was selected as the cathode material due to its robust crystal structure and facile de/insertion of alkali metal ions.[99] Similar to the cathode in MgABs, PBA used in ammonium aqueous batteries (AABs) was subjected to a process of replacement of pristine alkali ions by $NH_4^+$. With the aim to improve its rate performance, authors from the same group employed hydrated Bi-layered $V_2O_5$ to serve as a cathode to store ammonium ions.[100] In this work, the authors claimed the superior rate performance originated from a chemisorption-involved intercalation pseudocapacitance. Through comparison of different pseudocapacitive behavior of $NH_4^+$ and K$^+$, the authors suggested a 'monkey swinging' process in which the ammonium ion can twist to disconnect one of its trailing hydrogen bonds and form a new hydrogen bond with another oxygen atom during its migration in the crystal structure of hydrated bi-layer $V_2O_5$ (as shown in Figure 1.5h).

Apart from proton and ammonium, recently Wei *et al.* explored the potential of employing methyl viologen ($MV^{2+}$) as a dicationic charge carrier in an aromatic solid electrode material 3,4,9,10-perylenetetracarboxylic dianhydride (PTCDA).[101] $MV^{2+}$ is the largest insertion charge carrier (when non-solvated) ever reported for batteries. The interaction between MV and PTCDA is capable of offering a decent capacity ~100 mAh $g^{-1}$ and 60% rate capability retention from 100 mA $g^{-1}$ to 2000 mA $g^{-1}$. This result shows that the large charge carrier does not compromise the specific capacity and rate capability. Exploring novel multivalent ions, especially novel organic charge carriers, might be an interesting avenue for ABs.

Except for aqueous cation batteries, there are several reports on anionic non-metal-ion-based ABs.[15] For example, Chen *et al.* recently reported a chloride-ion AB in a NaCl solution with BiOCl anode and silver cathode.[15] The most extraordinary advantage of $Cl^-$ ion is its high abundance in the natural form, for example, NaCl solution, e.g., seawater. A stable reversible capacity of 92.1 mAh $g^{-1}$ was obtained. Other than chloride ions, there are also reports on the battery employing fluoride ion as charge carrier.[102] However, cycling stability was limited in the current halogen anion batteries. Apart from halogen anions, Jiang *et al.* reported the reversible insertion of nitrate into $Mn_3O_4$ for aqueous dual-ion batteries.[103] The insertion of $NO_3^-$ resulted in a capacity as high as 183 mAh $g^{-1}$ and more than 3,500 cycles at 1 A $g^{-1}$. These achievements imply that there is definitely a margin for developing anions as charge carriers for ABs.

## 1.4 Summary and Outlook

ABs are rising as the promising energy storage systems for intermittent energy utilization and sustainable large-scale applications. Benefiting from their low cost, abundant resources, easy assembling and recycling, environmental benignity, and above all safety, the advanced ABs have potential to replace conventional Li-ion, Ni-metal hydride (Ni/MH) and Pb-acid batteries for future automotive, aerial and scalable energy storage applications. In recent years we witnessed a rapid development of electrode materials with remarkable electrochemical performance and new electrochemical mechanisms. Although significant advances have been made in this area, unremitting efforts are still required, including pushing the energy/power densities and longstanding stability before meeting the requirements of practical applications. Next, we will summarize the advantages and disadvantages of some typical AB systems, and try to provide prospects for the next stage of development of ABs.

## 1.4.1 *Comparison of Different Systems*

Currently, LIBs, Ni/MH and lead-acid batteries remain as the mainstream energy storage systems in the world market of rechargeable batteries. LIBs have been the ubiquitous commercial power source for portable electronics, electric vehicles and backup power supplies as evidenced by their widespread usage in the past decades. The scarce abundance (20 ppm) and increasing high cost of Li (or Co) pose challenges for employing them for grid storage. In addition, some incidents have exposed serious safety concerns of LIBs with organic solvents and hindered their large-scale energy applications (see Table 1.1). At present, lead-acid ABs are the most recognizable aqueous-based batteries in our daily life and still hold a major proportion of the global battery market. Major market sectors for lead-acid batteries are Starting Lighting Ignition for cars, automotive including e-bikes, forklifts and other vehicles, and stationary industrial uses including Uninterruptible Power Supplies and grid-scale energy storage units. However, they are also known to have low energy density of ca. 35 Wh kg$^{-1}$, which is less than a quarter of the value of LIBs (see Figure 1.6a). They are also limited by problems of toxicity and low charge/discharge efficiency. Nickel-iron batteries have shown their environmental friendliness, longevity, and tolerance to electrical misuse compared with lead-acid batteries, while with the same problems of low energy density and CE (ca. 60%). Ni/MH battery remains a great success for use in hybrid electric vehicles (such as Prius, Toyota) due to its relatively high energy density, safety and wide temperature range performance.[104] The stagnation of the development can be ascribed to its limited energy density, usage of rare earth resources, and memory effects compared with LIBs (see Table 1.1). Recent improvements in alkaline Ni/Co-Zn and Mn-Zn batteries have focused on the advanced nanoengineering design with increased electrochemical reversibility and decreased formation of irreversible phases including $\gamma$-NiOOH, $Mn_3O_4$

**Table 1.1.** Comparison of different aqueous batteries and other commercialized electrochemical energy storage technologies.

| Representative battery type | Electrochemical reaction mechanism | Electrolyte | Working voltage (V) | Theoretical energy density (Wh kg$^{-1}$) | Practical energy density (Wh kg$^{-1}$) | Cost of electrolyte | Status | Advantages (A) vs. Disadvantages (D) |
|---|---|---|---|---|---|---|---|---|
| Li-ion battery | $LiC_6 + FePO_4 \leftrightarrow LiFePO_4 + 6C + e^-$ | 1 M LiPF$_6$ | 3.3 | 385 | ~145 (device scale) | Middle | Commercialized | A: High energy density, good overall performance. D: Safety risk, strict manufacture, limited low-temperature performance, limited Li/Co resources. |
| Lead-acid battery | $Pb + PbO_2 + 2H_2SO_4 \leftrightarrow 2PbSO_4 + 2H_2O + 2e^-$ | 5 M H$_2$SO$_4$ | 2.0 | 167 | ~35 (device scale) | Cheap | Commercialized | A: Good safety, low cost, low self-discharge. D: Low energy density, poor cyclability, environmental issue. |
| Ni-Fe battery | $Fe + 2NiOOH + 2H_2O \leftrightarrow Fe(OH)_2 + 2Ni(OH)_2 + 2e^-$ | 2 M KOH | 1.2 | 234 | ~40 (device scale) | Cheap | Commercialized | A: Good safety, low cost, long life, tolerance to electrical abuse. D: Low energy density, hard to maintain |

(Continued)

Table 1.1. (Continued)

| Representative battery type | Electrochemical reaction mechanism | Electrolyte | Working voltage (V) | Theoretical energy density (Wh kg$^{-1}$) | Practical energy density (Wh kg$^{-1}$) | Cost of electrolyte | Status | Advantages (A) vs. Disadvantages (D) |
|---|---|---|---|---|---|---|---|---|
| Ni-metal hydride battery | MH + NiOOH $\leftrightarrow$ M + Ni(OH)$_2$ + e$^-$ | 6 M KOH | 1.35 | 240 | ~100 (device scale) | Cheap | Commercialized | A: Good safety, wide temperature range, overcharge ability. D: Limited energy density, usage of rare earth resources, memory effect. |
| Alkaline Ni(Co)-Zn battery | Zn + 2NiOOH + 2H$_2$O $\leftrightarrow$ Zn(OH)$_2$ + 2Ni(OH)$_2$ + 2e$^-$ | 1 M KOH + 20 mM Zn(ac)$_2$ | 1.7 | 372 | ~90 (device scale) | Cheap | Commercialized | A: Good safety, low cost, good low-temperature performance. D: Limited energy density, limited cycling life. |
| Alkaline Mn-Zn battery (75) | Zn + 2MnO$_2$ + 2H$_2$O $\leftrightarrow$ Zn(OH)$_2$ + 2MnOOH + 2e$^-$ | 37 wt% KOH | 1.2 | 358 | ~100 (electrode scale) | Cheap | Bench-scale | A: Good safety, low cost. D: Limited energy density, poor high-DoD cyclability. |
| Li-ion aqueous battery (28) | C + LiBr + LiCl $\leftrightarrow$ C$_n$[BrCl] + 2Li$^+$ + 2e$^-$ | 21 M LiTFSL + 7 M LiOTf | 4.2 | 617 | ~460 (electrode scale) | Expensive | Bench-scale | A: Good safety, high energy density. D: High cost in electrolyte, limited rate capability. |

| Battery | Reaction | Electrolyte | Voltage | | Energy density | Cost | Scale | Advantages/Disadvantages |
|---|---|---|---|---|---|---|---|---|
| Na-ion aqueous battery (59) | $2Na_3MnTi(PO_4)_3 \leftrightarrow Na_4MnTi(PO_4)_3 + Na_2MnTi(PO_4)_3 + e^-$ | 1 M $Na_2SO_4$ | 1.4 | 41 | ~40 | Cheap | Bench-scale | A: Good safety, low cost. D: Low energy density, limited cyclability. |
| K-ion aqueous battery (12) | $KFeMnHCF + PTCDL \leftrightarrow K_xPTCDL + K_{1-x}FeMnHCF$ | 22 M $KCF_3SO_3$ | 1.2 | — | ~80 (electrode scale) | Expensive | Bench-scale | A: Good safety. D: High cost in electrolyte, low energy density. |
| Al-ion aqueous battery (101) | $Al_xMnO_2 \times nH_2O + (y-x)Al \leftrightarrow Al_yMnO_2 \times nH_3O$ | 5 M $Al(OTF)_3$ | 1.1 | 561 | — | Expensive | Bench-scale | A: Good safety. D: High cost in electrolyte, low energy density. |
| Zn-ion aqueous battery (85, 96) | $Zn + 2MnO_2 \leftrightarrow ZnMn_2O_4 + 2e^-$; $Zn + 2MnO_2 + 1/3ZnSO_4 + 11/3H_2O \leftrightarrow 2MnOOH + 1/3Zn_4SO_4(OH)_6 \cdot 5H_2O$ | 2 M $ZnSO_4$ + 0.1 M $MnSO_4$ | 1.35 | 302 | — | Cheap | Bench-scale | A: Good safety, low cost. D: Moderate energy density, limited rate and cycling performance. |
| | 1. $1Zn + Zn_{0.25}V_2O_5 \leftrightarrow Zn_{1.35}V_2O_5 + 2.2e^-$ | 1 M $ZnSO_4$ | 0.8 | 175 | — | Cheap | Bench-scale | A: Good safety, high capacity, low cost. D: Moderate energy density, low voltage. |
| Electrolytic battery (41) | $Zn + MnO_2 + 2H_2SO_4 \leftrightarrow ZnSO_4 + MnSO_4 + 2H_2O$ | 1 M $ZnSO_4$ and $MnSO_4$ + 0.1 M $H_2SO_4$ | 1.99 | 700 | ~409 (electrode scale) | Cheap | Bench-scale | A: Good safety, low cost, high energy density. D: Limited areal/volumetric capacity, Zn anode long-term sustainability. |

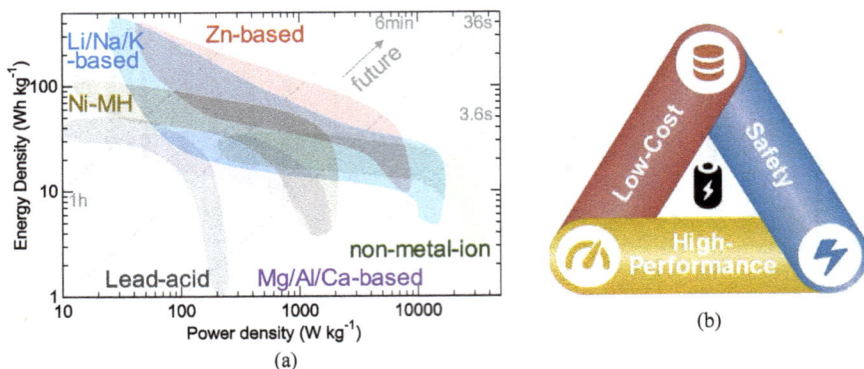

**Figure 1.6.** Overview of current and future developments in the area of ABs. (a) Energy/power density comparison between some conventional commercialized batteries, typical metal-ions (including Li+, Na+, K+, Zn2+, Mg2+, Al2+ and Ca2+) and non-metal-ion charge carrier-based ABs. (b) Proposed three criteria for the future road to commercialization of ABs.

and $ZnMn_2O_4$. These batteries have shown potential for applications as the next-generation power sources in low-speed electric vehicles, battery electric vehicles, hybrid electric vehicles and emerging grid-scale electrical energy storage systems.

The concept of aqueous 'Rocking-Chair' battery chemistry was inspired by intercalation electrodes used in organic solvent-based LIBs. The first-proposed $LiMn_2O_4//VO_2$ LiAB exhibited a practical energy density of ~55 Wh kg$^{-1}$,[51] which was competitive with that of the lead-acid battery. The major challenges faced by this chemistry are in their poor electrode stability and restricted ESW due to the problems of water electrolysis, protons or water intercalation, and electrode material dissolution. Na+ and K+ showing similar chemical behavior as Li+ are considered more attractive power sources for large-scale energy storage. However, their electrochemical performance is restricted due to the limited choices of host materials and low operation voltage (<1.2 V) and energy density. The energy density in Li/Na/K aqueous-based systems has been boosted since 2015 by expanding their output voltages via WISE or hydrate-melt electrolytes.[23a] There has been a jump in the energy density of aqueous LiABs from less than 100 Wh kg$^{-1}$ of traditional slat-in-water systems to the current ~200 Wh kg$^{-1}$. It should be noted that apart from the high cost of the commonly used salts in WISE, the high viscosity and volumetric weight of concentrated electrolyte put forward more rigorous requirements for the physical design and assembly process for practical applications. Exploration of low-cost salts such as sodium perchlorate ($NaClO_4$),[105] zinc chloride ($ZnCl_2$),[106] lithium acetate ($LiCH_2COO$) and potassium acetate ($KCH_3COO$)[66,107] should be vigorously explored.

Over the past five years, tremendous efforts have been made in near-neutral ZnABs ascribed to the high reversibility of $Zn/Zn^{2+}$ plating/striping in aqueous media compared with other multivalent metal ions of $Mg/Mg^{2+}$, $Ca/Ca^{2+}$ and $Al/Al^{3+}$, which remain at the primary stage of electrode materials exploration. The achieved energy and power densities of ZnABs have been approaching those of LiABs without using WISE. To achieve practically reliable ZnABs, several issues like limited output voltage, dissolution and irreversible by-product at cathode, and dendrite, corrosion and passivation at zinc anode should be moderated to further improve the longevity of ZnABs. The electrolytic Zn-Mn battery with two solid/solution redox pairs of $Zn/Zn^{2+}$ at anode and $MnO_2/Mn^{2+}$ at cathode was invented by our group showing a high output voltage of ~2 V and energy density of ~400 Wh kg$^{-1}$ (see Table 1.1 and Figure 1.6a), which has significantly extended the energy and power densities of Zn-based electrochemistry. Further judicial developments such as high mass-loading skeleton and optimized electrolyte with high Zn efficiency, are still expected to further boost the battery performance. As regards safety and cost, the 'Salt-in-Water' electrolyte, earth-abundant raw material resources, easy manufacturing, and high energy/power density of the electrolytic Zn-Mn batteries mean that the commercialization progress is on the way. Other types of electrolytic batteries such as Cu-Mn, Bi-Mn and Zn-I have also been reported by Zhi and coworkers,[108] Xia and coworkers[109] and Ji and coworkers,[106] respectively. It is still of particular interest to introduce more promising solid/solution redox pairs with electrolysis mechanism to practical applications.

### 1.4.2 *Principles and Road to Commercialization*

Over decades of development, ABs have been improved greatly to meet the selection criteria for next-generation commercial energy storage systems. Three criteria are recommended for the future road to commercialization of ABs, i.e., safety, low cost and high performance (see Figure 1.6b). The key aspect of any future battery technology in our daily life is safety. As can be found in Figure 1.6b, safety is the survival of ABs in comparison with current LIBs. Apart from the recent safety concerns that restricted the large-scale energy storage application of LIBs, low cost of ABs and associated abundant resources, simple manufacturing processes and facile auxiliary systems facilitate the fast-growing research on ABs. Regarding performance, compared with LIBs, the next-generation ABs should be long lasting and feature high energy and power densities.

While current ABs require further improvements in cyclability, some works report the possibility of improving the cycling stability by optimizing battery configuration, operation conditions, and electrolytes and electrode materials, for example Zn hybrid battery[110] and 'Rocking-Chair' Zn-ion full battery[111] designs, and host materials with an open framework[64] or organic materials.[56] Power density is a crucial factor for developing commercial ABs, especially for fast-charging or regenerative braking of electric vehicles. Novel intercalation electrode materials with tailored architecture design have been proved to be a prerequisite for providing short and fast migration channels and sustaining the structural integrity for long-term and high-rate cycling. In addition, pseudocapacitive behavior renders much faster charge transfer than volume lattice diffusion and thus can help to retain the capacity at high current rates of ABs.[14,32] The energy density by mass or volume has been regarded as a pivotal indicator for large-scale applications, which is the function of the voltage and capacity of the device. Generally, three strategies are effective to enhance the energy density of ABs, i.e., a) widen ESW and enlarge output voltage, b) explore new host materials or chemistries with more electron transfer but lower consumption of mass and/or volume, c) improve the utilization and prevent the loss of the electrolyte.

In addition to the aforementioned challenges, the self-discharge needs to be carefully evaluated before practically realizing a new AB technology. Recently, significant progress in electrode materials design has been made, as evidenced by a large spectrum of available systems for ABs with promising electrochemical performance. It is believed that the progress in electrode materials innovation will boost the performance of ABs in the coming years. Further developments in this area taking into account safety, low cost and high performance will surely advance the commercialization progress of high-performance ABs. While LIBs could be still the dominating power source for the next decade for consumable electronics, vehicles, drones, and even robots, the progress in post-LIBs toward more cost-effective (>50% cost reduction) and safer alternatives will be intensified in the near future. The battery that wins safety and low cost at the start would prevail later.

## Acknowledgements

**Funding**: This work was supported by the Australian Research Council Discovery Project (Australian Laureate Fellowship FL170100154). **Author contributions**: D.C. and S.-Z.Q. conceived the Review. All authors contributed to discussions of the content and conceived the topic of the Review. S.-Z.Q. edited and M.J. reviewed the article before submission. **Competing interests**: The authors declare that they have no competing interests. **Data and materials availability**: All data needed to evaluate the conclusions in the paper are present in the paper. Additional data related to this paper may be requested from the authors.

## References

[1] B. Dunn, H. Kamath, J. M. Tarascon, *Science* **2011**, 334, 928.

[2] K. Liu, Y. Liu, D. Lin, A. Pei, Y. Cui, *Sci. Adv.* **2018**, 4, eaas9820.

[3] a) O. Schmidt, A. Hawkes, A. Gambhir, I. Staffell, *Nat. Energy* **2017**, 2, 17110; b) D. Larcher, J. M. Tarascon, *Nat. Chem.* **2015**, 7, 19.

[4] a) D. Chao, B. Ouyang, P. Liang, T. T. T. Huong, G. Jia, H. Huang, X. Xia, R. S. Rawat, H. J. Fan, *Adv. Mater.* **2018**, 30 1804833; b) W. Zhang, Y. Liu, Z. Guo, *Sci. Adv.* **2019**, 5, eaav7412.

[5] H. Kim, J. Hong, K. Y. Park, H. Kim, S. W. Kim, K. Kang, *Chem. Rev.* **2014**, 114, 11788.

[6] G. J. May, A. Davidson, B. Monahov, *J. Energy Storage* **2018**, 15, 145.

[7] M. Huang, M. Li, C. J. Niu, Q. Li, L. Q. Mai, *Adv. Funct. Mater.* **2019**, 29, 1807847.

[8] J. Huang, Z. Guo, Y. Ma, D. Bin, Y. Wang, Y. Xia, *Small Methods* **2019**, 3, 1800272.

[9] H. Ao, Y. Zhao, J. Zhou, W. Cai, X. Zhang, Y. Zhu, Y. Qian, *J. Mater. Chem. A* **2019**, 7, 18708.

[10] L. Jiang, Y. Lu, C. Zhao, L. Liu, J. Zhang, Q. Zhang, X. Shen, J. Zhao, X. Yu, H. Li, X. Huang, L. Chen, Y.-S. Hu, *Nat. Energy* **2019**, 4, 495.

[11] J. Ming, J. Guo, C. Xia, W. Wang, H. N. Alshareef, *Mat. Sci. Eng. R* **2019**, 135, 58.

[12] M. Song, H. Tan, D. L. Chao, H. J. Fan, *Adv. Funct. Mater.* **2018**, 28, 1802564.

[13] a) W. Chen, G. Li, A. Pei, Y. Li, L. Liao, H. Wang, J. Wan, Z. Liang, G. Chen, H. Zhang, J. Wang, Y. Cui, *Nat. Energy* **2018**, 3, 428; b) X. Zhao, Z. Zhao-Karger, M. Fichtner, X. Shen, *Angew. Chem. Int. Ed.* **2020**, 59, 5902.

[14] D. Chao, H. J. Fan, *Chem* **2019**, 5, 1359.

[15] F. Chen, Z. Y. Leong, H. Y. Yang, *Energy Storage Mater.* **2017**, 7, 189.

[16] C. D. Wessells, S. V. Peddada, M. T. McDowell, R. A. Huggins, Y. Cui, *J. Electrochem. Soc.* **2011**, 159, A98.

[17] H. Wang, Z. Lu, S. Xu, D. Kong, J. J. Cha, G. Zheng, P. C. Hsu, K. Yan, D. Bradshaw, F. B. Prinz, Y. Cui, *Proc. Natl. Acad. Sci.* **2013**, 110, 19701.

[18] Y. Yin, J. Han, Y. Zhang, X. Zhang, P. Xu, Q. Yuan, L. Samad, X. Wang, Y. Wang, Z. Zhang, P. Zhang, X. Cao, B. Song, S. Jin, *J. Am. Chem. Soc.* **2016**, 138, 7965.

[19] H. Wang, H. W. Lee, Y. Deng, Z. Lu, P. C. Hsu, Y. Liu, D. Lin, Y. Cui, *Nat. Commun.* **2015**, 6, 7261.

[20] a) H. Li, L. Ma, C. Han, Z. Wang, Z. Liu, Z. Tang, C. Zhi, *Nano Energy* **2019**, 62, 550; b) R. Einerhand, W. Visscher, E. Barendrecht, *J. Appl. Electrochem.* **1988**, 18, 799.

[21] V. Verma, S. Kumar, W. Manalastas, R. Satish, M. Srinivasan, *Adv. Sustain. Syst.* **2019**, 3, 1800111.

[22] M. Yu, Y. Lu, H. Zheng, X. Lu, *Chemistry* **2018**, 24, 3639.

[23] a) L. Suo, O. Borodin, T. Gao, M. Olguin, J. Ho, X. Fan, C. Luo, C. Wang, K. Xu, *Science* **2015**, 350, 938; b) C. Yang, J. Chen, X. Ji, T. P. Pollard, X. Lu, C. J. Sun, S. Hou, Q. Liu, C. Liu, T. Qing, Y. Wang, O. Borodin, Y. Ren, K. Xu, C. Wang, *Nature* **2019**, 569, 245.

[24] L. Suo, O. Borodin, W. Sun, X. Fan, C. Yang, F. Wang, T. Gao, Z. Ma, M. Schroeder, A. v. Cresce, S. M. Russell, M. Armand, A. Angell, K. Xu, C. Wang, *Angew. Chem. Int. Ed.* **2016**, 55, 7136.

[25] Y. Yamada, K. Usui, K. Sodeyama, S. Ko, Y. Tateyama, A. Yamada, *Nat. Energy* **2016**, 1, 16129.

[26] F. Wang, Y. Lin, L. Suo, X. Fan, T. Gao, C. Yang, F. Han, Y. Qi, K. Xu, C. Wang, *Energy Environ. Sci.* **2016**, 9, 3666.

[27] Z. Hou, X. Zhang, X. Li, Y. Zhu, J. Liang, Y. Qian, *J. Mater. Chem. A* **2017**, 5, 730.

[28] J. Zhi, A. Z. Yazdi, G. Valappil, J. Haime, P. Chen, *Sci. Adv.* **2017**, 3, e1701010.

[29] S. B. Son, T. Gao, S. P. Harvey, K. X. Steirer, A. Stokes, A. Norman, C. Wang, A. Cresce, K. Xu, C. Ban, *Nat. Chem.* **2018**, 10, 532.

[30] Q. Zhao, M. J. Zachman, W. I. Al Sadat, J. Zheng, L. F. Kourkoutis, L. Archer, *Sci. Adv.* **2018**, 4, eaau8131.

[31] X. Shan, D. S. Charles, Y. Lei, R. Qiao, G. Wang, W. Yang, M. Feygenson, D. Su, X. Teng, *Nat. Commun.* **2016**, 7, 13370.

[32] D. Chao, C. R. Zhu, M. Song, P. Liang, X. Zhang, N. H. Tiep, H. Zhao, J. Wang, R. Wang, H. Zhang, H. J. Fan, *Adv. Mater.* **2018**, 30, 1803181.

[33]  C. Zhu, P. Yang, D. Chao, X. Wang, X. Zhang, S. Chen, B. K. Tay, H. Huang, H. Zhang, W. Mai, H. J. Fan, *Adv. Mater.* **2015**, 27, 4566.

[34]  M. R. Lukatskaya, O. Mashtalir, C. E. Ren, Y. Dall'Agnese, P. Rozier, P. L. Taberna, M. Naguib, P. Simon, M. W. Barsoum, Y. Gogotsi, *Science* **2013**, 341, 1502.

[35]  M. Yu, D. Lin, H. Feng, Y. Zeng, Y. Tong, X. Lu, *Angew. Chem. Int. Ed.* **2017**, 56, 5454.

[36]  D. L. Chao, W. H. Zhou, C. Ye, Q. H. Zhang, Y. G. Chen, L. Gu, K. Davey, S. Z. Qiao, *Angew. Chem. Int. Ed.* **2019**, 58, 7823.

[37]  S. E. Chun, B. Evanko, X. Wang, D. Vonlanthen, X. Ji, G. D. Stucky, S. W. Boettcher, *Nat. Commun.* **2015**, 6, 7818.

[38]  L. Chen, H. Shao, X. Zhou, G. Liu, J. Jiang, Z. Liu, *Nat. Commun.* **2016**, 7, 11982.

[39]  J. Yan, J. Wang, H. Liu, Z. Bakenov, D. Gosselink, P. Chen, *J. Power Sources* **2012**, 216, 222.

[40]  K. Zhao, C. Wang, Y. Yu, M. Yan, Q. Wei, P. He, Y. Dong, Z. Zhang, X. Wang, L. Mai, *Adv. Mater. Interfaces* **2018**, 5, 1800848.

[41]  J. H. Jo, Y.-K. Sun, S.-T. Myung, *J. Mater. Chem. A* **2017**, 5, 8367.

[42]  N. Zhang, F. Cheng, Y. Liu, Q. Zhao, K. Lei, C. Chen, X. Liu, J. Chen, *J. Am. Chem. Soc.* **2016**, 138, 12894.

[43]  N. Zhang, F. Cheng, J. Liu, L. Wang, X. Long, X. Liu, F. Li, J. Chen, *Nat. Commun.* **2017**, 8, 405.

[44]  F. Wang, O. Borodin, T. Gao, X. Fan, W. Sun, F. Han, A. Faraone, J. A. Dura, K. Xu, C. Wang, *Nat. Mater.* **2018**, 17, 543.

[45]  H. L. Pan, Y. Y. Shao, P. F. Yan, Y. W. Cheng, K. S. Han, Z. M. Nie, C. M. Wang, J. H. Yang, X. L. Li, P. Bhattacharya, K. T. Mueller, J. Liu, *Nat. Energy* **2016**, 1, 16039.

[46]  F. Wan, L. Zhang, X. Dai, X. Wang, Z. Niu, J. Chen, *Nat. Commun.* **2018**, 9, 1656.

[47]  a) J. F. Parker, C. N. Chervin, I. R. Pala, M. Machler, M. F. Burz, J. W. Long, D. R. Rolison, *Science* **2017**, 356, 415; b) J. F. Parker, C. N. Chervin, E. S. Nelson, D. R. Rolison, J. W. Long, *Energy Environ. Sci.* **2014**, 7, 1117; c) Z. Wang, J. Huang, Z. Guo, X. Dong, Y. Liu, Y. Wang, Y. Xia, *Joule* **2019**, 3, 1289.

[48]  Z. Yuan, Y. Duan, T. Liu, H. Zhang, X. Li, *iScience* **2018**, 3, 40.

[49]  Z. Zhao, X. Fan, J. Ding, W. Hu, C. Zhong, J. Lu, *ACS Energy Lett.* **2019**, 4, 2259.

[50]  W. Li, W. R. McKinnon, J. R. Dahn, *J. Electrochem. Soc.* **1994**, 141, 2310.

[51]  W. Li, J. R. Dahn, D. S. Wainwright, *Science* **1994**, 264, 1115.

[52]  D. Bin, Y. Wen, Y. Wang, Y. Xia, *J. Energy Chem.* **2018**, 27, 1521.

[53]  D. Sun, Y. Jiang, H. Wang, Y. Yao, G. Xu, K. He, S. Liu, Y. Tang, Y. Liu, X. Huang, *Sci. Rep.* **2015**, 5, 10733.

[54] W. Sun, L. Suo, F. Wang, N. Eidson, C. Yang, F. Han, Z. Ma, T. Gao, M. Zhu, C. Wang, *Electrochem. Commun.* **2017**, 82, 71.

[55] C. Yang, J. Chen, T. Qing, X. Fan, W. Sun, A. von Cresce, M. S. Ding, O. Borodin, J. Vatamanu, M. A. Schroeder, N. Eidson, C. Wang, K. Xu, *Joule* **2017**, 1, 122.

[56] Y. Liang, Y. Jing, S. Gheytani, K. Y. Lee, P. Liu, A. Facchetti, Y. Yao, *Nat. Mater.* **2017**, 16, 841.

[57] X. Dong, L. Chen, J. Liu, S. Haller, Y. Wang, Y. Xia, *Sci. Adv.* **2016**, 2, e1501038.

[58] L. Suo, O. Borodin, Y. Wang, X. Rong, W. Sun, X. Fan, S. Xu, M. A. Schroeder, A. V. Cresce, F. Wang, C. Yang, Y.-S. Hu, K. Xu, C. Wang, *Adv. Energy Mater.* **2017**, 7, 1701189.

[59] R.-S. Kühnel, D. Reber, C. Battaglia, *ACS Energy Lett.* **2017**, 2, 2005.

[60] M. H. Lee, S. J. Kim, D. Chang, J. Kim, S. Moon, K. Oh, K.-Y. Park, W. M. Seong, H. Park, G. Kwon, B. Lee, K. Kang, *Mater. Today* **2019**, 29, 26.

[61] D. Reber, R.-S. Kühnel, C. Battaglia, *ACS Mater. Lett.* **2019**, 1, 44.

[62] H. Gao, J. B. Goodenough, *Angew. Chem. Int. Ed.* **2016**, 55, 12768.

[63] C. D. Wessells, S. V. Peddada, R. A. Huggins, Y. Cui, *Nano Lett.* **2011**, 11, 5421.

[64] D. Su, A. McDonagh, S. Z. Qiao, G. Wang, *Adv. Mater.* **2017**, 29, 1604007.

[65] D. S. Charles, M. Feygenson, K. Page, J. Neuefeind, W. Xu, X. Teng, *Nat. Commun.* **2017**, 8, 15520.

[66] D. P. Leonard, Z. Wei, G. Chen, F. Du, X. Ji, *ACS Energy Lett.* **2018**, 3, 373.

[67] G. G. Yadav, J. W. Gallaway, D. E. Turney, M. Nyce, J. Huang, X. Wei, S. Banerjee, *Nat. Commun.* **2017**, 8, 14424.

[68] M. Gong, Y. Li, H. Zhang, B. Zhang, W. Zhou, J. Feng, H. Wang, Y. Liang, Z. Fan, J. Liu, H. Dai, *Energy Environ. Sci.* **2014**, 7, 2025.

[69] L. Zhou, X. Zhang, D. Zheng, W. Xu, J. Liu, X. Lu, *J. Mater. Chem. A* **2019**, 7, 10629.

[70] X. Wang, F. Wang, L. Wang, M. Li, Y. Wang, B. Chen, Y. Zhu, L. Fu, L. Zha, L. Zhang, Y. Wu, W. Huang, *Adv. Mater.* **2016**, 28, 4904.

[71] Y. Zeng, Z. Lai, Y. Han, H. Zhang, S. Xie, X. Lu, *Adv. Mater.* **2018**, DOI: 10.1002/adma.201802396e1802396.

[72] W. Chen, Y. Jin, J. Zhao, N. Liu, Y. Cui, *Proc. Natl. Acad. Sci.* **2018**, 115, 11694.

[73] T. Yamamoto, T. Shoji, *Inorg. Chim. Acta* **1986**, 117, L27.

[74] C. Xu, B. Li, H. Du, F. Kang, *Angew. Chem.* **2012**, 51, 933.

[75] J. Zheng, Q. Zhao, T. Tang, J. Yin, C. D. Quilty, G. D. Renderos, X. Liu, Y. Deng, L. Wang, D. C. Bock, C. Jaye, D. Zhang, E. S. Takeuchi, K. J. Takeuchi, A. C. Marschilok, L. A. Archer, *Science* **2019**, 366, 645.

[76] F. Wan, Z. Niu, *Angew. Chem. Int. Ed.* **2019**, 58, 2.

[77] M. Yan, P. He, Y. Chen, S. Wang, Q. Wei, K. Zhao, X. Xu, Q. An, Y. Shuang, Y. Shao, K. T. Mueller, L. Mai, J. Liu, J. Yang, *Adv. Mater.* **2018**, 30, 1703725.

[78] D. Kundu, B. D. Adams, V. Duffort, S. H. Vajargah, L. F. Nazar, *Nat. Energy* **2016**, 1, 16119.

[79] L. Wang, K.-W. Huang, J. Chen, J. Zheng, *Sci. Adv.* **2019**, 5, eaax4279.

[80] L. Zhang, L. Chen, X. Zhou, Z. Liu, *Adv. Energy Mater.* **2015**, 5, 1400930.

[81] F. Wan, L. Zhang, X. Wang, S. Bi, Z. Niu, J. Chen, *Adv. Funct. Mater.* **2018**, 28, 1804975.

[82] Q. Zhao, W. Huang, Z. Luo, L. Liu, Y. Lu, Y. Li, L. Li, J. Hu, H. Ma, J. Chen, *Sci. Adv.* **2018**, 4, eaao1761.

[83] Y. Fu, Q. Wei, G. Zhang, X. Wang, J. Zhang, Y. Hu, D. Wang, L. Zuin, T. Zhou, Y. Wu, S. Sun, *Adv. Energy Mater.* **2018**, 8, 1801445.

[84] Y. Zeng, X. Zhang, Y. Meng, M. Yu, J. Yi, Y. Wu, X. Lu, Y. Tong, *Adv. Mater.* **2017**, 29, 1702698.

[85] G. Z. Fang, C. Y. Zhu, M. H. Chen, J. Zhou, B. Y. Tang, X. X. Cao, X. S. Zheng, A. Q. Pan, S. Q. Liang, *Adv. Funct. Mater.* **2019**, 29, 1808375.

[86] W. Sun, F. Wang, S. Hou, C. Yang, X. Fan, Z. Ma, T. Gao, F. Han, R. Hu, M. Zhu, C. Wang, *J. Am. Chem. Soc.* **2017**, 139, 9775.

[87] S. Zhao, B. Han, D. Zhang, Q. Huang, L. Xiao, L. Chen, D. G. Ivey, Y. Deng, W. Wei, *J. Mater. Chem. A* **2018**, 6, 5733.

[88] B. Wu, G. Zhang, M. Yan, T. Xiong, P. He, L. He, X. Xu, L. Mai, *Small* **2018**, 14, 1703850.

[89] J. Huang, Z. Wang, M. Hou, X. Dong, Y. Liu, Y. Wang, Y. Xia, *Nat. Commun.* **2018**, 9, 2906.

[90] L. Chen, J. L. Bao, X. Dong, D. G. Truhlar, Y. Wang, C. Wang, Y. Xia, *ACS Energy Lett.* **2017**, 2, 1115.

[91] X. Q. Sun, V. Duffort, B. L. Mehdi, N. D. Browning, L. F. Nazar, *Chem. Mater.* **2016**, 28, 534.

[92] F. Wang, X. L. Fan, T. Gao, W. Sun, Z. H. Ma, C. Y. Yang, F. Han, K. Xu, C. S. Wang, *ACS Cent. Sci.* **2017**, 3, 1121.

[93] H. Y. Zhang, K. Ye, K. Zhu, R. B. Cang, J. Yan, K. Cheng, G. L. Wang, D. X. Cao, *Chem. Eur. J.* **2017**, 23, 17118.

[94] C. Wu, S. C. Gu, Q. H. Zhang, Y. Bai, M. Li, Y. F. Yuan, H. L. Wang, X. Y. Liu, Y. X. Yuan, N. Zhu, F. Wu, H. Li, L. Gu, J. Lu, *Nat. Commun.* **2019**, 10, 73.

[95] S. Kumar, R. Satish, V. Verma, H. Ren, P. Kidkhunthod, W. Manalastas, M. Srinivasan, *J. Power Sources* **2019**, 426, 151.

[96] Z. Rong, R. Malik, P. Canepa, G. Sai Gautam, M. Liu, A. Jain, K. Persson, G. Ceder, *Chem. Mater.* **2015**, 27, 6016.

[97] S. Gheytani, Y. Liang, F. Wu, Y. Jing, H. Dong, K. K. Rao, X. Chi, F. Fang, Y. Yao, *Adv. Sci.* **2017**, 4, 1700465.

[98] a) X. Wang, C. Bommier, Z. Jian, Z. Li, R. S. Chandrabose, I. A. Rodríguez-Pérez, P. A. Greaney, X. Ji, *Angew. Chem.* **2017**, 129, 2955; b) H. Jiang, J. J. Hong, X. Wu, T. W. Surta, Y. Qi, S. Dong, Z. Li, D. P. Leonard, J. J. Holoubek, J. C. Wong, J. J. Razink, X. Zhang, X. Ji, *J. Am. Chem. Soc.* **2018**, 140, 11556; c) X. Wu, J. J. Hong, W. Shin, L. Ma, T. Liu, X. Bi, Y. Yuan, Y. Qi, T. W. Surta, W. Huang, J. Neuefeind, T. Wu, P. A. Greaney, J. Lu, X. Ji, *Nat. Energy* **2019**, 4, 123.

[99] X. Y. Wu, Y. T. Qi, J. J. Hong, Z. F. Li, A. S. Hernandez, X. L. Ji, *Angew. Chem. Int. Ed.* **2017**, 56, 13026.

[100] S. Dong, W. Shin, H. Jiang, X. Wu, Z. Li, J. Holoubek, W. F. Stickle, B. Key, C. Liu, J. Lu, P. A. Greaney, X. Zhang, X. Ji, *Chem* **2019**, 5, 1537.

[101] Z. Wei, W. Shin, H. Jiang, X. Wu, W. F. Stickle, G. Chen, J. Lu, P. Alex Greaney, F. Du, X. Ji, *Nat. Commun.* **2019**, 10, 3227.

[102] X. Hou, Z. Zhang, K. Shen, S. Cheng, Q. He, Y. Shi, D. Y. W. Yu, C.-y. Su, L.-J. Li, F. Chen, *J. Electrochem. Soc.* **2019**, 166, A2419.

[103] H. Jiang, Z. Wei, L. Ma, Y. Yuan, J. J. Hong, X. Wu, D. P. Leonard, J. Holoubek, J. J. Razink, W. F. Stickle, F. Du, T. Wu, J. Lu, X. Ji, *Angew. Chem. Int. Ed.* **2019**, 58, 5286.

[104] C. Dongliang, Z. Chenglin, M. Zhewen, Y. Fei, W. Yucheng, Z. Ding, W. Chaoling, C. Yungui, *Int. J. Hydrogen Energy* **2012**, 37, 12375.

[105] K. Nakamoto, R. Sakamoto, M. Ito, A. Kitajou, S. Okada, *Electrochemistry* **2017**, 85, 179.

[106] J. J. Hong, L. Zhu, C. Chen, L. Tang, H. Jiang, B. Jin, T. C. Gallagher, Q. Guo, C. Fang, X. Ji, *Angew. Chem.* **2019**, 58, 15910.

[107] M. R. Lukatskaya, J. I. Feldblyum, D. G. Mackanic, F. Lissel, D. L. Michels, Y. Cui, Z. Bao, *Energy Environ. Sci.* **2018**, 11, 2876.

[108] G. Liang, F. Mo, H. Li, Z. Tang, Z. Liu, D. Wang, Q. Yang, L. Ma, C. Zhi, *Adv. Energy Mater.* **2019**, 9, 1901838.

[109] J. Huang, Z. Guo, X. Dong, D. Bin, Y. Wang, Y. Xia, *Sci. Bull.* **2019**, 64, 1780.

[110] X. Zeng, J. Hao, Z. Wang, J. Mao, Z. Guo, *Energy Storage Mater.* **2019**, 20, 410.

[111] W. Li, K. Wang, S. Cheng, K. Jiang, *Adv. Energy Mater.* **2019**, 9, 1900993.

Chapter 2

# Protocol in Evaluating Capacity of Zn-Mn Aqueous Batteries: A Clue of pH[1]

Hang Yang,[a,*] Tengsheng Zhang,[b,*] Duo Chen,[a,c] Yicheng Tan,[a]
Wanhai Zhou,[b] Li Li,[a] Wei Li,[b] Guangshe Li,[a] Wei Han,[a,†]
Hong Jin Fan,[d,†] Dongliang Chao[b,†]

[a]College of Physics, College of Chemistry, State Key Laboratory
of Inorganic Synthesis and Preparative Chemistry, International
Center of Future Science, Jilin University,
Changchun 130012, P. R. China

[b]Laboratory of Advanced Materials, Shanghai Key Laboratory
of Molecular Catalysis and Innovative Materials, and School of
Chemistry and Materials, Fudan University,
Shanghai 200433, P. R. China

[c]Jiangsu Key Laboratory of Electrochemical Energy Storage
Technologies, College of Materials Science and Technology,
Nanjing University of Aeronautics and Astronautics,
Nanjing 210016, P. R. China

[d]School of Physical and Mathematical Sciences,
Nanyang Technological University,
Singapore 637371, Singapore

---

* These authors contributed equally.

† Corresponding author: whan@jlu.edu.cn; fanhj@ntu.edu.sg; chaod@fudan.edu.cn

[1] Adapted with permission from H. Yang, T. Zhang, D. Chen, Y. Tan, W. Zhou, L. Li, W. Li, G. Li, W. Han, H. J. Fan, D. Chao, *Adv. Mater.* 2023, 35, 2300053.

In literature, Zn-Mn aqueous batteries (ZMABs) confront abnormal capacity behavior, such as capacity fluctuation and diverse "unprecedented performances." Because of the electrolyte additive-induced complexes, various charge/discharge behaviors associated with different mechanisms are being reported. However, the current performance assessment remains unregulated, and only the electrode or electrolyte is considered. The lack of a comprehensive and impartial performance evaluation protocol for ZMABs hinders forward research and commercialization. In this chapter, we first propose a pH clue (proton-coupled reaction) to understand different mechanisms and normalize the capacity contribution. Then, a series of performance metrics, including rated capacity ($C_r$) and electrolyte contribution ratio from $Mn^{2+}$ (*CfM*), are systematically discussed based on diverse energy storage mechanisms. The relationship between Mn (II) $\leftrightarrow$ Mn (III) $\leftrightarrow$ Mn (IV) conversion chemistry and proton consumption/production is subsequently established. Finally, we propose concrete design concepts of a tunable $H^+/Zn^{2+}/Mn^{2+}$ storage system for customized application scenarios, opening the door for next-generation high-safety and reliable energy storage systems.

## 2.1 Speaking at the Front

The general trend of current energy storage technology is to achieve storage devices with high energy and high safety.[1–3] Among various energy storage devices, Zn-Mn aqueous batteries (ZMABs) have shared the limelight due to their low cost, high safety, decent voltage platform, and considerable theory capacity.[4] Up to now, a variety of high-energy and long-term stable ZMABs have been extensively developed, which confers great potential for application purposes. In parallel to the progress in battery performance, the reaction chemistry of ZMABs has been complex and remains controversial.[5–6]

Since the first report of introducing $Mn^{2+}$ additive into the electrolyte in 2016, the $Mn^{2+}$ oxidation behavior in aqueous batteries has attracted burgeoning attention (Figure 2.1a).[5,7–19] In 2018, the oxidation of $Mn^{2+} \rightarrow$ Mn(III)-based oxides was found in ZMABs for contributing extra capacities.[13,20–21] In the numerous subsequent studies, it has been unveiled that the abnormal attenuation is triggered by the inert Zn/Mn oxides (ZMO) aggregation ("dead Mn") on the electrode surface.[15,19,22–24] In 2019, Chao *et al.* proposed the latent high-voltage $Mn^{2+} \leftrightarrow MnO_2$ (IV) electrolysis process

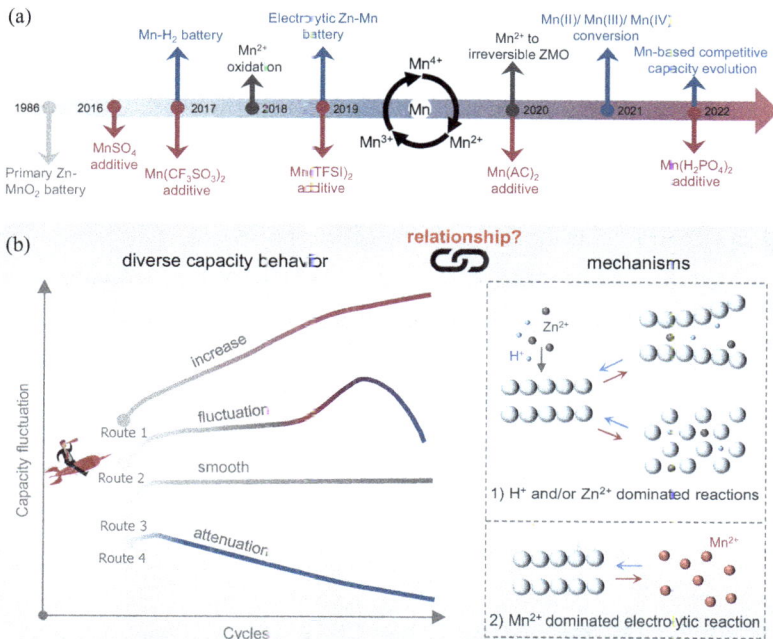

**Figure 2.1.** Summary of charge/discharge mechanisms and abnormal behaviors for ZMABs. (a) Main progress for ZMABs. Light gray, dark red, gray, and dark blue arrows represent the prototype of the primary Zn-MnO$_2$ battery, iteration of types of Mn$^{2+}$ ion additives, side reactions of Mn$^{2+}$ ion, and new battery systems associated with Mn$^{2+}$ ions, respectively. (b) Schematics of abnormal capacity behavior and possible reaction mechanisms for ZMABs. The red part of the curve corresponds to the rapid capacity increase from the electrolyte. Gray represents reversible ion (de)intercalation. Dark blue indicates capacity decay due to irreversible phase transitions.

in the conventional ZMABs, which exhibits not only high capacity *ca.* 600 mAh g$^{-1}$ but also high voltage plateau near 2 V.[14] The generally recognized ZMAB energy storage mechanisms can be summarized into two: (1) H$^+$- and/or Zn$^{2+}$-dominated reactions;[25–26] (2) Mn$^{2+}$-dominated electrolytic reaction (Mn$^{2+}$ to Mn$^{n+}$, 2 < n ≤ 4),[14,27–32] as depicted in Figure 2.1b. These mechanisms or their combination are proposed in some cases. They have also been extended to explain the charge-discharge behavior of other manganese oxides, like MnO$_2$,[11] ZnMn$_2$O$_4$,[33] and MnO.[34]

However, the controlling factors behind these storage mechanisms remain unclear due to the variable valences, different phases of manganese oxide, and various electrolytes.[1,4]

By summarizing the literature, we propose that the pH associated with proton-coupled reactions (PCRs) mainly dominates the storage mechanism, which couples with the $Mn^{2+}$ in the electrolyte to affect the capacity evolution of the ZMABs.[19,24] It is known that the $Mn^{2+}$ additives introduced to the electrolyte can suppress cathode dissolution caused by the Jahn–Teller effect. While this additive has been commonly applied in almost every ZMAB, it contributes to capacity unexpectedly.[7,13] In the meantime, it brings up a disturbance to the charge-discharge behavior. For example, various unusual fluctuation phenomena caused by $Mn^{2+}$ additive have been observed, including unnatural increase,[13,35] attenuation,[15–16,22] combined fluctuation,[19,36–38] and perfectly smooth trend of the cyclic curve (Figure 2.1a).[39–41] In most reports, a single electrochemical reaction process is declared in the whole cycle calendar. However, electrochemistry usually comprises multiple processes that dominate at different stages.[19,23,42] Such a combination of several phenomena has not been systematically analyzed, and the capacity calculation remains unregulated in current reports. The complexity of reaction mechanisms and the lack of unicity call for a deep re-examination of the ZMABs in a more practical way.

In previous reviews, the charge/discharge mechanisms are classified usually based on the crystallography and morphography characteristics of manganese oxides. This chapter aims to unravel the Zn-Mn aqueous chemistry *via* a pH clue. The relationship between Mn (II) ↔ Mn (III) ↔ Mn (IV) conversion chemistry and proton consumption/production is established here. We believe the new perspective would be a solid supplement to the current electrode-materials-classified mechanisms, and should shield light on the practical performances tailoring from the ion/molecule/cluster view.

## 2.2 Convergence is on the Way

In relatively acidic solutions (empirically defined as pH < 3) with high proton reactivity, the $Mn^{2+}$ and manganese oxides would undergo a dominant reversible $Mn^{2+}/MnO_2$ dissolution/deposition process.[1,14] Therefore, the capacity contribution can be easily estimated. However, most reported ZMABs operate in near-neutral environments (pH = 3~6), where $Mn^{2+}$ additive in the aqueous electrolytes was employed. The concomitant complex reaction mechanisms varying with the pH value and PCRs make it hard to evaluate the capacity contribution from the electrolyte and/or electrode.

For example, converting $Mn^{2+}$ to manganese oxides could provide extra capacity to the electrode,[19,27] resulting in "fake" high-performance ZMABs, especially for cathode materials. Therefore, the evaluation protocol for cathode materials should be scrutinized. In addition, the overall charge-discharge mechanism needs a broader conclusion than individual ones. These are the first steps in the convergence process. In particular, there are three major controversies about the electrochemical behavior in near-neutral aqueous solutions: 1) What is the role of $H^+/Zn^{2+}$ ions in the electrochemical process: intercalating in the crystal structure of manganese oxides or participating in the dissolution reaction? 2) What is the main reason for the capacity fluctuation? 3) What are the final forms of manganese after cycles? There may be no general answers to these questions, but more insights into the reaction mechanisms will lead to better reliability in peer comparison.

We first discuss what drives the Mn deposition during the charging process. According to the Nernst equation (Figure 2.2a), the $H^+$ concentration is strongly related to Mn deposition potential. Fewer $H^+$ ions would decrease the energy requirement of Mn deposition. That is to say, the formation of manganese oxide can be more easily triggered at relatively higher pH values. For instance, the near-neutral pH environment of typical electrolytes, such as $ZnSO_4$, $Zn(OTF)_2$, $Zn(Ac)_2$, $Zn(NO_3)_2$, and $ZnCl_2$, are suitable to promote the Mn deposition reaction at a lower potential of $E_{c2}$ (Figure 2.2a).[43–45] The conceivable capacity evolution curve manipulated by pH is illustrated in Figure 2.2b.[19] Specifically, the curve can be divided into four regions: the activation region, the $H^+/Zn^{2+}$-dominated region, the $Mn^{2+}$-dominated region, and the "dead Mn" oxide aggregation region.[19] The corresponding dominating electrochemical reaction evolves with pH in each region (Figure 2.2c). Notably, in the $H^+/Zn^{2+}$-dominated region, the main contribution of the capacity is ion de-intercalation because the deposition potential of Mn at this region is high. The corresponding capacity evolution curve is relatively steady, which can be ascribed to the fixed number of ion storage sites in the electrode material. As the cycle proceeds, the pH value and capacity increase continuously with the dominant process changes to $Mn^{2+}$ deposition. Then the capacity attenuation occurs with the pH increase further accompanied by the formation of abundant inert "dead Mn". Such a dynamic model integrates various electrochemical mechanisms of ZMABs, which may contain, generalize and account for most of the present electrochemical behaviors of different manganese oxide-based cathodes.

(a)

$Mn^{2+} + 2H_2O \rightleftharpoons MnO_2 + 4H^+ + 2e^-$

cathode 1

$0.5Zn^{2+} + Mn^{2+} + 2H_2O \rightleftharpoons 0.5ZnMn_2O_4 + 4H^+ + e^-$

cathode 2

anode

$Zn^{2+} + 2e^- \rightarrow Zn$

Voltage (V vs. $Zn^{2+}/Zn$)

$Mn^{2+}$  $Zn^{2+}$  $H^+$

$E_{c1}$

$E_{c2}$

$E_a$

$\Delta E = E_c - E_a$
$= (E_c^\theta - E_a^\theta) + \dfrac{RT}{nF}\ln(\dfrac{C_H^4}{C_{Mn^{2+}}})$

(b)

Capacity

activation

(1) $H^+/Zn^{2+}$-dominated

(2) $Mn^{2+}$-dominated

(3) "dead Mn"

pH evolution

Cycles

(c)

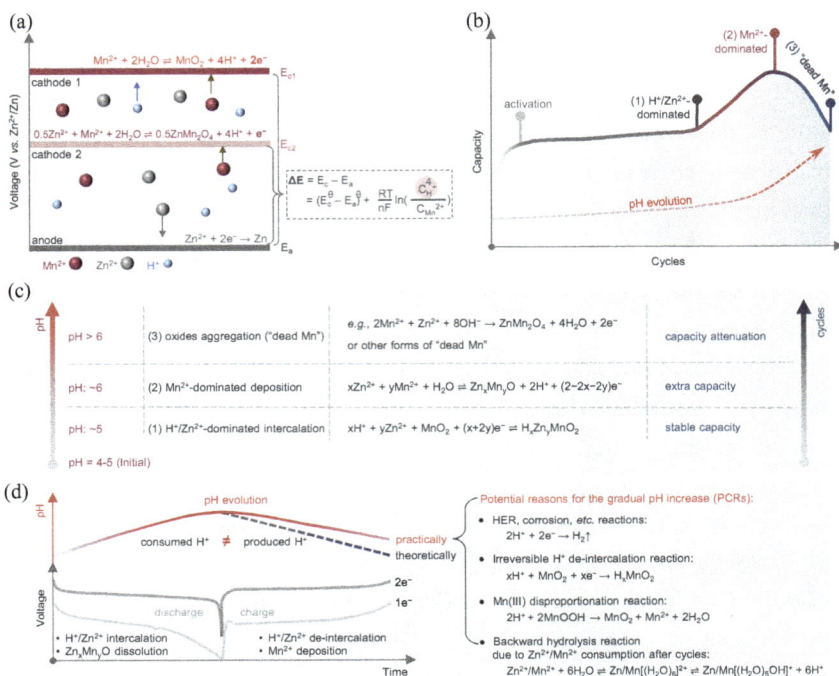

| pH | | reaction | | |
|---|---|---|---|---|
| pH > 6 | (3) oxides aggregation ("dead Mn") | e.g., $2Mn^{2+} + Zn^{2+} + 8OH^- \rightarrow ZnMn_2O_4 + 4H_2O + 2e^-$ or other forms of "dead Mn" | capacity attenuation | cycles |
| pH: ~6 | (2) $Mn^{2+}$-dominated deposition | $xZn^{2+} + yMn^{2+} + H_2O \rightleftharpoons Zn_xMn_yO + 2H^+ + (2-2x-2y)e^-$ | extra capacity | |
| pH: ~5 | (1) $H^+/Zn^{2+}$-dominated intercalation | $xH^+ + yZn^{2+} + MnO_2 + (x+2y)e^- \rightleftharpoons H_xZn_yMnO_2$ | stable capacity | |
| pH = 4-5 (Initial) | | | | |

(d)

pH evolution

consumed $H^+$ $\neq$ produced $H^+$

practically

theoretically

$2e^-$

$1e^-$

Voltage

discharge    charge

• $H^+/Zn^{2+}$ intercalation
• $Zn_xMn_yO$ dissolution

• $H^+/Zn^{2+}$ de-intercalation
• $Mn^{2+}$ deposition

Time

Potential reasons for the gradual pH increase (PCRs):

• HER, corrosion, etc. reactions:
  $2H^+ + 2e^- \rightarrow H_2\uparrow$
• Irreversible $H^+$ de-intercalation reaction:
  $xH^+ + MnO_2 + xe^- \rightarrow H_xMnO_2$
• Mn(III) disproportionation reaction:
  $2H^+ + 2MnOOH \rightarrow MnO_2 + Mn^{2+} + 2H_2O$
• Backward hydrolysis reaction due to $Zn^{2+}/Mn^{2+}$ consumption after cycles:
  $Zn^{2+}/Mn^{2+} + 6H_2O \rightleftharpoons Zn/Mn[(H_2O)_6]^{2+} \rightleftharpoons Zn/Mn[(H_2O)_5OH]^+ + 6H^+$

**Figure 2.2.** Zn-Mn aqueous chemistry *via* a pH clue and its associated PCRs. (a) The relationship between pH values and the potentials of $Mn^{2+}$ deposition based on the Nernst equation. (b) The evolution of capacity and charge storage mechanisms during cycles associated with pH change, and (c) the dominant reactions determined by the pH environment (pH < 7). (d) Summary of potential reasons for the PCRs and gradual pH increase.

As shown in Figure 2.2d, theoretically, there should be a perfect pH balance during discharge ($H^+$ consumption in the processes of $H^+/Zn^{2+}$ intercalation and $Zn_xMn_yO$ deposits dissolution) and charge ($H^+$ production in the processes of $H^+/Zn^{2+}$ de-intercalation and $Mn^{2+}$ deposition). However, practically, the consumed $H^+$ cannot be fully reproduced in a single charge/discharge cycle. Possible factors (PCRs) that cause the gradual pH increase are (see also Figure 2.2d): 1) side reactions such as hydrogen evolution reaction (HER), corrosion, *etc.*; 2) irreversible $H^+$ de-intercalation reaction; 3) Mn(III) disproportionation reaction; and 4) backward hydrolysis reaction due to $Zn^{2+}/Mn^{2+}$ consumption after cycles. In the typical electrolytes ($ZnSO_4$, $Zn(OTF)_2$, and $Zn(Ac)_2$, *etc.*) of ZMABs with pH values less than 7, oxygen evolution reaction ($2H_2O \rightarrow O_2 + 4H^+ + 4e^-$) would be restrained,[14,18] and HER ($2H^+ + 2e^- \rightarrow H_2$)[46]

and possible corrosion at the metallic Zn anode would consume the protons in the electrolyte. In addition, irreversible proton de-intercalation from the host and Mn(III) disproportionation reactions can also cause the rising of pH.[45,47–49] Furthermore, in the deep cycles, a large number of charge carriers ($Zn^{2+}/Mn^{2+}$) are deposited on the electrode surface, forming irreversible "dead Mn" oxide aggregation, which leads to the irreversible loss of $Zn^{2+}/Mn^{2+}$ in the electrolyte. So the original hydrolysis of $Zn^{2+}/Mn^{2+}$ is inhibited ($Zn^{2+}/Mn^{2+} + 6H_2O \leftrightarrow Zn/Mn[(H_2O)_6]^{2+} \leftrightarrow Zn/Mn[(H_2O)_5OH]^+ + 6H^+$), resulting in the backward hydrolysis reaction and further consumption of protons.[50–53] These reactions would inevitably result in irreversible proton circulation. As shown in Figure 2.2b and 2.2d, the pH cannot be restored to the initial state in a single cycle, eventually leading to an increase in pH as the number of loops continues. It can be inferred that these factors play different roles in systems with different electrolyte and electrode metrics, including pH, solvation, crystallinities, phases, and morphologies.

During the discharge process, the electrochemical behavior of ZMABs becomes much more complex. For ZMABs with a low-pH (< 3) electrolyte,[10,14,28] the high proton reactivity is conducive to facilitating electrolytic reaction activity and promoting electron transfer dynamics, which tends to double the electron transfer number from $MnO_2$ to $Mn^{2+}$. Therefore, the $2e^-$ electrolytic $Mn^{2+}$-dominated deposition behavior would be magnified (Figure 2.3a).[14,18,28,54] For most ZMABs in the mild pH environment, the dissolution for high-valence manganese (IV) oxides proceeds first with conversion to manganese (III), then disproportionate to low-valence manganese (II).[55–57] To ensure the charge neutrality in the crystal during the valence reduction, the $H^+$ and/or $Zn^{2+}$ intercalation would occur with high probability during the dissolution process. Notably, intermediate valence manganese (III) is characterized by disproportionated risk, which triggers the recycling of by-products (Figure 2.3b).[17,24,58] The disproportionation reaction might be inhibited by adding $Mn^{2+}$ ions in the electrolyte, which can balance the dissolution of $MnO_2$.[18] In addition, a thorough conversion of Mn (IV, solid) to Mn (II, liquid) requires a sufficient concentration of protons at voltages above 1.6 V or even above 2.0 V.[14,18,58–60] Many manganese oxides that carry zinc elements could be detected on the electrodes at low potentials, which might be the inexhaustive form originating from the first step of $Zn^{2+}$ intercalation or disproportionation.[17,59–61] On the other hand, the confirmation of the unstable MnOOH formed with intercalated $H^+$ ions requires, for example, using advanced *in situ* detection methods.[30] In addition to the current indirect explanation through the formation of basic zinc sulfate (BZS), atomic-level electron microscopy has been developed to demonstrate the intercalation of $H^+$ ions into manganese dioxide.[62]

In short, the decrease in the Mn valence might be the parallel result of $Zn^{2+}$ and/or $H^+$ intercalation in the mild pH environment. Partial dissolution behavior would be induced during this period. So the $H^+/Zn^{2+}$-dominated intercalation behavior develops (Figure 2.3b).[50,58] Specifically, some manganese dissolution may be triggered at the cathode-electrolyte interface due to the intercalation of $H^+$ and/or $Zn^{2+}$ in the initial cycle.[62–63] In the meantime, the nanostructured surface of the electrode material ensures close contact with the electrolyte to maintain a dynamic balance between the

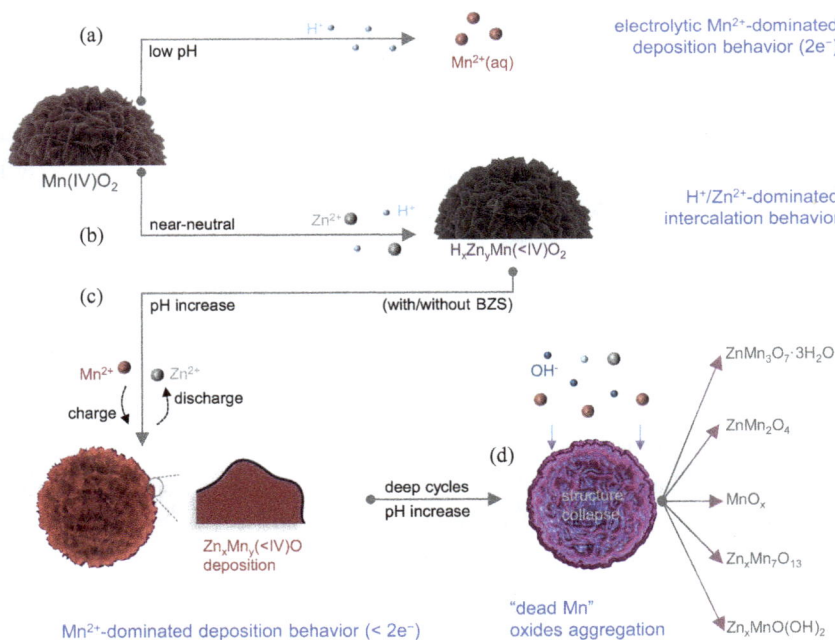

**Figure 2.3.** Illustration of the $H^+/Zn^{2+}/Mn^{2+}$-dominated electrochemical process at different pH environments. (a) Schematic diagram of electrolytic $Mn^{2+}$-dominated electrochemical behavior at low pH. (b) Schematic diagram of $H^+/Zn^{2+}$-dominated electrochemical behavior at near-neutral pH. (c) Schematic diagram of $Mn^{2+}$-dominated deposition at near-neutral pH. (d) Schematic diagram of "dead Mn" oxide aggregation and the morphological deterioration at near-neutral pH.

$H^+/Zn^{2+}$-dominated intercalation behavior and $Mn^{2+}$-dominated deposition behavior (Figure 2.3c). As a result, the capacity evolution could still satisfy a flat or mildly ascending trend, as observed in many reports.[41,64–66] It should be noted that the discharge process is usually accompanied by the formation of a large number of BZS (such as $Zn_4(OH)_6SO_4 \cdot 5H_2O$) by-products in the near-neutral electrolytes of $ZnSO_4$, $Zn(OTF)_2$, and $Zn(Ac)_2$, etc. During the charging process, the additional protons produced by $Mn^{2+}$ deposition react with BZS without causing a significant decrease in pH at the interface, accounting for the experimentally observed gradual BZS dissolution.[52]

As the cycles go on, the dynamic balance would be perturbated. Electrode structure would be gradually reshaped (Figure 2.3c) after booming dissolution/re-deposition.[27,63] The capacity would boost rapidly with the increased pH environment. Notably, due to different charge voltages and/or electrolyte environments during $Mn^{2+}$ deposition ($Mn^{2+}$-dominated deposition behavior, < 2e⁻), various forms of manganese depositions (noted as $Zn_xMn_yO$) would be formed, such as $ZnMn_2O_4$,[63] $ZnMn_3O_7 \cdot 3H_2O$,[42] $Zn_xMn_7O_{13}$,[67] $Zn_xMnO(OH)_2$,[27] and $MnO_x$.[22,68] As the thickness of the $Mn^{2+}$ deposition layer increases with cycles, the electrode conductivity declines because of the non-conductive nature of these oxides. The incomplete dissolution and the further raised pH environment during the deep cycles result in "dead Mn" oxide aggregation and poor contact between inner active materials and the electrolyte. The "dead Mn"-induced capacity attenuation and even structure collapse occur in the deep cycling profiles (Figure 2.3d).[19,48,56,63,69–70] Consequently, the capacity fluctuation with a rapid increase and decay becomes rationalized due to pH evolution in near-neutral pH electrolytes.

Re-analysis of the dynamic reactions can inspire new classification of the current complex behaviors for ZMABs, which would hopefully facilitate a re-understanding of the $H^+/Zn^{2+}/Mn^{2+}$-dominated charge-discharge behavior in various pH environments.

## 2.3 Quantitative Analyses and Proposed Metrics

Mn deposition provides new opportunities to boost the overall capacity, but quantifying its capacity contribution remains an unsolved issue. As discussed in the previous section, it has been observed in numerous reports that the capacity evolution curve exhibits rising or flat regions owing to the complex electrochemical reactions. As the pH environment evolves, abnormal capacity fluctuation is triggered by the competitive capacity contribution of the electrolyte and cathode material. Therefore, it is imperative to establish a more universal and systematic evaluation protocol to accurately evaluate the origins of capacity and rationally re-assess these unusual phenomena.

## 2.3.1 Protocol for Cathode-leading Zn-Mn Aqueous Batteries

Since the capacity rise originates from the electrolyte, monitoring the $Mn^{2+}$ concentration change would be the most directive approach. With the help of inductively coupled plasma-optical emission spectroscopy,[19,35] the electrolyte contribution ratio from $Mn^{2+}$ (*CfM*) can be determined *via* Eq. (2.1), also shown in Figure 2.4a:

$$CfM = \frac{(C_1 \times V_1 - C_2 \times V_2) \times nF}{3.6 \times m_0 \times C_s} \quad (2.1)$$

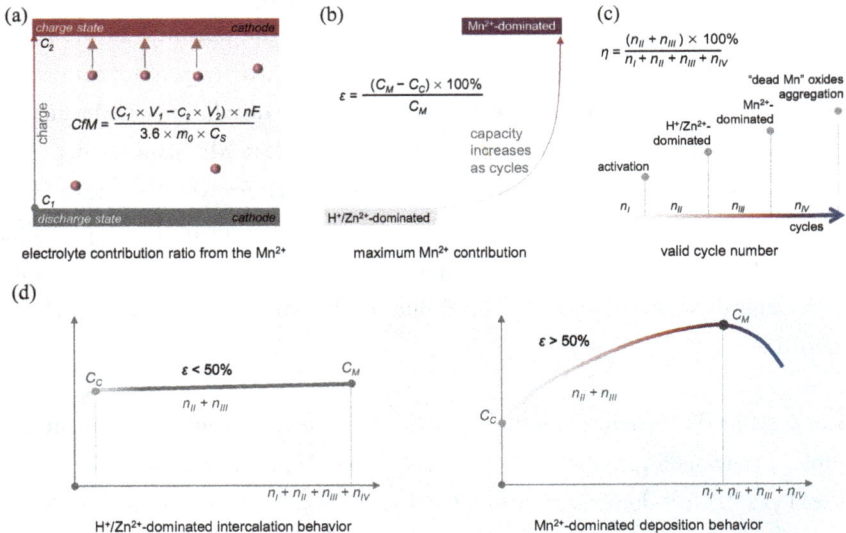

**Figure 2.4.** Evaluation protocol for the *cathode-leading* ZMABs. (a) The metrics of electrolyte contribution ratio from the $Mn^{2+}$ (*CfM*). (b) The metrics of the maximum $Mn^{2+}$ contribution $\varepsilon$. (c) The metrics of the valid cycle number $\eta$. (d) The evaluation methods for $\varepsilon$ and $\eta$ in the "stable capacity" (left) and "extra capacity" (right) situations. $n_I$, $n_{II}$, $n_{III}$, and $n_{IV}$ represent the cycle numbers of the activation region, $H^+$/ $Zn^{2+}$-dominated region, $Mn^{2+}$-dominated region, and "dead Mn" oxide aggregation region, respectively.

where $C_1$ and $C_2$ (mol L$^{-1}$) represent the initial and detected concentration, respectively; $V_1$ and $V_2$ (L) correspond to the volume of electrolyte at the initial and detected states, respectively; $F$ is Faraday constant; $m_0$ (mg) is the mass of the active material; and $C_s$ stands for the overall specific capacity corresponding to cycle. Despite different deposition forms of Mn$^{2+}$ (Zn$_x$Mn$_y$O), such as MnO$_2$,[23] ZnMn$_2$O$_4$,[19] Zn$_x$MnO(OH)$_2$,[27,71] and Zn$_x$Mn$_3$O$_7$,[67] the extra capacity contribution is all from the electrolyte, which can be evaluated by $CfM$ using different electron transfer numbers $n$. This method allows one to explicitly quantify the actual capacity of the Mn-based cathode material, avoiding the "fake" high capacity caused by the electrolyte.

To simplify the test method and calculation, the maximum Mn$^{2+}$ contribution, $\varepsilon$, is used to estimate the contribution of Mn$^{2+}$ to the total capacity (Figure 2.4b):

$$\varepsilon = \frac{(C_M - C_C) \times 100\%}{C_M} \tag{2.2}$$

where the capacity ($C_C$) after activation is defined as the initial capacity with low Mn$^{2+}$ contributions. $C_M$ represents the maximum capacity. A small $\varepsilon$ indicates low capacity involving the Mn$^{2+}$ deposition and good cathode performance, suggesting good ionic intercalation capability of the original cathode.

In association with the different stages, the parameter of valid cycle number, $\eta$, can be used to indicate the valid operation for an electrode during cycling, in which the abnormal cycles in the activation and "dead Mn" oxide aggregation regions are excluded (Figure 2.4c)[19]:

$$\eta = \frac{(n_{II} + n_{III}) \times 100\%}{n_I + n_{II} + n_{III} + n_{IV}} \tag{2.3}$$

where $n_I$, $n_{II}$, $n_{III}$ and $n_{IV}$ refer to the corresponding cycling numbers in the four regions, respectively. The evaluation methods for the new metrics of $\varepsilon$ and $\eta$, with "stable capacity" (left) and "extra capacity" (right) situations, are shown schematically in Figure 2.4d.

The combinatory or preferential electrochemical behavior, including H$^+$/Zn$^{2+}$-dominated and Mn$^{2+}$-dominated fluctuations, are prevalent in the reported ZMABs at near-neutral electrolytes.[8-9,13-15,17-20,22-23,26-27,35-37,39-41,48,65-66,68,72-75] In particular, the

H+/Zn²⁺-dominated electrochemical behavior with flat or mildly ascending characteristics usually exhibits low Mn contribution ($\varepsilon < 50\%$). Such a system can be identified as *cathode-leading* ZMABs, of which the Mn²⁺-dominated region is generally absent or very short. Conversely, Mn²⁺-dominated materials typically present steeply rising capacity curves ($\varepsilon > 50\%$) with short or absent H+/Zn²⁺-dominated regions. It should be noted that the "electrolytic Mn²⁺/MnO₂ system" with Mn²⁺ electrolyte at low pH ($< 3$) can be considered as an electrolytic Mn²⁺-dominated 2e⁻ electrochemical process.[10,14,17–18,27–28,59]

## 2.3.2 *Protocol for Electrolyte-leading Zn-Mn Aqueous Batteries*

In addition to the metrics described above to re-evaluate the electrode performance, the performance of electrolyte that provides capacity through conversion reactions should also be effectively evaluated. For the electrolyte in an assembled battery, we introduce a rated performance index, such as rated capacity $C_r$:

$$C_r = \frac{x \times y \times nF}{3.6} \tag{2.4}$$

where $x$ is Mn²⁺ concentration (mol L⁻¹), and $y$ corresponds to electrolyte volume (L). For a given volume ($V$, L), electrolyte mass ($m$, kg), and electrode area ($s$, cm²), the corresponding formulas for the rated volume capacity $C_{VC}$, rated area capacity $C_{AC}$, rated volumetric energy density $C_{VED}$, and rated gravimetric energy density $C_{GED}$ are depicted in Figure 2.5a. In particular cases of 0.1~3 mol L⁻¹ Mn²⁺ additive, the calculation results of rated performance are shown in Table 2.1.

The voltage platform for one- and two-electron transfer can be determined as *ca.* 2.0 V and 1.4 V, respectively. Taking "0.1 mol L⁻¹, 100 μL Mn²⁺ additive for one-electron transfer" as an example for the calculation of $C_{VED}$:

$$C_{VED} = \frac{0.1 \times 100 \times 10^{-6} \times 96320 \times 1.4}{3.6 \times 100 \times 10^{-6}} = 3746\ \text{Wh L}^{-1} \tag{2.5}$$

Furthermore, the percentage of the actual performance to the theoretical value represents electrolyte conversion efficiency, which depicts the effective utilization of the electrolyte (Figure 2.5b).

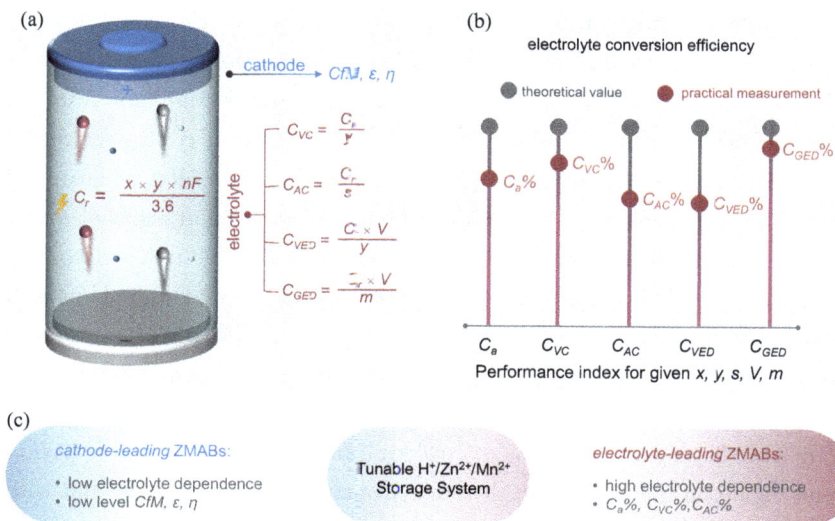

(a)

$$C_r = \frac{x \times y \times nF}{3.6}$$

cathode $\longrightarrow$ C$fM$, $\varepsilon$, $\eta$

$$C_{VC} = \frac{C_r}{y}$$

$$C_{AC} = \frac{C_r}{s}$$

$$C_{VED} = \frac{C_r \times V}{y}$$

$$C_{GED} = \frac{C_r \times V}{m}$$

(b) electrolyte conversion efficiency

⬤ theoretical value　⬤ practical measurement

$C_a\%$　$C_{VC}\%$　$C_{AC}\%$　$C_{VED}\%$　$C_{GED}\%$

$C_a$　$C_{VC}$　$C_{AC}$　$C_{VED}$　$C_{GED}$

Performance index for given $x$, $y$, $s$, $V$, $m$

(c)

cathode-leading ZMABs:
- low electrolyte dependence
- low level C$fM$, $\varepsilon$, $\eta$

Tunable H$^+$/Zn$^{2+}$/Mn$^{2+}$ Storage System

electrolyte-leading ZMABs:
- high electrolyte dependence
- $C_a\%$, $C_{VC}\%$, $C_{AC}\%$

**Figure 2.5.** Evaluation protocol for the *electrolyte-leading* ZMABs. (a) The metrics of the rated capacity $C_a$, rated volume capacity $C_{VC}$, rated area capacity $C_{AC}$, rated volumetric energy density $C_{VED}$, and rated gravimetric energy density $C_{GED}$. (b) The performance $C_a$, $C_{VC}$, $C_{AC}$, $C_{VED}$, and $C_{GED}$. (c) Comparison of the electrolyte performance in *cathode-leading* ZMABs and *electrolyte-leading* ZMABs.

**Table 2.1.** Rated performance induced by Mn$^{2+}$ additive.

| Mn$^{2+}$ Additive Concentration (mol L$^{-1}$) | $n$ (e$^-$) | $C_r$ (mAh) | $C_{VC}$ (mAh L$^{-1}$) | $C_{AC}$ (mAh cm$^{-2}$) | $C_{VED}$ (Wh L$^{-1}$) | $C_{GED}$ (Wh kg$^{-1}$) |
|---|---|---|---|---|---|---|
| 0.1 | 1 e$^-$ | 2676y | 2676 | 2676y/s | 3746 | 3746y/m |
|  | 2 e$^-$ | 5351y | 5351 | 5351y/s | 10702 | 10702y/m |
| 0.2 | 1 e$^-$ | 5351y | 5351 | 5351y/s | 7492 | 7492y/m |
|  | 2 e$^-$ | 10702y | 10702 | 10702y/s | 21404 | 21404y/m |
| 0.5 | 1 e$^-$ | 13378y | 13378 | 13378y/s | 18729 | 18729y/m |
|  | 2 e$^-$ | 26756y | 26756 | 26756y/s | 53511 | 53511y/m |
| 1 | 1 e$^-$ | 26756y | 26756 | 26756y/s | 37458 | 37458y/m |
|  | 2 e$^-$ | 53511y | 53511 | 53511y/s | 107022 | 107022y/m |
| 2 | 1 e$^-$ | 53511y | 53511 | 53511y/s | 74916 | 74916y/m |
|  | 2 e$^-$ | 107022y | 107022 | 107022y/s | 214044 | 214044y/m |
| 3 | 1 e$^-$ | 80267y | 80267 | 80267y/s | 112373 | 112373y/m |
|  | 2 e$^-$ | 160533y | 160533 | 160533y/s | 321067 | 321067y/m |

*Note*: Data at the upper and lower parts correspond to one-electron transfer and two-electron transfer, respectively.

Consequently, a universal and systematic protocol can be established according to the different electrochemical behaviors of ZMABs. For the *cathode-leading* ZMABs (with $H^+/Zn^{2+}$-dominated charge storage process), the contribution ratio from the $Mn^{2+}$ (*CfM*), maximum $Mn^{2+}$ contribution ratio ($\varepsilon$), and effective cycling percentage ($\eta$) can be adopted to quantify the capacity contribution of the electrode material with an $H^+/Zn^{2+}$-dominated and/or $Mn^{2+}$-dominated deposition behavior in the mild pH environment. Typically, the *cathode-leading* ZMABs exhibit a low electrolyte dependence, so the discharge capacity and output voltage may be limited but with relatively stable cyclability. For the *electrolyte-leading* ZMABs (with $Mn^{2+}$-dominated deposition process), $C_a$, $C_{VC}$, $C_{AC}$, $C_{VED}$, and $C_{GED}$ can be employed to describe the $Mn^{2+}$ contribution in acidic and near-neutral electrolytes. In contrast, the *electrolyte-leading* ZMABs, especially for the electrolytic $Mn^{2+}/MnO_2$ deposition behavior, depend highly on the electrolyte; they may display high energy output but with inadequate cyclability. Ideally, if a system that combines these two features (*i.e.*, $H^+/Zn^{2+}$ and $Mn^{2+}$ contributions) can be realized, the advantages from both the electrode and electrolyte would be fully utilized. Such a tunable $H^+/Zn^{2+}/Mn^{2+}$ storage system (Figure 2.5c) will render considerable and stable capacity and high energy density.

## 2.4 Overall Conclusions and Suggestions

The current consensus focuses on the energy storage mechanisms of ZMABs, including ion intercalation and deposition reactions. We suggest that the above reaction mechanisms usually coexist and evolve alternately with a strong dependence on the pH environment or proton reactivity (Figure 2.6a and 2.6b). In mild pH (3–6) electrolytes, the $Mn^{2+}$ deposition/dissolution (usually single-electron transfer) would mix with zinc ion de-insertion/insertion, and the corresponding dominant reaction changes with the variation of pH value and PCRs, leading to abnormal capacity fluctuations. While electrolytes with low pH (< 3) benefit the dissolution of the deposited manganese or the original $MnO_2$ cathode material, an enhanced dissolution behavior occurs, corresponding to the favorable two-electron transfer reaction.[10,14,43] Hence, in evaluating the electrochemistry of ZMABs, and probably other aqueous zinc batteries, one needs to go down to the atomic scale and correlate the microscopic charge storage process to the macroscopic electrochemical performances.

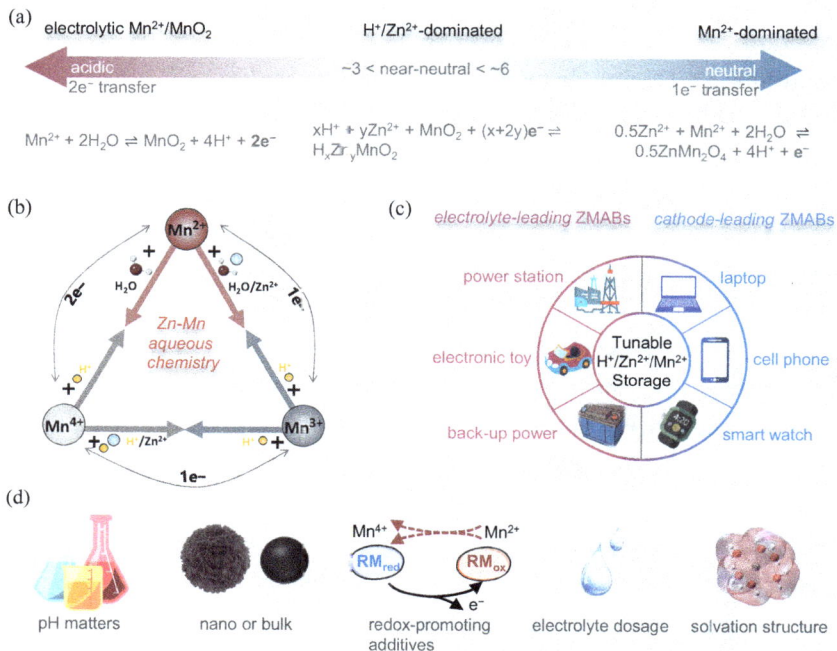

**Figure 2.6.** Overall conclusion and perspective toward future commercialization. (a) The summary of pH-dependent ZMABs. (b) Zn-Mn aqueous chemistry with the relationship between Mn (II) ↔ Mn (III) ↔ Mn (IV) conversion and proton consumption/production. (c) The potential device application scenarios of ZMABs. (d) Proposed future research directions in regulating ZMABs.

When it comes to ZMABs, it always does not suffice to consider only the mass of active cathode material in calculating specific capacity and energy density. Especially when the mass of the electrolyte is much larger than that of the cathode material, an artificial high capacity is inevitable. Therefore, we put forward new metrics to re-evaluate the capacity of ZMABs, including the following: the electrolyte contribution ratio from $Mn^{2+}$ (*CfM*), maximum $Mn^{2+}$ contribution ratio ($\varepsilon$), effective cycling percentage ($\eta$), the rated capacity $C_a$, rated volume capacity $C_{VC}$, rated area capacity $C_{AC}$, rated volumetric energy density $C_{VED}$, and rated gravimetric energy density $C_{GED}$.

In assessing the performance of battery devices in practical applications, the total mass and volume of the cell need to be considered (Figure 2.6c). Accordingly, ZMABs with different mechanisms may fit different application scenarios. For example, the *cathode-leading* ZMABs display a low demand for electrolyte content, which can power portable devices with a long lifespan. On the other hand, the *electrolyte-leading* ZMABs, including the electrolytic Zn-Mn battery, can be adopted for mass-insensitive devices, such as electronic toys, backup power, and power stations.

To boost the progress of ZMABs as a competitive energy storage technology, we would like to highlight the following points for future research (Figure 2.6d):

(1) *pH matters*. It is known that the pH value controls the PCRs and electrochemical behavior of the cathode part and $Mn^{2+}$ electrolyte additives. A mild pH value (3-6) is conducive to the $Mn^{2+}$ deposition, but an excess of hydroxide ions is not beneficial to Mn dissolution from the point of view of the reaction equilibrium.[43] That is why electrolytic Zn-Mn batteries usually work in a high proton concentration environment (pH < 3) to ensure complete dissolution ($MnO_2 \rightarrow Mn^{2+}$ dissolution reaction at above 1.6 V or even 2.0 V[59–60]). In other words, an increased pH would lower the deposition/dissolution capacity contribution, shifting the electrochemical mechanism to the ion intercalation region.[60] Hence, controlling the pH environment and associated PCRs in ZMABs is critical to achieving desired performances.

(2) *Cathode nano-engineering*. There are several advantages of nano-design for the cathode. First, a nano-engineered crystal structure (for example, cation or molecular pillaring effect) is favorable for $H^+/Zn^{2+}$ ion intercalation, which can prevent unwanted dissolution and volume expansion of the cathode material during charging/discharging. Consequently, the $H^+/Zn^{2+}$-dominated intercalation and/or $Mn^{2+}$-dominated deposition mechanisms can be balanced. [11,39–40,64–65,76–77] Conversely, a bulky crystal structure suppresses the intercalation due to weakened nanoscale effects, leading to an increase in the proportion of $Mn^{2+}$-dominated behavior.[13,22] Second, a nanostructure with a high specific surface area and porosity is preferable because it improves the contact between electrode and electrolyte and provides sufficient anchoring sites for both $H^+/Zn^{2+}$ intercalation and $MnO_2$ nucleation. In comparison, on a low-surface-area surface, the ion-embedding behavior is restricted. And a thick Mn oxide layer

might be formed outside the active sites, which is detrimental to the $Mn^{2+}$-dominated behavior due to their poor conductivity.[14,17,70,78–79]

(3) *Redox-promoting additives.* At present, anionic chemistry has been demonstrated as an effective method to achieve reversible $Mn^{2+}/MnO_2$ conversion with high efficiency and without "dead Mn". For instance, the addition of $Ac^-$ can reduce the barrier for both the dissolution and deposition process.[43,69] $PO_4^{3-}$ can act as a proton reservoir to maintain the pH value for reversible $Mn^{2+}/MnO_2$ conversion.[18] Moreover, catalysts such as poly(vinylpyrrolidone),[80] $I_2$,[9,81] $Br_2$,[82–83], and $Ni^{2+}$[28] have been employed to boost the $Mn^{2+}/MnO_2$ redox chemistry. However, when redox mediator additives are introduced, their effects on the zinc stripping/plating on the anode side should also be carefully investigated, as the reversibility of Zn plating/striping can be the major factor in determining the full battery cycle life.

(4) *Electrolyte dosage.* Most of the ZMABs were evaluated with excess electrolyte, which stabilizes the pH values. Under flooded conditions, the effect of micro-environmental change could be minimized due to a sufficient supply of electrolytes, which may correspond to fewer side reactions.[84] That might be the reason for the reported "unprecedented performance" of Mn-based cathodes. However, with limited electrolyte, the local pH value would inevitably increase due to proton storage and/or side reactions such as HER and Zn corrosion. It is easy to speculate that limited electrolyte dosage would deteriorate the electrochemical performances of the full cell. Therefore, it is necessary to consider the amount of electrolyte through the proposed metrics ($CfM$, $C_a$, $C_{GED}$, etc.) for a fair evaluation of battery performance.

(5) *Solvation structure.* In recent years, regulating $Zn^{2+}$ solvation shells of electrolytes has proven effective in suppressing dendrite growth and $H_2O$-induced parasitic side reactions at the solid-liquid interface.[85–86] However, many studies focus only on the Zn anode. Limited attention has been paid to the possible influence of the solvation structure on the cathode side, for example, Mn deposition behavior and the intercalation/de-intercalation kinetics of solvated Zn ions. These fundamental issues should be considered in future research to understand the effect of electrolyte additives on full battery performances. Fortunately, preliminary investigations have been conducted on other cathode materials, such

as Prussian blue[87] and vanadium-based cathodes.[88] Given the high sensitivity of ZMABs to the electrolyte, exploring the $Mn^{2+}$ solvation structure and even cathode-electrolyte behaviors should be of great significance.

The above discussions imply that the tunable $H^+/Zn^{2+}/Mn^{2+}$ storage system is promising in both scientific exploration and industrial applications. This tunable $H^+/Zn^{2+}/Mn^{2+}$ storage system can achieve either high-efficiency electrolyte conversion or high capacity from electrode materials. As a result, different application scenarios, such as intelligent electronics, flexible and wearable devices, electric vehicles, household energy storage, and power stations, can be unlocked. However, the trade-off between inhibiting "dead Mn" formation at the cathode side and HER/corrosion at the zinc anode side remains challenging. Generally, electrolytes with low pH help achieve high energy, but the stability of zinc anodes needs extra attention. Ingenious electrolyte activity regulation and metal anode interface engineering would be essential in constructing a reliable $H^+/Zn^{2+}/Mn^{2+}$ storage system.[18,89–90] Despite the increasing number of publications and commercial prospects, ZMABs are still in the laboratory stage. Therefore, a reliable and systematic evaluation protocol is indispensable before industrial applications. We hope this chapter can shed light on the development of ZMABs.

## Acknowledgements

The authors sincerely acknowledge financial support from the National Natural Science Foundation of China (NSFC Grant No. 21571080, 22109029, and 22279023), Natural Science Foundation of Shanghai (22ZR1403600), International Center of Future Science, Jilin University, Changchun, P. R. China (ICFS Seed Funding for Young Researchers), and the Singapore Ministry of Education by Academic Research Fund Tier 2 (MOE-T2EP50121-0006).

## References

[1] D. Chao, W. Zhou, F. Xie, C. Ye, H. Li, M. Jaroniec, S.-Z. Qiao, *Sci. Adv.* **2020**, 6, eaba4098.

[2] P. Ruan, S. Liang, B. Lu, H. J. Fan, J. Zhou, *Angew. Chem. Int. Ed.* **2022**, 61, e202200598.

[3] Y. Liang, Y. Yao, *Nat. Rev. Mater.* **2022**, 8, 109–122.

[4] D. Chen, M. Lu, D. Cai, H. Yang, W. Han, *J. Energy Chem.* **2021**, 54, 712.

[5] T. Yamamoto, T. Shoji, *Inorganica Chimica Acta* **1986**, 117, L27.

[6] V. Mathew, B. Sambandam, S. Kim, S. Kim, S. Park, S. Lee, M. H. Alfaruqi, V. Soundharrajan, S. Islam, D. Y. Putro, J.-Y. Hwang, Y.-K. Sun, J. Kim, *ACS Energy Lett.* **2020**, 5, 2376.

[7] H. Pan, Y. Shao, P. Yan, Y. Cheng, K. S. Han, Z. Nie, C. Wang, J. Yang, X. Li, P. Bhattacharya, K. T. Mueller, J. Liu, *Nat. Energy* **2016**, 1, 16039.

[8] Y. Jin, L. Zou, L. Liu, M. H. Engelhard, R. L. Patel, Z. Nie, K. S. Han, Y. Shao, C. Wang, J. Zhu, H. Pan, J. Liu, *Adv. Mater.* **2019**, 31, e1900567.

[9] C. Xie, T. Li, C. Deng, Y. Song, H. Zhang, X. Li, *Energy Environ. Sci.* **2020**, 13, 135.

[10] W. Chen, G. Li, A. Pei, Y. Li, L. Liao, H. Wang, J. Wan, Z. Liang, G. Chen, H. Zhang, J. Wang, Y. Cui, *Nat. Energy* **2018**, 3, 428.

[11] N. Zhang, F. Cheng, J. Liu, L. Wang, X. Long, X. Liu, F. Li, J. Chen, *Nat. Commun.* **2017**, 8, 405.

[12] B. Lee, H. R. Lee, H. Kim, K. Y. Chung, B. W. Cho, S. H. Oh, *Chem. Commun.* **2015**, 51, 9265.

[13] Y. Fu, Q. Wei, G. Zhang, X. Wang, J. Zhang, Y. Hu, D. Wang, L. Zuin, T. Zhou, Y. Wu, S. Sun, *Adv. Energy Mater.* **2018**, 8, 1801445.

[14] D. Chao, W. Zhou, C. Ye, Q. Zhang, Y. Chen, L. Gu, K. Davey, S. Z. Qiao, *Angew. Chem. Int. Ed.* **2019**, 58, 7823.

[15] C. Qiu, X. Zhu, L. Xue, M. Ni, Y. Zhao, B. Liu, H. Xia, *Electrochim. Acta* **2020**, 351, 136445.

[16] X. Gao, H. Wu, W. Li, Y. Tian, Y. Zhang, H. Wu, L. Yang, G. Zou, H. Hou, X. Ji, *Small* **2020**, 16, e1905842.

[17] X. Shen, X. Wang, Y. Zhou, Y. Shi, L. Zhao, H. Jin, J. Di, Q. Li, *Adv. Funct. Mater.* **2021**, 31, 2101579.

[18] Y. Liu, Z. Qin, X. Yang, J. Liu, X.-X. Liu, X. Sun, *ACS Energy Lett.* **2022**, 7, 1814.

[19] H. Yang, W. Zhou, D. Chen, J. Liu, Z. Yuan, M. Lu, L. Shen, V. Shulga, W. Han, D. Chao, *Energy Environ. Sci.* **2022**, 15, 1106.

[20] B. Wu, G. Zhang, M. Yan, T. Xiong, P. He, L. He, X. Xu, L. Mai, *Small* **2018**, 14, e1703850.

[21] M. Chamoun, W. R. Brant, C.-W. Tai, G. Karlsson, D. Noréus, *Energy Stor. Mater.* **2018**, 15, 351.

[22] V. Soundharrajan, B. Sambandam, S. Kim, S. Islam, J. Jo, S. Kim, V. Mathew, Y.-k. Sun, J. Kim, *Energy Stor. Mater.* **2020**, 28, 407.

[23] Y. Liao, H.-C. Chen, C. Yang, R. Liu, Z. Peng, H. Cao, K. Wang, *Energy Stor. Mater.* **2022**, 44, 508.

[24] X. Ye, D. Han, G. Jiang, C. Cui, Y. Guo, Y. Wang, Z. Zhang, Z. Weng, Q. Yang, *Energy Environ. Sci.* **2023**, 16, 1016–1023.

[25] W. Sun, F. Wang, S. Hou, C. Yang, X. Fan, Z. Ma, T. Gao, F. Han, R. Hu, M. Zhu, C. Wang, *J. Am. Chem. Soc.* **2017**, 139, 9775.

[26] Q. Zhao, X. Chen, Z. Wang, L. Yang, R. Qin, J. Yang, Y. Song, S. Ding, M. Weng, W. Huang, J. Liu, W. Zhao, G. Qian, K. Yang, Y. Cui, H. Chen, F. Pan, *Small* **2019**, 15, e1904545.

[27] H. Chen, C. Dai, F. Xiao, Q. Yang, S. Cai, M. Xu, H. J. Fan, S. J. Bao, *Adv. Mater.* **2022**, 34, e2109092.

[28] D. Chao, C. Ye, F. Xie, W. Zhou, Q. Zhang, Q. Gu, K. Davey, L. Gu, S. Z. Qiao, *Adv. Mater.* **2020**, 32, e2001894.

[29] M. Han, L. Qin, Z. Liu, L. Zhang, X. Li, B. Lu, J. Huang, S. Liang, J. Zhou, *Mater. Today Energy* **2021**, 20, 100626.

[30] B. Sambandam, V. Mathew, S. Kim, S. Lee, S. Kim, J. Y. Hwang, H. J. Fan, J. Kim, *Chem* **2022**, 8, 924.

[31] J. Huang, Z. Guo, X. Dong, D. Bin, Y. Wang, Y. Xia, *Sci. Bull.* **2019**, 64, 1780.

[32] D. Wu, L. M. Housel, S. T. King, Z. R. Mansley, N. Sadique, Y. Zhu, L. Ma, S. N. Ehrlich, H. Zhong, E. S. Takeuchi, A. C. Marschilok, D. C. Bock, L. Wang, K. J. Takeuchi, *J. Am. Chem. Soc.* **2022**, 144, 23405.

[33] N. Zhang, F. Cheng, Y. Liu, Q. Zhao, K. Lei, C. Chen, X. Liu, J. Chen, *J. Am. Chem. Soc.* **2016**, 138, 12894.

[34] C. Zhu, G. Fang, S. Liang, Z. Chen, Z. Wang, J. Ma, H. Wang, B. Tang, X. Zheng, J. Zhou, *Energy Stor. Mater.* **2020**, 24, 394.

[35] J. Huang, J. Zeng, K. Zhu, R. Zhang, J. Liu, *Nano-micro Lett.* **2020**, 12, 110.

[36] X. Z. Zhai, J. Qu, S. M. Hao, Y. Q. Jing, W. Chang, J. Wang, W. Li, Y. Abdelkrim, H. Yuan, Z. Z. Yu, *Nano-micro Lett.* **2020**, 12, 56.

[37] Z. Yang, X. Pan, Y. Shen, R. Chen, T. Li, L. Xu, L. Mai, *Small* **2022**, 18, e2107743.

[38] Q. Xie, G. Cheng, T. Xue, L. Huang, S. Chen, Y. Sun, M. Sun, H. Wang, L. Yu, *Mater. Today Energy* **2022**, 24, 100934.

[39] J. Huang, Z. Wang, M. Hou, X. Dong, Y. Liu, Y. Wang, Y. Xia, *Nat. Commun.* **2018**, 9, 2906.

[40] Y. Zhao, P. Zhang, J. Liang, X. Xia, L. Ren, L. Song, W. Liu, X. Sun, *Energy Stor. Mater.* **2022**, 47, 424.

[41] X.-Z. Zhai, J. Qu, J. Wang, W. Chang, H.-J. Liu, Y.-H. Liu, H. Yuan, X. Li, Z.-Z. Yu, *Energy Stor. Mater.* **2021**, 42, 753.

[42]  S. Islam, M. H. Alfaruqi, D. Y. Putro, S. Park, S. Kim, S. Lee, M. S. Ahmed, V. Mathew, Y. K. Sun, J. Y. Hwang, J. Kim, *Adv. Sci.* **2021**, 8, 2002636.

[43]  X. Zeng, J. Liu, J. Mao, J. Hao, Z. Wang, S. Zhou, C. D. Ling, Z. Guo, *Adv. Energy Mater.* **2020**, 10, 1904163.

[44]  A. S. Poyraz, J. Laughlin, Z. Zec, *Electrochim. Acta* **2019**, 305, 423.

[45]  P. Oberholzer, E. Tervoort, A. Bouzid, A. Pasquarello, D. Kundu, *ACS Appl. Mater. Interfaces* **2019**, 11, 674.

[46]  Y. Kim, Y. Park, M. Kim, J. Lee, K. J. Kim, J. W. Choi, *Nat. Commun.* **2022**, 13, 2371.

[47]  B. Lee, H. R. Seo, H. R. Lee, C. S. Yoon, J. H. Kim, K. Y. Chung, B. W. Cho, S. H. Oh, *ChemSusChem* **2016**, 9, 2948.

[48]  L. Li, T. K. A. Hoang, J. Zhi, M. Han, S. Li, P. Chen, *ACS Appl. Mater. Interfaces* **2020**, 12, 12834.

[49]  R. C. Paul Mulvaney, Franz Grieser and Dan Meise, *J. Phys. Chem.* **1990**, 94, 8339–8345.

[50]  O. Fitz, C. Bischoff, M. Bauer, H. Gentischer, K. P. Birke, H. M. Henning, D. Biro, *ChemElectroChem* **2021**, 8, 3555.

[51]  I. Aguilar, P. Lemaire, N. Ayouni, E. Bendadesse, A. V. Morozov, O. Sel, V. Balland, B. Limoges, A. M. Abakumov, E. Raymundo-Piñero, A. Slodczyk, A. Canizarès, D. Larcher, J.-M. Tarascon, *Energy Stor. Mater.* **2022**, 53, 238.

[52]  I. A. Rodríguez-Pérez, H. J. Chang, M. Fayette, B. M. Sivakumar, D. Choi, X. Li, D. Reed, *J. Mater. Chem. A* **2021**, 9, 20766.

[53]  M. Mateos, N. Makivic, Y. S. Kim, B. Limoges, V. Balland, *Adv. Energy Mater.* **2020**, 10.

[54]  C. J. Clarke, G. J. Browning, S. W. Donne, *Electrochim. Acta* **2006**, 51, 5773.

[55]  J. Yang, J. Cao, Y. Peng, W. Yang, S. Barg, Z. Liu, I. A. Kinloch, M. A. Bissett, R. A. W. Dryfe, *ChemSusChem* **2020**, 13, 4103.

[56]  S. Zhao, B. Han, D. Zhang, Q. Huang, L. Xiao, L. Chen, D. G. Ivey, Y. Deng, W. Wei, *J. Mater. Chem. A* **2018**, 6, 5733.

[57]  H. Moon, K. H. Ha, Y. Park, J. Lee, M. S. Kwon, J. Lim, M. H. Lee, D. H. Kim, J. H. Choi, J. H. Choi, K. T. Lee, *Adv. Sci.* **2021**, 8, 2003714.

[58]  H. Lv, Y. Song, Z. Qin, M. Zhang, D. Yang, Q. Pan, Z. Wang, X. Mu, J. Meng, X. Sun, X.-X. Liu, *Chem. Eng. J.* **2022**, 430, 133064.

[59]  G. Li, W. Chen, H. Zhang, Y. Gong, F. Shi, J. Wang, R. Zhang, G. Chen, Y. Jin, T. Wu, Z. Tang, Y. Cui, *Adv. Energy Mater.* **2020**, 10, 1902085.

[60]  Z. Liu, Y. Yang, S. Liang, B. Lu, J. Zhou, *Small Struct.* **2021**, 2, 2100119.

[61]  M. H. Alfaruqi, V. Mathew, J. Gim, S. Kim, J. Song, J. P. Baboo, S. H. Choi, J. Kim, *Chem. Mater.* **2015**, 27, 3609.

[62] Y. Yuan, R. Sharpe, K. He, C. Li, M. T. Saray, T. Liu, W. Yao, M. Cheng, H. Jin, S. Wang, K. Amine, R. Shahbazian-Yassar, M. S. Islam, J. Lu, *Nat. Sustain.* **2022**, 5, 890.

[63] S. J. Kim, D. Wu, N. Sadique, C. D. Quilty, L. Wu, A. C. Marschilok, K. J. Takeuchi, E. S. Takeuchi, Y. Zhu, *Small* **2020**, 16, e2005406.

[64] D. Wang, L. Wang, G. Liang, H. Li, Z. Liu, Z. Tang, J. Liang, C. Zhi, *ACS Nano* **2019**, 13, 10643.

[65] T. Sun, Q. Nian, S. Zheng, J. Shi, Z. Tao, *Small* **2020**, 16, e2000597.

[66] Q. Gou, H. Luo, Y. Zheng, Q. Zhang, C. Li, J. Wang, O. Odunmbaku, J. Zheng, J. Xue, K. Sun, M. Li, *Small* **2022**, e2201732.

[67] H. Li, H. Yao, X. Sun, C. Sheng, W. Zhao, J. Wu, S. Chu, Z. Liu, S. Guo, H. Zhou, *Chem. Eng. J.* **2022**, 446, 137205.

[68] R. Liang, J. Fu, Y.-P. Deng, Y. Pei, M. Zhang, A. Yu, Z. Chen, *Energy Stor. Mater.* **2021**, 36, 478.

[69] Z. Zhong, J. Li, L. Li, X. Xi, Z. Luo, G. Fang, S. Liang, X. Wang, *Energy Stor. Mater.* **2022**, 46, 165.

[70] J. Lei, Y. Yao, Z. Wang, Y.-C. Lu, *Energy Environ. Sci.* **2021**, 14, 4418.

[71] H. Chen, S. Cai, Y. Wu, W. Wang, M. Xu, S. J. Bao, *Mater. Today Energy* **2021**, 20, 100646.

[72] W. Li, X. Gao, Z. Chen, R. Guo, G. Zou, H. Hou, W. Deng, X. Ji, J. Zhao, *Chem. Eng. J.* **2020**, 402, 125509.

[73] M. H. Alfaruqi, S. Islam, D. Y. Putro, V. Mathew, S. Kim, J. Jo, S. Kim, Y.-K. Sun, K. Kim, J. Kim, *Electrochim. Acta* **2018**, 276, 1.

[74] G. Wang, Y. Wang, B. Guan, J. Liu, Y. Zhang, X. Shi, C. Tang, G. Li, Y. Li, X. Wang, L. Li, *Small* **2021**, 17, e2104557.

[75] J. Wang, J.-G. Wang, H. Liu, Z. You, C. Wei, F. Kang, *J. Power Sources* **2019**, 438, 226951.

[76] H. Peng, H. Fan, C. Yang, Y. Tian, C. Wang, J. Sui, *RSC Adv.* **2020**, 10, 17702.

[77] M. Shi, B. Wang, C. Chen, J. Lang, C. Yan, X. Yan, *J. Mater. Chem. A* **2020**, 8, 24635.

[78] X. Guo, J. Zhou, C. Bai, X. Li, G. Fang, S. Liang, *Mater. Today Energy* **2020**, 16, 100396.

[79] M. Zhang, T. Hu, X. Wang, P. Chang, Z. Jin, L. Pan, H. Mei, L. Cheng, L. Zhang, *J. Mater. Chem. A* **2022**, 10, 7195.

[80] M. Chuai, J. Yang, R. Tan, Z. Liu, Y. Yuan, Y. Xu, J. Sun, M. Wang, X. Zheng, N. Chen, W. Chen, *Adv. Mater.* **2022**, 34, e2203249.

[81] X. Zheng, R. Luo, T. Ahmad, J. Sun, S. Liu, N. Chen, M. Wang, Y. Yuan, M. Chuai, Y. Xu, T. Jiang, W. Chen, *Energy Environ. Mater.* **2022**, 0, e12433.

[82]  T. Zhang, Q. Chen, X. Li, J. Liu, W. Zhou, B. Wang, Z. Zhao, W. Li, D. Chao, D. Zhao, *CCS Chem.* **2022**, 4, 2874.

[83]  X. Zheng, Y. Wang, Y. Xu, T. Ahmad, Y. Yuan, J. Sun, R. Luo, M. Wang, M. Chuai, N. Chen, T. Jiang, S. Liu, W. Chen, *Nano Lett.* **2021**, 21, 8863.

[84]  G. Zampardi, F. La Mantia, *Nat. Commun.* **2022**, 13, 687.

[85]  J. Cao, D. Zhang, X. Zhang, Z. Zeng, J. Qin, Y. Huang, *Energy Environ. Sci.* **2022**, 15, 499.

[86]  P. Sun, L. Ma, W. Zhou, M. Qiu, Z. Wang, D. Chao, W. Mai, *Angew. Chem. Int. Ed.* **2021**, 60, 18247.

[87]  L. Chen, W. Sun, K. Xu, Q. Dong, L. Zheng, J. Wang, D. Lu, Y. Shen, J. Zhang, F. Fu, H. Kong, J. Qin, H. Chen, *ACS Energy Lett.* **2022**, 7, 1672.

[88]  W. Chen, S. Guo, L. Qin, L. Li, X. Cao, J. Zhou, Z. Luo, G. Fang, S. Liang, *Adv. Funct. Mater.* **2022**, 32, 2112609.

[89]  J. Sun, X. Zheng, K. Li, G. Ma, T. Dai, B. Ban, Y. Yuan, M. Wang, M. Chuai, Y. Xu, Z. Liu, T. Jiang, Z. Zhu, J. Chen, H. Hu, W. Chen, *Energy Stor. Mater.* **2023**, 54, 570.

[90]  S. Liu, J. P. Vongsvivut, Y. Wang, R. Zhang, F. Yang, S. Zhang, K. Davey, J. Mao, Z. Guo, *Angew. Chem. Int. Ed.* **2023**, 62, e202215600.

https://doi.org/10.1142/9789811278327_0003

Chapter 3

# Open Challenges and Good Experimental Practices in the Research Field of Aqueous Zn-ion Batteries[1]

Giorgia Zampardi,[a,*] Fabio La Mantia[b,*]

[a]*University of Bremen, Energy Storage and Energy Conversion Systems, Bibliothekstraße 1, 28359 Bremen, Germany*
[b]*Fraunhofer Institute for Manufacturing Technology and Advanced Materials — IFAM*
*Wiener Str. 12, 28359, Bremen, Germany*

Aqueous zinc-ion batteries are realistic candidates as stationary storage systems for power-grid applications. However, to accelerate their commercialization, some important challenges must be specifically tackled, and appropriate experimental practices need to be embraced to align academic research efforts with realistic industrial working conditions for stationary storage. Within this chapter, both open challenges and good experimental practices are discussed in relation to their impact on the future development of the aqueous Zn-ion technology.

## 3.1 Introduction

In response to the increasing awareness regarding climate change, many countries have set the goal to increase the share of renewable energy within their total energy production.[1] Therefore, the development of cost-effective and safe energy storage technologies to accumulate the electrical energy harvested from renewable resources before putting it into the power grid has become a critical necessity of our society.[2,3]

---

* Corresponding author: zampardi@uni-bremen.de; lamantia@uni-bremen.de
[1] Adapted with permission from G. Zampardi, F. L. Mantia, *Nat. Commun.* **2022**, *13*, 687.

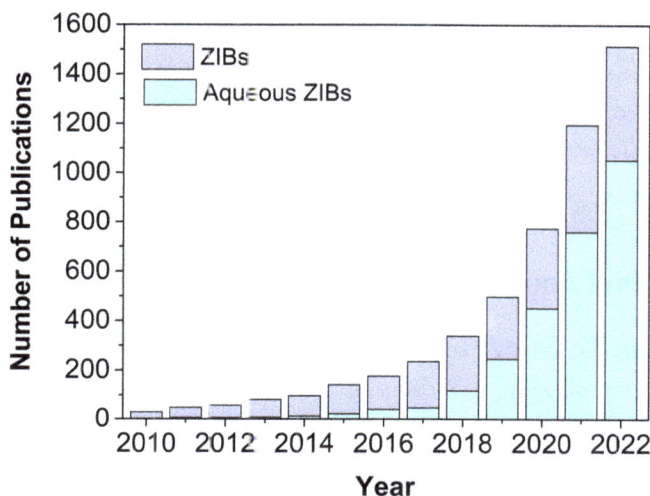

**Figure 3.1.** Total number of scientific publications on ZIBs (organic- plus aqueous-based ZIBs) and specifically on aqueous ZIBs published in recent years (data collected in April 2023 from Clarivate Web of Science. Keywords used: "zinc-ion batteries", "aqueous zinc-ion batteries").

In contrast to the already established Li-ion batteries, mild acidic aqueous Zn-ion batteries (ZIBs) operating in a pH range of ca. 4–5.5[4–6] are excellent candidates as storage systems for power grid applications. The reasons for this lay in their intrinsic high safety and environmental friendliness, high specific power and high reversibility, non-toxicity, and most importantly, low costs and abundance of metallic zinc.[4,7,8] Indeed, the ZIB field has recently received great attention, as evidenced by the exponential growth of related publications in the last 10 years (Figure 3.1).

The Zn-ion concept usually consists of a Zn-based negative electrode, onto/from which metallic zinc is electrodeposited and dissolved, and a positive electrode deinserting and inserting $Zn^{2+}$ cations from/within its lattice during the battery cycling.[4,8,9] In mild acidic ZIBs, typical insertion materials for the positive electrodes are based on manganese oxide,[10] vanadium oxides[10] and Prussian blue analogues.[4]

Although the Zn-ion concept can also be implemented with organic-based electrolytes, a realistic power-grid application of the Zn-ion technology implies the use of water-based electrolyte solutions to ensure high safety levels and keep costs of the final devices at a minimum.

In order to boost the commercialization of aqueous ZIBs as cheap and safe storage devices for the stationary grid, it is worth highlighting the challenges that remain yet to be addressed, together with the adoption of good experimental practices needed to align academic research efforts with industrial working conditions envisaged by a practical application of this battery technology.

## 3.2 Current Aqueous ZIB Limitations

### 3.2.1 Specific Energy and Utilization of the Zinc Anode

Despite the clear advantages that aqueous ZIBs offer in terms of high safety and low costs, the current technology does not attain high enough specific energies (e.g., >40 Wh/kg on full-cell scale, including passive elements and casing) that are needed to access the stationary energy market.

The specific energy of a battery system is a function of the specific capacity of both the positive and negative electrodes (often referred to as cathode and anode, respectively) together with the average discharge voltage of the cell. Although for a proper evaluation of the specific energy of the battery, the inactive components of the cell (such as contacts, current collector, case, etc.) must also be taken into account. Eq. (3.1) gives an acceptable first estimation of the battery specific energy:

$$E = \frac{Q_{cat} \times Q_{an}}{Q_{cat} + Q_{an}} \times \Delta V_{cell} \tag{3.1}$$

where $Q_{cat}$, $Q_{an}$ are the specific capacities of the active materials constituting the positive and the negative electrodes, respectively, and $\Delta V_{cell}$ is the difference between the average operating potentials of the positive and negative electrodes, which defines the average discharge voltage of the cell.

In order to ensure low costs of the final aqueous ZIB, the use of a negative electrode based on metallic zinc is the obvious choice due to the high availability of metallic Zn and its low price (ca. 3.5 $/kg).[11]

Current academic research efforts to increase the energy content of aqueous ZIBs generally focus on developing new cathode materials with high specific capacity

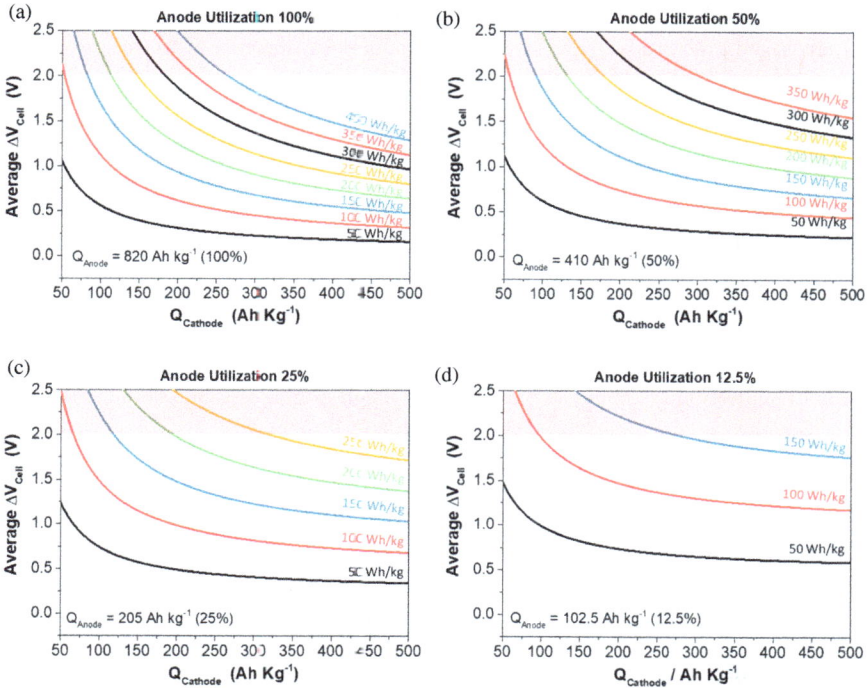

**Figure 3.2.** Equi-energy curves of a ZIB calculated from Eq. (3.1) as a function of the specific charge capacity of a generic cathode. Each graph represents a different extent of utilization of a metallic zinc anode, namely: (a) 100%, (b) 50%, (c) 25% and (d) 12.5%. Shaded in red is the region in which water splitting occurs, an inaccessible operating regime for ZIBs.

(e.g., >200 Ah/kg).[9] At the same time, however, the role of the utilization of the zinc-based negative electrode and of its Coulombic efficiency on the electrochemical performance of the battery is widely ignored, despite its heavy influence on the specific energy of a full Zn-ion cell (Eq. (3.1)). The Zn-based anodes routinely employed consist of Zn foils, metallic zinc deposited on a substrate, or composite electrodes with Zn particles within their formulation.[7-9] In order to understand the great challenges that hide behind the Zn-based anode utilization (often referred to as depth of discharge), it is helpful to consider the equi-energy curves, which are a graphical representation of Eq. (3.1). Such curves can be calculated as a function of the specific capacity of a generic cathode, as shown in Figure 3.2. In each graph, a different utilization (or depth of discharge) of the zinc anode has been taken into consideration. Figure 3.2a represents the ideal scenario, in which the zinc anode is fully used (100% utilization or depth of discharge) and perfectly matched in terms of charge with the cathode. Figure 3.2b is the envisaged practical scenario in which the anode is utilized at 50%, meaning that it is two times oversized with respect to the cathode. This means that the total mass of the active material within the electrode formulation (metallic Zn in the case of the anode) is two times more than the one needed to balance the capacity of the cathode. Figure 3.2c and 3.2d represent the more common, worse scenarios in which only 25 and 12.5% of the anode is used, corresponding to four times and eight times oversized anode with respect to the cathode, respectively.

Eq. (3.1) is also helpful when new cathode materials are developed. As can be seen in its graphical representation in Figure 3.2, developing high specific capacity cathode materials ($Q_{cat} > 100$ Ah/kg) has no practical significance if their operating potential is not high enough to lead to an average discharge voltage of the Zn-ion cell at least higher than 1.0–1.2 V. To make an example, a cathode material having 300 Ah/kg with a working potential of 0.6 V vs $Zn/Zn^{2+}$ tested with a Zn anode utilized at 50% (practical envisaged scenario) would lead to a Zn-ion cell with a specific energy lower than 100 Wh/kg (Figure 3.2b). Higher specific energy could be reached with a cathode material having only 100 Ah/kg but with a working potential of 1.5 V vs $Zn/Zn^{2+}$, keeping the same utilization of the zinc anode (50%). This effect is evident in the case of the lower zinc anode utilization (Figure 3.2d), as the performance of the ZIB is dominated by the Zn anode itself.

Unfortunately, the operational conditions that are mostly used in academic research employ an anode even more than eight times oversized with respect to the cathode, as the positive electrode is usually tested with low active mass loadings of around 1–2 mg/cm$^2$ while a metallic Zn foil anode usually has a Zn loading up to 1000 mg/cm$^2$.[9]

## 3.2.2 Parasitic Hydrogen Evolution and Irreversible Zn-ion Losses

Another important challenge that is often ignored within the aqueous ZIB literature concerns the main parasitic reaction occurring within a Zn-ion cell, namely: hydrogen evolution occurring at the zinc-based anode.

The unwanted, although thermodynamically favored, H$_2$ evolution reaction occurs mainly during the zinc deposition step. Due to the reduction of the hydronium ion

H$^+$(aq.) to gaseous H$_2$, the pH in the proximity of the Zn-based anode locally increases up to values of ca. 6–7.5.[12] In this pH range, the formation and stabilization of inactive and/or poorly conductive zinc passivation species and clay-like layered double hydroxides are strongly favored.[13–17] This not only causes an irreversible loss of Zn$^{2+}$ from the electrolyte, but also a decrease of the Coulombic efficiency of the Zn anode and thus of the whole Zn-ion cell.

It is worth mentioning that due to the local increase of pH caused by occurrence of the H$_2$ evolution reaction, evolution of the morphology of the Zn-based anode during the cycles often shows the appearance of non-homogeneous lamellar deposits, which are typical of the poorly conductive/inactive layered double hydroxides,[13–15] and less commonly of fine dendrites.[18]

## 3.3 Good Experimental Practices

### 3.3.1 Realistic Current Rates (C-rates) During the Electrochemical Tests

The current rate (C-rate) represents the inverse of the time in hours needed by the battery to complete a full discharge or charge. It is very important to perform power rate experiments in order to evaluate the efficiency and capacity losses of the electrodes when subjected to different cycling rates, as this represents a realistic working scenario for a battery intended for power grid applications.

However, it is of no practical interest to perform cycle-life duration experiments at C-rates higher than 10 C (meaning full charge or discharge of the battery occurring in 6 min), as the operational currents required by a storage device for the power grid are around 0.5–2 C and do not exceed 10 C.[19] Long cycle life ($\geq$1,000 cycles) of the cathode materials is very often claimed in the literature based on C-rates much higher than 10 C, up to ca. 100 C (full charge or discharge of the electrodes in 36 s), where capacitive effects of the double layer often mask the ones related to the Faradic reactions. Moreover, the high current rates strongly underestimate the chemical degradation of the materials, simply because degradation processes depend on time and not on the number of cycles.

However, it is generally known that the higher the C-rate, the slower the aging processes of an insertion material.[9,20,21] In order to collect non-misleading and meaningful results on a realistic estimation of the cycle life of anode and cathode materials, it is critical to choose a 1 C rate for the aging test, as this mimics the rate required for aqueous ZIBs intended to be used as storage devices for the power grid.

### 3.3.2 Mass Loading of the Active Materials

The relative mass loading of the active materials for both the anode and the cathode is a very important aspect, which strongly affects both the cycle life of the electrodes and the specific energy of the Zn-ion cell. Despite its clear importance on the cell's performance, the active mass loading of the electrodes is not generally given proper consideration in the literature.

The electrodes employed for lab-scale measurements often contain a low active material mass loading of ca. 1 mg/cm². When such electrodes are electrochemically tested, they experience negligible diffusion limitations and increased electrical connectivity among the different particles, which results in negligible polarization effects. This behavior artificially improves the performance of the material under investigation.[9,20] A realistic evaluation of ZIBs, and in particular of aqueous ZIBs, requires electrodes with an active material mass loading of at least ca. 7–10 mg/cm².[9,20] The mass loading should be high enough that an areal capacity of 2–4 mAh/cm² is reached.

### 3.3.3 Electrodes Balancing and Amount of Zn Ions Involved During Cycling

As already pointed out in the previous section, the relative mass loading of both anode and cathode should be balanced in terms of charge.

As clearly demonstrated by Eq. (3.1) and Figure 3.2, the smaller the mass of the active material within the cathode with respect to the active mass of the anode, the smaller the utilization (depth of discharge) of the anode. Consequently, the more unbalanced the electrodes, the lower the specific energy of the Zn-ion cell. Unfortunately, it is a very common practice to work with cathodes having active material loadings of around hundred-folds smaller than the active mass of the anode (e.g., cathode active material around 1–2 mg/cm², coupled with an anode active mass of ca. 150–300 mg/cm²). Unbalanced electrodes are unrealistic as they not only yield artificially high electrochemical performance and Coulombic efficiency of both electrodes (though this effect is more drastic for the zinc electrodeposition/dissolution process at the anode) but also drastically reduce the specific energy of the Zn-ion cell and vastly increase its costs.

## 3.3.4 Reproducibility of the Electrochemical Tests and Data Representation

In order to demonstrate the reproducibility of the electrochemical results disclosed in a scientific report or publication, the defining cycling performance characteristics of a material (such as specific capacity, capacity retention, Coulombic efficiency, etc.) should be always reported with mean values and standard deviations averaged over at least three different experiments.[21] In this way, the reproducibility of the acquired measurements would be clearly demonstrated and not only claimed.

Moreover, quantities such as the Coulombic efficiency should be correctly displayed with an appropriate resolution, implying the use of a scale with appropriate dimensions that do not span from 0 to >120%. Only in this way, the data are represented clearly enough to distinguish the possible fluctuation of efficiencies that could be related to important physical phenomena occurring at the electrodes (such as passivation of the surface of the electrode, occurrence of parasitic reactions, degradation of additives in the electrode/electrolyte, etc.).

Last but not least, the electrochemical analysis of ZIB electrodes should be carried out not only in flooded cells containing excess electrolyte (usually around 10 mL, corresponding to ca. 5 mL/cm² with respect to the geometric area of the electrode), but also in compact ones, where "electrolyte starving" conditions are attained (around 50–100 μL/cm² with respect to the geometric area of the electrode).[9] During the experiments carried out in cells with electrolyte starving conditions the detrimental effect of the parasitic reactions, leading to the loss of active $Zn^{2+}$ due to the formation of zinc oxide/hydroxide-based inactive compounds, is amplified due to the smaller amount of $Zn^{2+}$-containing electrolyte. This would then result in lower cycle life and lower Coulombic efficiencies with respect to the case where electrolyte-flooded conditions are used for the ZIB electrodes analysis.[9]

## 3.4 Summary and Outlook

Due to their promising characteristics, mild aqueous ZIBs represent a viable, green, and cost-effective energy storage technology for stationary grid applications. In order to quickly and efficiently reach this goal, it is worth focusing on the remaining challenges: increase of the specific energy of the full Zn-ion cell, and prevention of the parasitic $H_2$ evolution reaction occurring during the Zn electrodeposition step.

In order to do so, particular attention must be taken in cycling a fair amount of $Zn^{2+}$ contained within the Zn-based anode with respect to its total mass and in developing cathode materials resulting in an average discharge voltage of the cell of at least $\geq 1.0$–$1.2$ V. Moreover, research efforts should also address the optimization of the electrolyte composition (e.g., additives), towards the development of future water-based electrolytes with the aim to hinder the parasitic hydrogen evolution reaction at the anode.

In order to produce meaningful experimental results, which represent realistic ZIB working conditions, the electrochemical tests should be standardized. In particular: (i) the C-rates employed for the cycle-life assessment of the materials should reflect the need for a stationary power grid (i.e., long-term cycling tests at around 1 C must be carried out); (ii) the electrodes employed during the electrochemical tests should have a realistic amount of active material mass loading (e.g., in the 7–15 mg/cm² range); (iii) both electrodes (anode and cathode) of the electrochemical cell should be balanced in terms of charge. Moreover, attention should be paid to testing ZIB electrodes and full cells not only in flooded conditions, but also in cells where "electrolyte starving" conditions are attained, as this would mimic the realistic working conditions of the battery.

On the other hand, in order to demonstrate the robustness and reproducibility of research findings, experimental data (such as Coulombic efficiency, specific capacity, specific capacity retention, etc.) should always be averaged over at least three different measurements.

Only through focusing on the main issues that are currently limiting the practical application of aqueous ZIBs and on the adoption of good experimental practice in the academic research, which reflects the real working conditions envisaged for stationary power grid applications, researchers will be able to push forward the development of the aqueous Zn-ion technology, allowing it to permeate the stationary energy market.

## Acknowledgments

The financial support of the Federal Ministry of Education and Research (BMBF) in the framework of the project "ZIB" (FKZ 03XP0204A) is gratefully acknowledged.

## References

[1] Eurostat, Datasets, https://ec.europa.eu/eurostat/web/products-datasets/-/nrg_ind_ren (Last update: Apr 2023).

[2] Battery 2030+: A long-term roadmap for forward-looking battery research in Europe, **2021**; https://battery2030.eu/research/roadmap/ (accessed: March 2021).

[3] V. Blay, R. E. Galian, L. M. Muresan, D. Pankratov, P. Pinyou, G. Zampardi, *Adv. Sustain. Syst.* **2020**, 4, 1900145.

[4] G. Zampardi, F. La Mantia, *Curr. Opin. Electrochem.* **2020**, 21, 84.

[5] G. Kasiri, J. Glenneberg, R. Kun, G. Zampardi, F. La Mantia, *ChemElectroChem* **2020**, 7, 3301.

[6] J. Lim, G. Kasiri, R. Sahu, K. Schweinar, K. Hengge, D. Raabe, F. La Mantia, C. Scheu, *Chem. Eur. J.* **2020**, 26, 4917.

[7] G. Zampardi, R. G. Compton, *J. Solid State Electrochem.* **2020**, 24, 2695.

[8] J. Shin, J. Lee, Y. Park, J. W. Choi, *Chem. Sci.* **2020**, 11, 2028.

[9] Y. Li, B. Liu, J. Ding, X. Han, Y. Deng, T. Wu, K. Amine, W. Hu, C. Zhong, J. Lu, *Batteries & Supercaps* **2021**, 4, 60.

[10] J. Ming, J. Guo, C. Xia, W. Wang, H. N. Alshareef, *Mater. Sci. Eng. R Rep.* **2019**, 135, 58.

[11] Metalle IfsEus. Base metals prices in December 2020, **2021**; https://en.institut-seltene-erden.de/prices-for-base-metals-in-december/ (accessed: October 2021).

[12] F. Argoul, A. Kuhn, *J. Electroanal. Chem.* **1993**, 359, 81.

[13] A. Bani Hashemi, G. Kasiri, J. Glenneberg, F. Langer, R. Kun, F. La Mantia, *ChemElectroChem* **2018**, 5, 2073.

[14] A. Bani Hashemi, G. Kasiri, F. La Mantia, *Electrochim. Acta* **2017**, 258, 703.

[15] M. A. González, R. Trócoli, I. Pavlovic, C. Barriga, F. La Mantia, *Electrochem. Commun.* **2016**, 68, 1.

[16] P. Oberholzer, E. Tervoort, A. Bouzid, A. Pasquarello, D. Kundu, *ACS Appl. Mater. Interfaces* **2019**, 11, 674.

[17] A. Moezzi, M. B. Cortie, A. M. McDonagh, *Dalton Trans.* **2013**, 42, 14432.

[18] Q. Yang, G. Liang, Y. Guo, Z. Liu, B. Yan, D. Wang, Z. Huang, X. Li, J. Fan, C. Zhi, *Adv. Mater.* **2019**, 31, 1903778.

[19] H. C. Hesse, M. Schimpe, D. Kucevic, A. Jossen, *Energies* **2017**, 10, 2107.

[20] Z. Lin, T. Liu, X. Ai, C. Liang, *Nat. Commun.* **2018**, 9, 5262.

[21] G. Zampardi, M. Warnecke, M. Tribbia, J. Glenneberg, C. Santos, F. La Mantia, *Electrochem. Commun.* **2021**, 126, 107030.

Chapter 4

# Design Strategies for High-Energy-Density Aqueous Zinc Batteries[1]

Pengchao Ruan,[a] Shuquan Liang,[a,*] Bingan Lu,[b]
Hong Jin Fan,[c,*] Jiang Zhou[a,*]

[a]*School of Materials Science and Engineering, Key Laboratory
of Electronic Packaging and Advanced Functional Materials
of Hunan Province, Central South University,
Changsha, 410083 P. R. China*
[b]*School of Physics and Electronics, Hunan University,
Changsha, 410082 P. R. China*
[c]*School of Physical and Mathematical Sciences,
Nanyang Technological University, 637371 Singapore*

In recent years, the increasing demand for high-capacity and safe energy storage has focused attention on zinc batteries featuring high voltage, high capacity, or both. Despite extensive research progress, achieving high-energy-density zinc batteries remains challenging and requires the synergistic regulation of multiple factors including reaction mechanisms, electrodes, and electrolytes. In this chapter, we comprehensively summarize the rational design strategies of high-energy-density zinc batteries and critically analyze the positive effects and potential issues of these strategies in optimizing the electrochemistry, cathode materials, electrolytes, and device architecture. Finally, the challenges and perspectives for the further development of high-energy-density zinc batteries are outlined to guide research towards new-generation batteries for household appliances, low-speed electric vehicles, and large-scale energy storage systems.

* Corresponding authors: lsq@csu.edu.cn; fanhj@ntu.edu.sg; zhou_jiang@csu.edu.cn
[1] Adapted with permission from P. Ruan, S. Liang, B. Lu, H. J. Fan, J. Zhou, *Angew. Chem. Int. Ed.* **2022**, 61, e202200598.

# 4.1 Introduction

The energy crisis, greenhouse effect, and air pollution have driven research on new energy storage systems.[1] Aqueous rechargeable batteries are among the promising candidates for grid-scale energy storage, owing to their high safety, low cost, and environmental friendliness.[2] The advantages of aqueous rechargeable zinc batteries are their high safety, the rich reserves of zinc, simple preparation, low anode potential (−0.763 V vs. standard hydrogen electrode (SHE)), and the high theoretical capacity of zinc (820 mA h g$^{-1}$), which have stimulated growing interest towards new-generation safe batteries.[3] In spite of this, in contrast to organic rechargeable batteries, the energy density determining the future prospects of zinc batteries remains far from the target, suffering from the narrow electrochemically stable potential windows (ESPW), limited discharging capacity, complicated side reactions, and low working voltages (Figure 4.1a).[4] Hence, rationally addressing these deficiencies is pivotal to the development of high-energy-density batteries.[5]

## The issues for achieving high-energy-density zinc batteries

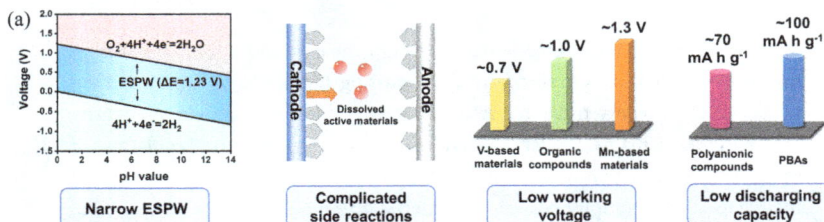

## Recent developments in high-energy-density zinc batteries

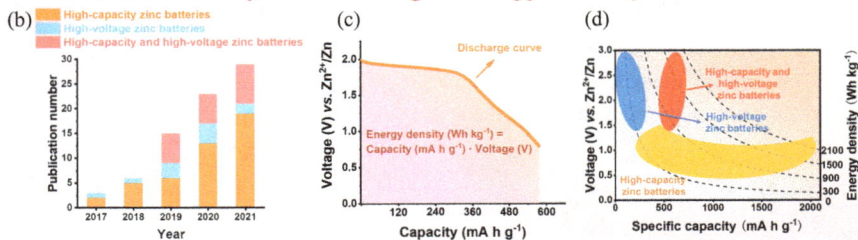

**Figure 4.1.** (a) A summary of issues for achieving high-energy-density zinc batteries. (b) Plot of the number of publications on high-energy-density zinc batteries in recent years. (c) Relationship between energy density, voltage, and capacity. (d) Comparison of the specific capacities and average output voltages for high-capacity (yellow), high-voltage (blue), and both high-capacity and high-voltage (red) zinc batteries.

Generations of electrochemical batteries (e.g., lead-acid batteries,[6] nickel-hydrogen batteries,[7] lithium batteries,[8] zinc batteries,[9] etc.) have been continuously developed to meet the increasing demand for energy storage. Nevertheless, as the energy density continues to increase, so do the safety hazards of the batteries. To mitigate this, aqueous-based high-safety zinc batteries have become the focus of battery research, for which the priority is to boost the energy density (Figure 4.1b). Fundamentally, the energy density is the product of the specific capacity and the voltage of the battery (Figure 4.1c).[10] Accordingly, there are three design strategies toward high-energy-density zinc batteries with mild or acidic aqueous electrolytes: achieving high voltage, high capacity, or both high capacity and high voltage. The design strategy for high-capacity batteries focuses on the cathode materials,[11] and that for high-voltage batteries focuses on modification of the electrolytes for high-voltage redox reactions (Figure 4.2).[12] In particular, the optimization of the $MnO_2/Mn^{2+}$ conversion reaction mechanism is a crucial strategy to realize both high-capacity and high-voltage zinc batteries (Figure 4.2).[13] Thanks to these strategies, the capacity, voltage, and energy density of zinc batteries can now reach up to $2000\ mAh\,g^{-1}$, 2.9 V, and $2372\ Wh\,kg^{-1}$, respectively (based on the mass of active materials) (Figure 4.1d).

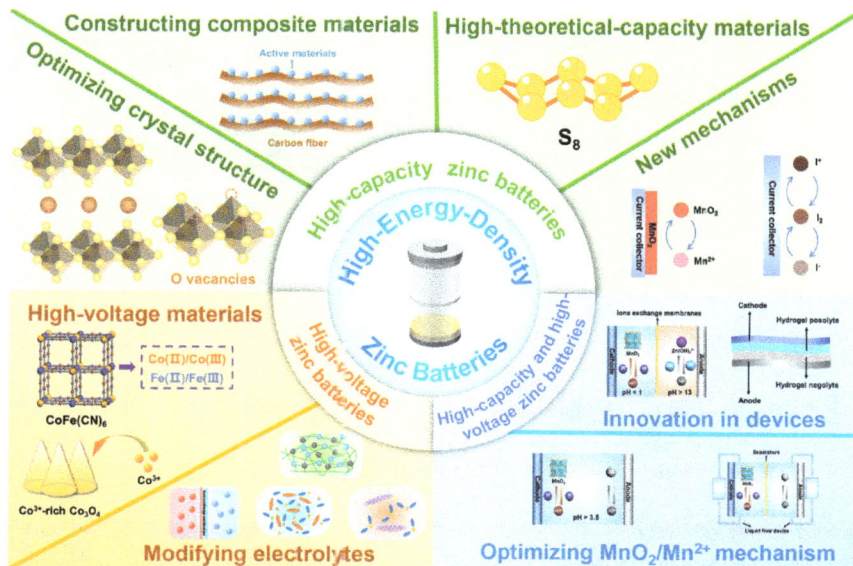

**Figure 4.2.** A summary of the design strategies towards high-energy-density zinc batteries.

Although extensive research has been carried out in this field, a comprehensive summary of the design strategies of high-energy-density zinc batteries has not been available. In this chapter, by surveying the recent literature, we highlight the challenges and analyze the design strategies for high-energy-density zinc batteries with mild or acidic aqueous electrolytes, and then provide a perspective for future directions. We put our discussion in a broader context beyond zinc batteries and the strategies summarized here will also apply to other aqueous-based batteries.

## 4.2 Design Strategies for High-Capacity Aqueous Zinc Batteries

### 4.2.1 Exploiting Cathode Materials with High Theoretical Capacity

The theoretical capacity of an electrode material is determined by the active material mass and electron transfer number, which is the guideline in looking for high-theoretical-capacity cathode materials (Figure 4.3a and 4.3b). Currently, common intercalation-type materials, including manganese-based and vanadium-based oxides,

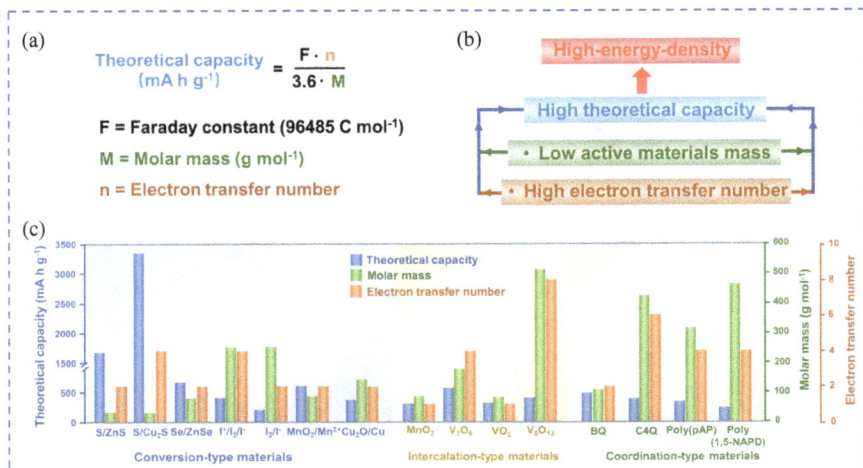

**Figure 4.3.** (a) Formula for calculating the theoretical capacity of an electrode material and (b) the corresponding design guidelines for high-theoretical-capacity materials. (c) Comparison of the theoretical capacity, relative molecular mass, and electron transfer number for various high-capacity zinc battery cathode materials.

have theoretical capacities in the range of 308–589 mAh g$^{-1}$, with an ultimate energy density of about 410 Wh kg$^{-1}$ (Figure 4.3c).[14] However, these values are based on only active materials; reducing the mass of the inactive fraction of the electrodes while ensuring structural stability is extremely challenging, and is usually accompanied by a reduction in the valence state of the active element. Accordingly, increasing the number of redox reaction centers, i.e., improving the electron transfer number, has become the pivotal design strategy for high-capacity materials.[15] In particular, *in situ* electrochemical oxidation has been an effective strategy to oxidize low-valence vanadium-based compounds to high-valence vanadium oxides (such as $V_2O_{5-x}$) in the first charging process, thereby endowing the cathodes with higher capacities (over 600 mAh g$^{-1}$).[16] It is noted that these high-capacity oxides formed *in situ* still require surface encapsulation by carbon to maintain chemical and structural stability.

Compared with inorganic compounds, the theoretical capacity of coordination-type organic compounds tends to be inferior owing to the large fraction of inactive carbon in the framework (Figure 4.3c).[17] Fortunately, the structural flexibility of carbon-based organic compounds is controllable, which allows the introduction of active coordination groups on the carbon skeleton. This, together with their low mass, make coordination-type materials ideal for high-theoretical-capacity batteries.[18] For instance, redox-active quinone groups can be introduced into 1,4,5,8,9,12-hexaazatriphenylene-based covalent organic frameworks (HAQ-COFs) to provide higher redox activity and more active coordination sites, resulting in an enhancement of the specific capacity (344 mAh g$^{-1}$) and output voltage (0.84 V).[19] Furthermore, the hierarchical configuration of poly(1,5-naphthalenediamine) and poly(*para*-aminophenol) on nanoporous carbon can facilitate the complete activation of the coordination centers (C=O and C=N) for $Zn^{2+}$.[20] On the other hand, despite the high theoretical capacity of small organic compounds such as benzoquinone (BQ), their electrochemical stability is poor due to the lack of sufficient covalent structure. Therefore, simply pursuing high capacity in coordination-type organic compounds may be counterproductive. The aim is to achieve a good balance between structure (carbon skeleton and active groups) and electrochemical performance (cycling stability and capacity).

In contrast with intercalation-type or coordination-type materials, conversion-type materials dominate the high-theoretical-capacity group. Representatively, sulfur, which has served as the cathode of zinc batteries, has various redox reaction pairs with ultrahigh theoretical capacity, such as S/ZnS ($\approx$1675 mAhg$^{-1}_s$)[21] and S/Cu$_2$S ($\approx$3350 mAhg$^{-1}_s$)[22] (Figure 4.3c). However, issues including sluggish kinetics, irreversible oxygenated by-products, and stringent pH requirements deteriorate the capacity and the actual electrode potential of aqueous Zn//S batteries. It has been observed that the electrolyte additive I$_2$ can act as a medium for Zn$^{2+}$ while reducing the reaction barriers, so it is beneficial for achieving low-voltage hysteresis and high energy efficiency of Zn//S batteries.[21] Furthermore, the formation of a S–Se infinite solid solution can optimize the electron density difference, band structure, and reaction energy of S, resulting in great enhancement in conductivity and reactivity.[23] In addition to S, zinc batteries employing Se as cathode materials have been proposed, which can provide an impressive capacity of 611 mAhg$^{-1}$.[24] Despite the natural advantages of elemental S and Se in terms of mass and electron transfer number, the low operating voltage (0.455–1.01 V) remains an insurmountable obstacle. Matching with high-voltage redox reactions is one of the solutions, such as S/Cu$_2$S (working voltage: 1.15 V).

## 4.2.2 Optimizing the Actual Capacity

The deterioration of the actual capacity of active materials can be caused by adverse intrinsic conductivity, poor selectivity to Zn$^{2+}$ charge carriers, and blocked active sites (Figure 4.4). Therefore, it is crucial to optimize the actual capacity of the active material, for which various strategies have been proposed so far, including engineering the crystal structure and combining with highly conductive materials.

In the conventional intercalation energy-storage mechanism, the strong electrostatic repulsion between the host lattice oxygen and the zinc ions results in the poor utilization and low capacity of the active materials.[25] Pre-intercalating some molecules[26] or ions[27] into layer-like materials can lower this electrostatic interaction and expand the layer spacing to enhance the storage capacity for Zn$^{2+}$ ions (425–550 mAhg$^{-1}$). Note that the intercalation of inactive guest substances will decrease the proportion and valence of active elements in the host materials. Another pitfall of these semiconductor oxides is the inadequate electron transfer capability, which impairs the capacity. Generally, reducing the band gap of oxides via doping with heteroatoms is favorable to enhance the electrical conductivity. For instance, a La–Ca co-doped ε-MnO$_2$ presented low charge transfer resistance (29.8 vs. 198.1 Ω) and higher capacity

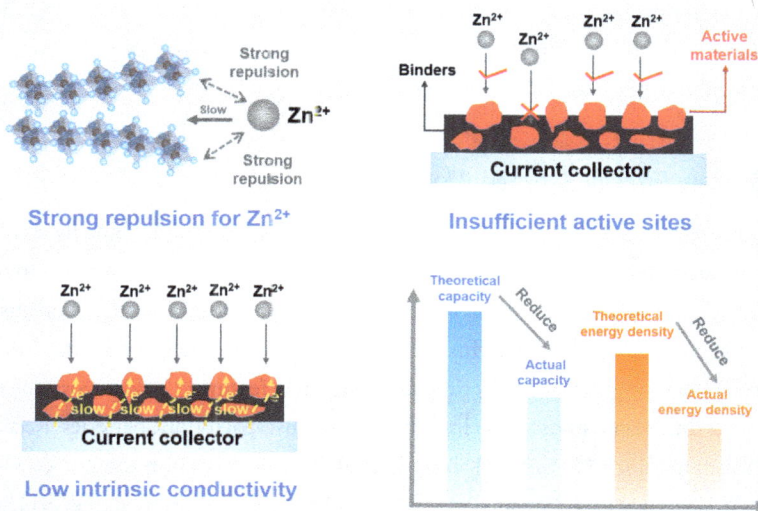

**Figure 4.4.** Illustration of adverse effects due to intrinsic conductivity, poor selectivity to $Zn^{2+}$ charge carriers, and blocked active sites in active materials.

(297 vs. 199 mAh g$^{-1}$) compared with pristine $MnO_2$.[28] Furthermore, as another effective approach for improving capacity, defect engineering can be effective in increasing carrier concentration and reducing electrostatic repulsion.[29] For this purpose, it is necessary but sometimes challenging to accurately control and characterize the defects.

Carbon-containing materials or lightweight carbon materials (graphene, carbon fiber, carbon nanotubes, MXene, etc.) featuring high electrical conductivity and good accessibility to the electrolyte are often employed to facilitate the contact of ions with the active materials and improve the charge transfer rate, which thus effectively increases the capacity of the cathodes.[30] Apart from the intrinsic properties of active materials, extrinsic factors also affect the actual capacity. Very often, part of the effective area of the electrode material is enveloped by binders, which hinders direct contact between charge carriers and the electrode.[31] Functional groups on the surface of carbon substrates can obviate the binders and allow for binder-free loading of active materials and maximum exposure of the active sites. For example, after combination with a carbon nanosheet array, the capacity of $Co$-$Mn_3O_4$ can increase from 220 to 362 $mAh\,g^{-1}$.[32] And N-doped carbon cloth is able to firmly anchor benzoquinone, thereby inhibiting its dissolution and stabilizing its capacity.[33]

### 4.2.3 New Energy Storage Mechanisms

To date, the recognized energy storage mechanisms of zinc batteries are broadly classified into three categories: (de)intercalation, coordination, and conversion reactions. In most cases, the charge carriers of the (de)intercalation and coordination mechanisms are $Zn^{2+}$ and $H^+$. However, the low diffusion rate and high charge density of $Zn^{2+}$ ions are not ideal for active materials. Selecting charge carriers featuring low corrosion, small size, and fast diffusion is particularly important for high-capacity zinc batteries. Recently, a mechanism in which $H^+$ and $NH_4^+$ ions co-insert was developed in the $Zn//\alpha$-$MnO_2$ battery.[34] The stable hydrogen bonds formed between the intercalated $NH_4^+$ and $Mn$–$O$ stabilize the $Mn^{3+}O_6$ octahedra, and the electron density of oxygen atoms in $Mn$–$O$ will be increased as a result of the co-insertion of $H^+$. Therefore, with the synergistic effect of the co-insertion of $H^+$ and $NH_4^+$, the $Zn//\alpha$-$MnO_2$ battery exhibited high discharging capacity and energy density (365 $mAh\,g^{-1}$ and 486 $Wh\,kg^{-1}$, respectively). Furthermore, in the intensive study of the energy storage mechanism of manganese-based zinc batteries, it has been commonly observed that additional $MnO_2$ will deposit on the electrode owing to the existence of $Mn^{2+}$ ions in the electrolyte, resulting in capacities (300–400 $mAh\,g^{-1}$) exceeding the theoretical value and unexpectedly high energy densities of 400–630 $Wh\,kg^{-1}$.[4b,35] It is worth mentioning that the $MnO_2$ deposition as the primary source of exorbitant extra capacity requires operating at low current densities, which is closely correlated with the electrochemical polarization and variation of the pH value of the system. More recently, the distinct $MnO_2$ dissolution/deposition mechanism has been adopted for multifarious $Zn//MnO_2$ batteries with exceptional energy density (discussed in Section 4.2).[36]

The advantages of the conversion reaction energy storage mechanism are high charge carrier selectivity, practical universality, and high versatility of the reaction system, implying that more redox reaction systems can be employed in aqueous zinc batteries. In the design of high-capacity energy storage mechanisms, one also should consider the mass of active materials and the number of transferred electrons. Therefore, it is feasible to tailor the continuous conversion reactions with a larger number of transferred electrons. For instance, the unique mechanism of continuous $I^+/I_2/I^-$ conversion reaction based on four-electron transfer has been elaborated, which pushes the zinc-iodine batteries to a superior level with a high capacity of 594 mAh g$^{-1}$.[37] Moreover, the universal nature of the conversion reaction mechanism also makes it versatile for a range of similar active materials towards high-capacity reaction systems. Very recently, the CuI/Cu conversion process has been discussed, and other Cu-based materials ($Cu_2O$, $Cu_2S$, $Cu_3N$, etc.) show impressive electrochemical properties by this mechanism by virtue of their own unique structures.[38] While the conversion reaction is a seemingly simple system, the compatibility of the charge carriers with the anode and the electrolyte as well as the additional electrolyte mass should be taken into account along with the reaction stability.

The design of high-capacity zinc batteries has been significantly competitive in achieving high-energy-density zinc batteries. High capacities of 300–2000 mAh g$^{-1}$ and energy densities of 300–2372 Wh kg$^{-1}$ have been achieved (Table 4.1). However, the high ratio of the Gibbs free energy change ($\Delta G$) to the electron transfer number for the electrochemical reaction system at the cathode results in low operating voltages (0.45–1.35 V) of zinc batteries. Hence, high-capacity performance alone is not enough.

**Table 4.1.** The performance of high-capacity aqueous zinc batteries.

| Cathode Material | Electrolyte | Working Voltage (V) | Capacity (mA h g$^{-1}$)/Energy Density (Wh kg$^{-1}$) | Mechanism | Ref. |
|---|---|---|---|---|---|
| $VN_{0.9}O_{0.15}$ | 3 M $Zn(CF_3SO_3)_2$ | 0.70 | 603/421 | $Zn^{2+}$ intercalation | [16e] |
| a-$V_2O_5$@C | 3 M $Zn(CF_3SO_3)_2$ | 0.77 | 620/475 | $Zn^{2+}$ intercalation | [39] |
| $V_2O_3$ | 2 M $Zn(CF_3SO_3)_2$ | 0.65 | 625/406 | $Zn^{2+}$ intercalation | [16b] |
| O-VN | 3 M $ZnSO_4$ | 0.77 | 705/~541 | $Zn^{2+}$ intercalation | [40] |
| $VO_2$ | 2 M $Zn(CF_3SO_3)_2$ | 0.70 | 610/426 | $H^+/Zn^{2+}$ co-intercalation | [16c] |
| MnVO@C | 3 M $Zn(CF_3SO_3)_2$ | 0.77 | 610.2/469 | $Zn^{2+}$ intercalation | [16d] |
| $V_6O_{13}$/CC | 3 M $ZnSO_4$ | 0.98 | 520/~511 | $Zn^{2+}$ intercalation | [15] |
| PANI-V | 2 M $ZnSO_4$ | 0.88 | 490/430 | $Zn^{2+}$ intercalation | [26a] |
| PA-$VOPO_4 \cdot 2H_2O$ | 2 M $Zn(CF_3SO_3)_2$ | 1.22 | 268.2/328 | $Zn^{2+}$ intercalation | [41] |
| $V_2O_5 \cdot 4VO_2 \cdot 2.72H_2O$ | 3 M $Zn(CF_3SO_3)_2$ | 0.66 | 567/375 | $Zn^{2+}$ intercalation | [26d] |

(*Continued*)

**Table 4.1.** (*Continued*)

| Cathode Material | Electrolyte | Working Voltage (V) | Capacity (mA h g$^{-1}$)/Energy Density (Wh kg$^{-1}$) | Mechanism | Ref. |
|---|---|---|---|---|---|
| O$_d$-MnO$_2$ | 1 M ZnSO$_4$ + 0.2 M MnSO$_4$ | 1.36 | 345/470 | Zn$^{2+}$ intercalation | [29] |
| Ca$_2$Mn$_3$O$_8$ | 3 M Zn(CF$_3$SO$_3$)$_2$ + 0.3 M Mn(CF$_3$SO$_3$)$_2$ | 1.39 | 368/512 | Zn$^{2+}$ intercalation | [42] |
| KMO | 2 M ZnSO$_4$ + 0.2 M MnSO$_4$ | 1.25 | 425/503 | H$^+$/Zn$^{2+}$ co-intercalation | [27b] |
| Na:MnO$_2$/GCF | 2 M ZnSO$_4$ + 0.1 M MnSO$_4$ | 1.34 | 382/512 | H$^+$/Zn$^{2+}$ co-intercalation | [30a] |
| Co-Mn$_3$O$_4$/CNA | 2 M ZnSO$_4$ + 0.2 M MnSO$_4$ | 1.28 | 362/463 | H$^+$/Zn$^{2+}$ co-intercalation | [32] |
| La–Ca co-doped ε-MnO$_2$ | 1 M ZnSO$_4$ + 0.4 M MnSO$_4$ | 1.35 | 297/401 | H$^+$/Zn$^{2+}$ co-intercalation | [28] |
| MnO$_x$@N-C | 2 M ZnSO$_4$ + 0.1 M MnSO$_4$ | 1.34 | 385/515 | Zn$^{2+}$ intercalation | [31b] |
| MnO$_2$@PEDOT | 2 M ZnCl$_2$ + 0.4 M MnSO$_4$ | 1.32 | 367/484 | Zn$^{2+}$ intercalation | [31a] |
| N-CC@MnO$_2$ | 2 M ZnCl$_2$, 0.4 M MnSO$_4$ | 1.33 | 353/469 | Zn$^{2+}$ intercalation | [43] |
| MnO$_2$ | 1 M (NH$_4$)$_2$SO$_4$ + 0.1 M MnSO$_4$ | 1.35 | 365/493 | H$^+$/NH$_4^+$ co-intercalation | [34] |
| C4Q | 3 M Zn(CF$_3$SO$_3$)$_2$ | 1.0 | 335/335 | Zn$^{2+}$ coordination | [18b] |
| HAQ-COF | 2 M ZnSO$_4$ | 0.84 | 344/289 | H$^+$/Zn$^{2+}$ coordination | [19] |
| C@multi-layer polymer | 2 M ZnSO$_4$ | 0.79 | 348/275 | Zn$^{2+}$ coordination | [20] |
| BQ-NCC | 3 M ZnSO$_4$ | 1.1 | 489/538 | Zn$^{2+}$ coordination | [33] |
| S@C | Cathode: gelatin/ CuSO$_4$ Anode: gelatin/ ZnSO$_4$ | 1.15 | 2063/2372 | S/Cu$_2$S conversion | [22a] |

Table 4.1. (Continued)

| Cathode Material | Electrolyte | Working Voltage (V) | Capacity (mA h g$^{-1}$)/Energy Density (Wh kg$^{-1}$) | Mechanism | Ref. |
|---|---|---|---|---|---|
| S/KB | Cathode: 0.5 M CuSO$_4$ Anode: 0.5 M ZnSO$_4$ Middle: 0.5 M Na$_2$SO$_4$ | 1.04 | 1990/2065 | S/Cu$_2$S conversion | [22c] |
| S@CNTs-50 | 1 M Zn(CH$_3$COO)$_2$ + 0.05 wt% I$_2$ | 0.455 | 1105/502 | Zn/ZnS and I$_5^-$/I$_2$ conversion | [21] |
| SeS$_{5.76}$@3D-NPCF | 3 M ZnSO$_4$ + 0.1 wt% I$_2$ | 0.71 | 1222/868 | S/ZnS and Se/ZnSe conversion | [23] |
| Se/CMK-3 | 2 M ZnTFSI/PEG$^+$ water | 1.01 | 611/751 | Se/ZnSe conversion | [24] |
| PAC/I$_2$ | ZnCl$_2$ + LiCl + CH$_3$CN | 1.26 | 594/750 | I$^+$/I$_2$/I$^-$ conversion | [37] |

# 4.3 Design Strategies for High-Voltage Aqueous Zinc Batteries

## 4.3.1 Exploiting Cathode Materials with High Working Voltage

In contrast to high-capacity zinc batteries, high-voltage zinc batteries rely on both high-voltage electrode materials and electrolytes with wide ESPW, which are mutually constrained. Despite various cathode materials with high discharging platforms, such as Prussian blue analogues (PBAs),[44] it is practically necessary that these electrodes based on high-voltage materials contain a low percentage of inactive matter. Essentially, this unbalanced molecular structure severely reduces the theoretical capacity of active materials, but it is also indispensable for stabilizing structure and high-voltage redox reactions (Figure 4.5a). To address the drawback of the low capacity of PBAs caused

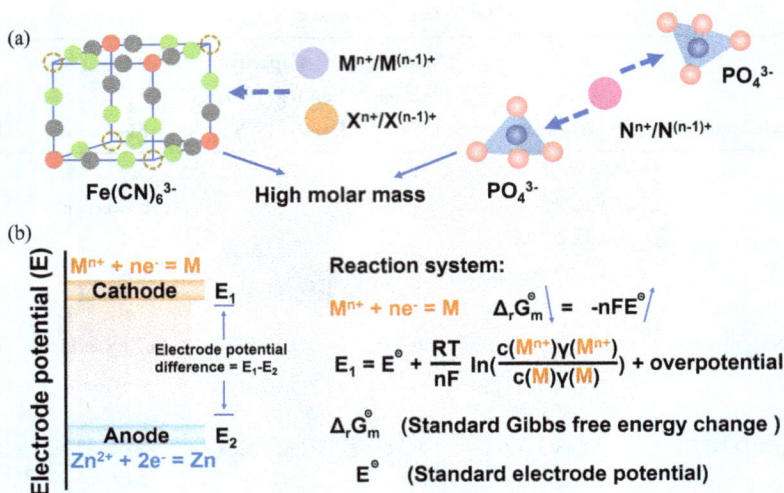

**Figure 4.5.** (a) Schematic illustration of high-voltage cathode materials with high molar mass. (b) Design principle of high-voltage conversion reactions.

by the single redox-active center, PBAs can be designed for with multiple high-voltage redox reactions, such as $CoFe(CN)_6$ with high-voltage $Co^{2+}/Co^{3+}$ and $Fe^{2+}/Fe^{3+}$ redox reactions. The introduced Co species can endow the $CoFe(CN)_6$ with an appropriate capacity of 173.4 mA h g$^{-1}$ while ensuring a high working voltage of 1.75 V.[45] However, the design of high-voltage intercalation-type materials with suitable capacity does have challenges, considering that high-voltage redox-active centers need to bond with specific anions (such as $Fe(CN)_6^{3-}$, $PO_4^{3-}$, $PO_4F^{4-}$). From a thermodynamic point of view, there is a connection between the standard electrode potential ($E^\ominus$) and the standard Gibbs free energy ($\Delta_r G^\ominus$) of the reaction: $\Delta_r G^\ominus = -nE^\ominus F$, where $n$ is the electron transfer number and $F$ is the Faraday constant (Figure 4.5b). Therefore, the high-voltage conversion reaction adapted to the ESPW can be designed according to the aforesaid relationship. A $Co^{3+}$-rich $Co_3O_4$ was first proposed for aqueous zinc batteries with a mild electrolyte. Compared with the alkaline electrolyte, this $Co^{3+}$-rich $Co_3O_4$ has a higher working voltage (1.81 V vs. 1.54 V) and capacity of 200 mA h g$^{-1}$.[46] Nevertheless, because of the narrow ESPW of aqueous batteries, it is not enough to consider only the high-potential cathode materials. To meet the demand of high-voltage zinc batteries, rational electrolyte engineering to expand the ESPW is pivotal.

## 4.3.2 *Electrolyte Optimization Strategy*

In contrast to organic electrolytes, the ESPW of aqueous electrolytes is controlled by the oxygen and hydrogen evolution reactions (OER and HER) on the cathode and anode, respectively (Figure 4.6a).[47] The theoretical decomposition potential of water is very low (1.23 V vs. SHE), and the actual operating voltage hardly exceeds 2.0 V. Therefore, such a narrow ESPW is the biggest challenge for high-voltage zinc batteries. A core strategy for broadening the ESPW is to reduce the contact between water molecules and electrodes. For example, sodium dodecyl sulfate (SDS, ESPW >2.5 V) and 21 M lithium bis(trifluoromethane)sulfonimide (LiTFSI, water-in-salt, ESPW >2.6 V) can build a hydrophobic layer on the electrode surface to effectively block the passage of water molecules (Figure 4.6b).[48] LiTFSI can also form a F-based solid-electrolyte interphase (SEI) on the anode surface, but the high concentration also

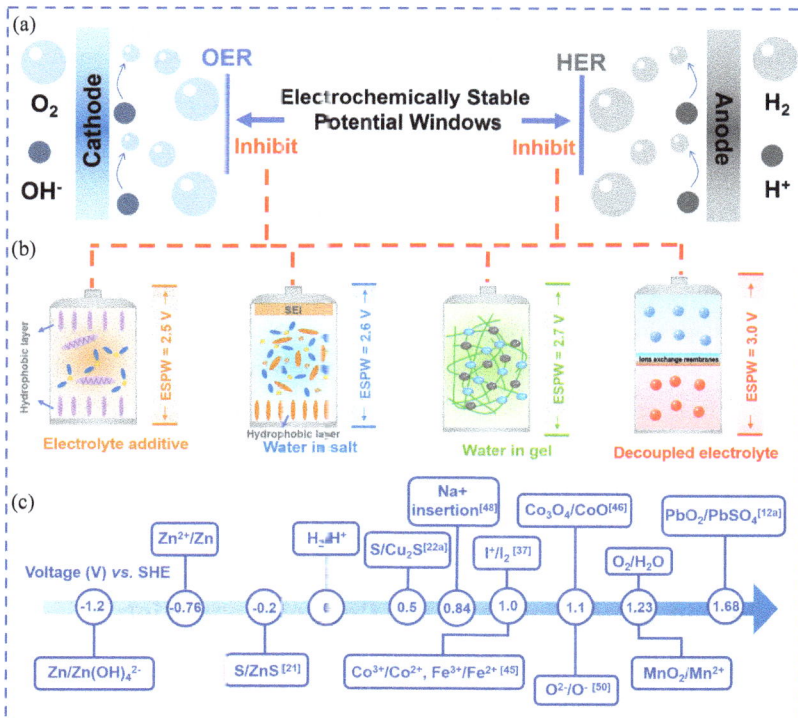

**Figure 4.6.** Strategies to expand the ESPW for high-voltage zinc batteries. (a, b) Schematic illustration of ESPW in different electrolytes; (c) various redox reaction pairs.

reduces the ionic conductivity. Although low-concentration SDS can effectively circumvent this issue, the hydrophobic layer also restricts the diffusion of charge carriers on the electrode surface. In view of this, it has been proposed to use the polar groups of sodium alginate hydrogels (water-in-gel) to capture water molecules and restrict their movement by hydrogen bonding (ESPW >2.7 V, Figure 4.6b).[49] Also the carboxylate groups in the gel can provide special channels for ion transport to ensure sufficiently high ionic conductivity. Furthermore, ion-exchange membrane engineering (decoupled electrolyte, ESPW >3.0 V) can decouple the cathodic and anodic electrolyte to inhibit OER and HER by high concentrations of $H^+$ and $OH^-$, respectively (Figure 4.6b). Various high-voltage reactions can be realized only if the ESPW is large enough (Figure 4.6c). For example, $VOPO_4$ exhibits additional $O^{2-}/O^-$ high-voltage reactions in water-in-salt electrolytes.[50] And the intercalation of $Na^+$ and the de-intercalation of $ZnCl_4^{2-}$ at high potentials are also dependent on SDS and water-in-gel electrolytes.[48,49] Notably, some of the high-voltage reactions need to be matched to specific electrolytes, especially those reactions with $H^+$ participation. For instance, the $PbO_2/PbSO_4$ conversion reaction (2.92 V vs. $Zn/Zn(OH)_4^{2-}$) should be applied in the decoupled electrolyte battery.[12a] Meanwhile, the cost and environmental issues associated with the use of ion-exchange membranes and highly concentrated electrolytes should not be underestimated.

At present, compared to the wide application of high-capacity zinc batteries, only a few high-voltage zinc batteries can achieve a high energy density of 200–440 Wh kg$^{-1}$ due to their low theoretical capacity and limited voltage window (Table 4.2). Obviously, there are difficulties in making major breakthroughs in either high-capacity or high-voltage zinc batteries to meet the high energy density due to their intrinsic drawbacks. Therefore, research on zinc batteries that show both high voltage and high capacity is imperative.

## 4.4 Design Strategies for High-Capacity and High-Voltage Aqueous Zinc Batteries

### 4.4.1 High-Capacity and High-Voltage Cathode Materials

In recent years, some cathode materials (Mn- and V-based materials) with both relatively high capacity and high voltage have been proposed, such as $Co_{0.247}V_2O_5 \cdot 0.944 H_2O$,[51]

**Table 4.2.** The performance of high-voltage aqueous zinc batteries.

| Cathode Material | Electrolyte | Working Voltage (V) | Capacity (mA h g⁻¹)/Energy Density (Wh kg⁻¹) | Mechanism | Ref. |
|---|---|---|---|---|---|
| $CoFe(CN)_6$ | 4 M $Zn(CF_3SO_3)_2$ | 1.75 | 173.4/303 | $Zn^{2+}$ intercalation | [45] |
| $Co^{3+}$-rich $Co_3O_4$ | 2 M $ZnSO_4$ + 0.2 M $CoSO_4$ | 1.80 | 200/360.8 | $Co_3O_4$/CoO conversion | [46] |
| $LiVPO_4F$ | 21 M LiTFSI + 2 M $Zn(OTf)_2$ | 1.65 | 140/230 | $Li^+$ intercalation | [47] |
| $Na_2MnFe(CN)_6$ | 1 M $Na_2SO_4$ + 1 M $ZnSO_4$ + 0.8 mM SDS | 1.6 | 145/235 | $Na^+$ intercalation | [48] |
| $VOPO_4$ | 21 M LiTFSI + 1 M $Zn(Tr)_2$ | 1.56 | 139/216 | $O^{2-}/O^-$ conversion and $Zn^{2+}$ intercalation | [50] |
| CuHCF-CNT | WIG-NaCl/$ZnSO_4$/SA | 1.7 | 260/440 | $Na^+/Zn^{2+}$ intercalation and $ZnCl_4^{2-}$ deintercalation | [49] |
| $PbO_2$ | Cathode: 4 M $H_2SO_4$ + 0.5 M $K_2SO_4$ Anode: 6 M KOH + 0.8 M $Zn(CH_3COO)_2$ | 2.92 | 86/252.39 | $PbO_2$/$PbSO_4$ conversion | [12a] |

$MnV_2O_6$,[14b] and $\alpha$-$MnO_2$[52] (see Table 4.3). However, they are still not ideal for ultrahigh-energy-density zinc batteries. These zinc batteries employ conventional intercalation mechanisms, so the capacity is limited by the structure of the cathode materials. First, the low electron transfer number and the high mass of active material lead to a low theoretical capacity. Second, the limited redox-active sites, slow reaction kinetics, and unsatisfactory electron transfer capability in the structure further lower the actual capacity. Furthermore, although the zinc anode possesses an ideal electrode potential of –0.763 V, the cathode with a low electrode potential due to the intrinsic high $\Delta G$ of the redox reaction may be the bottleneck. Therefore, it has been suggested that finding or further optimizing new storage mechanisms with high electrode potential, multiple electron transfer, and low active material mass might be an effective strategy to achieve ultrahigh-energy-density zinc batteries.

Table 4.3. The performance of both high-capacity and high-voltage aqueous zinc batteries.

| Cathode Material | Battery Configuration | Electrolyte | Working Voltage (V) | Capacity (mA h g⁻¹)/ Energy Density (Wh kg⁻¹) | Mechanism | Ref. |
|---|---|---|---|---|---|---|
| $Co_{0.247}V_2O_5 \cdot 0.944H_2O$ | Coin-type battery | 20 M LiTFSI + 1 M Zn (TFSI)2 | 1.15 | 432/497 | $Zn^{2+}$ intercalation | [51] |
| MVO·CNT | Coin-type battery | 2 M $ZnSO_4$ + 0.1 M $MnSO_4$ | 1.21 | 487/589 | $Zn^{2+}$ intercalation | [14b] |
| $\alpha\text{-}MnO_2$ | Coin-type battery | 1 M $Al(CF_3SO_3)_3$ + 1 M $Zn(CF_3SO_3)_2$ | 1.5 | 265/395 | $Al^{3+}/Zn^{2+}$ intercalation | [52] |
| $MnO_2$ | Home-made battery (open system) | 1 M $MnSO_4$ + 1 M $ZnSO_4$ + 0.1 M $H_2SO_4$ | 1.93 | 570/1100 | $MnO_2/Mn^{2+}$ conversion reaction | [36] |
| $MnO_2$ | Coin-type battery | 1 M $Zn(CH_3COO)_2$ + 0.4 M $Mn(CH_3COO)_2$ | 1.43 | 556/793 | $MnO_2/Mn^{2+}$ conversion | [53] |
| CMO | Coin-type battery | 0.5 M $MnSO_4$ + 1 M $ZnSO_4$ | 2.00 | 616/1232 | $MnO_2/Mn^{2+}$ conversion and $H^+/Zn^{2+}$ intercalation | [54] |
| $MnO_2$ | Coin-type battery | 0.005 M $MnSO_4$ + 2 M $ZnSO_4$ | 1.41 | 430/602 | $MnO_2/Mn^{2+}$ conversion and $H^+/Zn^{2+}$ intercalation | [55] |
| $MnO_2$ | Flow battery | 0.5 M $Mn(Ac)_2$ + 0.5 M $Zn(Ac)_2$ + 2 M KCl | 1.58 | 616/973 | $MnO_2/Mn^{2+}$ conversion | [56] |
| $MnO_2$ | Flow battery | 1 M $MnSO_4$ + 1 M $ZnSO_4$ | 1.72 | — | $MnO_2/Mn^{2+}$ conversion and $H^+/Zn^{2+}$ intercalation | [57] |

| | | | | | | |
|---|---|---|---|---|---|---|
| $MnO_2$ | Decoupled battery | Cathode: 3 M $H_2SO_4$ + 0.1 M $MnSO_4$<br>Anode: 6 M KOH + 0.2 M ZnO + 5 mM vanillin<br>Middle chamber: 0.1 M $K_2SO_4$ | 2.71 | 609/1650 | $MnO_2/Mn^{2+}$ conversion | [13b] |
| $MnO_2$ | Decoupled battery | Cathode: 3 M $MnSO_4$ + 0.3 M $H_2SO_4$ + 0.06 M $NiSO_4$<br>Anode: 3 M NaOH + 0.3 M ZnO | 2.44 | 616/1503 | $MnO_2/Mn^{2+}$ conversion | [13a] |
| $MnO_2$ | Decoupled battery | Cathode: 1 M $MnSO_4$ + 1 M $H_2SO_4$/PAM<br>Anode: 2 M $ZnSO_4$/PAM | 1.82 | — | $MnO_2/Mn^{2+}$ conversion | [58] |
| $MnO_2$ | Decoupled battery | Cathode: 1 M $MnSO_4$ or 0.5 M $KMnO_4$ + 0.5 M $H_2SO_4$<br>Anode: 45 wt% KOH/PAA | 2.45 or 2.8 | 616/1509 or 1724 | $MnO_2/Mn^{2+}$ or $MnO_4^-/Mn^{2+}$ conversion | [59] |
| $MnO_2$ | Decoupled battery | Cathode: 1 M $MnSO_4$ + 1 M $H_2SO_4$ + 0.3 M $CuSO_4$<br>BPE: Cu<br>Anode: 2.4 M KOH + 0.1 M $Zn(Ac)_2$ | 1.79 | 616/1100 | $MnO_2/Mn^{2+}$ conversion | [60] |
| Graphite felt | Decoupled battery | Cathode: 1 M KI<br>BPE: Ag<br>Anode: 2.4 M KOH + 0.1 M $Zn(Ac)_2$ | 2.31 | — | $I_2/I^-$ conversion reaction | [60] |
| Graphite felt | Decoupled battery | Cathode: 1 M NaBr<br>BPE: Ag<br>Anode: 2.4 M KOH + 0.1 M $Zn(Ac)_2$ | 2.53 | — | $Br_2/Br^-$ conversion reaction | [60] |

(Continued)

**Table 4.3.** (*Continued*)

| Cathode Material | Battery Configuration | Electrolyte | Working Voltage (V) | Capacity (mA h g$^{-1}$)/ Energy Density (Wh kg$^{-1}$) | Mechanism | Ref. |
|---|---|---|---|---|---|---|
| Carbon felt | Flow and decoupled battery | Cathode: 1.7 M V$^{3.5+}$ + 4 M H$_2$SO$_4$<br>Anode: 3 M ZnBr$_2$ + 1 M ZnCl$_2$ + 1:1 M MEP/MEM | 1.7 | 318/559.68 | VO$^{2+}$/VO$_2^+$ conversion reaction | [61] |
| Carbon felt | Flow and decoupled battery | Cathode: 0.5 M Ce(ClO$_4$)$_3$ + 0.5 M HClO$_4$<br>Anode: 3 M NaOH + 0.5 M Na$_2$[Zn(OH)$_4$]<br>Middle chamber: 4 M NaClO$_4$ | 2.8 | 191/536 | Ce$_2$O$^{6+}$/Ce$^{3+}$ conversion reaction | [62] |
| Carbon felt | Flow and decoupled battery | Cathode: 1 M FeCl$_2$ +1 M HCl<br>Anode: 4 M NaOH + 0.5 M Na$_2$[Zn(OH)$_4$]<br>Middle chamber: 3 M NaCl | 1.76 | 359/632 | Fe$^{2+}$/Fe$^{3+}$ conversion reaction | [63] |
| Porous carbon | Flow and decoupled battery | Cathode: 1 M KBr + Br$_2$ + 0.5 M H$_2$SO$_4$<br>Anode: 2 M KOH + 0.02 M Zn(CH$_3$COO)$_2$ | 1.98 | 392/775 | Br$_3^-$/Br$^-$ conversion reaction | [64] |
| MnO$_2$ | Flow and decoupled battery | Cathode: 1 M MnSO$_4$ + 0.5 M H$_2$SO$_4$<br>Anode: 2.4 M KOH + 0.1 M Zn(CH$_3$COO)$_2$ | 2.44 | 616/1503 | MnO$_2$/Mn$^{2+}$ conversion | [65] |

## 4.4.2 *Optimizing the MnO$_2$/Mn$^{2+}$ Conversion Reaction Mechanism*

In 2019, the proposal of a novel Zn//MnO$_2$ battery containing a high concentration of acid (pH < 1) based on the MnO$_2$/Mn$^{2+}$ conversion mechanism created a new avenue to increase the energy density.[35] This new energy storage mechanism [as shown in Eqs. (4.1)–(4.3) and Figure 4.7] benefits from the high theoretical potential (1.991 V vs. Zn$^{2+}$/Zn) and theoretical capacity of 616 mAh g$^{-1}$, which endows this Zn//MnO$_2$ battery with a high output voltage of 1.95 V, a remarkable capacity of ≈570 mAh g$^{-1}$, and unprecedented energy density of ≈1100 Wh kg$^{-1}$.

Cathode:

$$Mn^{2+} + 2H_2O = MnO_2 + 4H^+ + 2e^- \qquad E^\ominus = 1.228 \text{ V vs. SHE} \qquad (4.1)$$

Anode:

$$Zn^{2+} + 2e^- = Zn \qquad E^\ominus = -0.763 \text{ V vs. SHE} \qquad (4.2)$$

Overall:

$$Mn^{2+} + 2H_2O + Zn^{2+} = MnO_2 + 4H^+ + Zn \quad E = 1.991 \text{ V vs. } Zn^{2+}/Zn \qquad (4.3)$$

However, MnO$_2$ dissolution/deposition is a solid-liquid reaction with a phase interface, which is controlled by diffusion.[35] MnO$_2$ dissolution/deposition consists of two elementary reactions from Mn$^{2+}$ to Mn$^{3+}$ to MnO$_2$ (Figure 4.7). Therefore, the specific

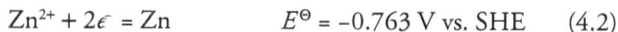
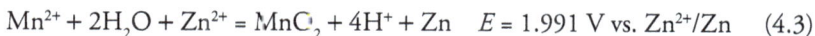
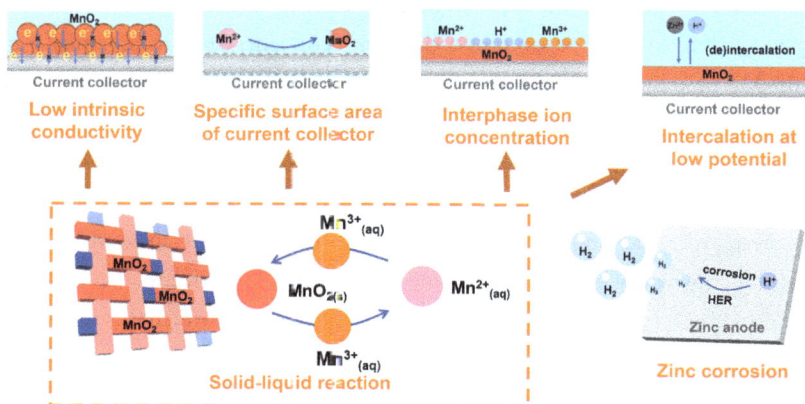

**Figure 4.7.** Summary of existing key issues in Zn//MnO$_2$ batteries based on the MnO$_2$ dissolution/deposition mechanism.

surface area and reaction redox-active sites of the current collector determine the reversible $MnO_2$ deposition/dissolution capacity. Once the $MnO_2$ layer on the electrode surface becomes too thick, the reaction kinetics will slow down greatly, in addition to the poor reaction reversibility due to the low intrinsic conductivity. Slow reaction kinetics also leads to increased overpotential as well as a decrease in energy efficiency. Furthermore, aggregation and diffusion of ions at the interface including $Mn^{3+}$, $Mn^{2+}$, and $H^+$ will profoundly affect the discharge voltage and plateau of $MnO_2$, during which ion intercalation may occur at low potentials. Each of the above factors can be detrimental to the Coulomb efficiency, energy efficiency, and reaction kinetics of Zn// $MnO_2$ batteries. Moreover, hydrogen may be produced on the zinc anode in a strong acid environment.[36,65] Therefore, in contrast to the conventional coin or pouch cells, this type of Zn//$MnO_2$ battery is usually in a home-made configuration, which inevitably increases the mass and volume of the battery. Immediately after the pioneering work, researchers proposed many strategies to tackle these issues, including designing mild electrolytes and decoupled or liquid flow devices. As a result, the output voltage has been pushed beyond 2.0 V and ultrahigh energy densities of 600–1700 $Wh\,kg^{-1}$ have been achieved (Table 4.3).

In comparison to strongly acidic and alkaline electrolytes, mild electrolytes with low corrosiveness can endow the battery with higher durability (Figure 4.8a). However, the low acidity of the electrolyte also means that the OER will compete with the deposition of $MnO_2$ during the charging process, giving rise to low Coulombic efficiency. One solution is to adjust the $Mn^{2+}$ concentration to a critical range (0.005 M) to achieve the reversible $MnO_2$/$Mn^{2+}$ reaction while suppressing the occurrence of OER.[55] Afterward, a $Mn^{2+}$-rich multivalent manganese oxide layer on carbon cloth (CMO) was proposed. It is worth mentioning that the oxygen vacancies in CMO can reduce the local oxygen mobility to inhibit the OER and the low-valence manganese oxides can endow CMO with more active electron transfer dynamics.[54] In addition, the type of manganese and zinc salts also affects the electrochemical process in the Zn//$MnO_2$ batteries. Different from the $MnSO_4$ electrolyte, in the mild acetate-based electrolyte, $Mn^{2+}$ ions are directly transformed into $MnO_2$ without disproportionation to $Mn^{3+}$ during the charging process, due to the coordination effect of $CH_3COO^-$, which effectively promotes the reversibility of the $MnO_2$/$Mn^{2+}$ conversion reaction (Figure 4.8b).[56] Meanwhile, the mild acetate-based electrolyte

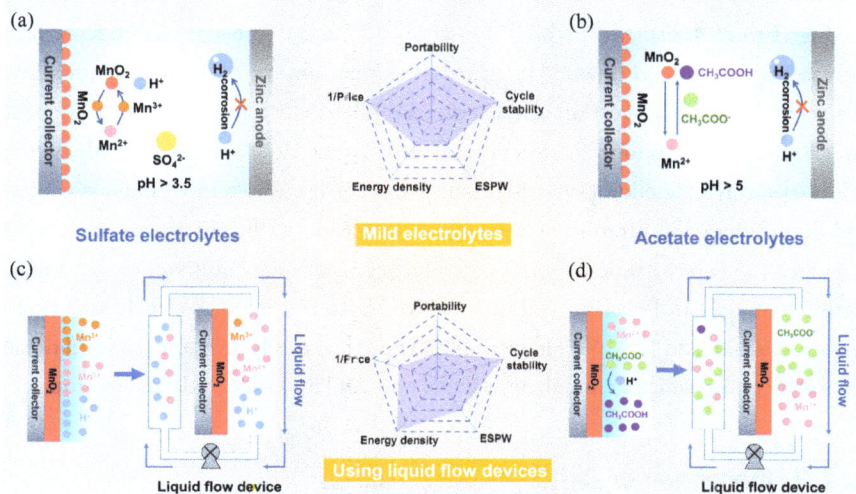

**Figure 4.8.** Zn//MnO$_2$ batteries with (a) mild sulfate and (b) acetate electrolytes along with their various property indices. Zn//MnO$_2$ flow batteries with (c) sulfate and (d) acetate electrolytes along with their various property indices.

has excellent compatibility and stability for the zinc anode.[53,67] Furthermore, a KI mediator can promote the dissolution of residual MnO$_2$ in this system, which undoubtedly opens a new direction towards high-area-capacity Zn//MnO$_2$ batteries.[68] On the other hand, acetate ions will reduce the overall reaction voltage of the Zn//MnO$_2$ battery. Hence, the concentration of acetate ions should be carefully adjusted.

We elaborate on the critical effect of the Mn$^{3+}$ intermediate in the Zn//MnO$_2$ battery on the reversibility of MnO$_2$ dissolution/deposition. Especially in sulfate electrolytes, Mn$^{3+}$ as an intermediate of the dissolution/deposition of MnO$_2$ determines the rate and extent of the reaction, while the stability of Mn$^{3+}$ in solution improves with increasing acidity (Figure 4.8c). Consequently, if the specific surface area and reaction sites of the carbon-based current collector are not sufficient to support the Mn$^{3+}$ disproportionation, the reversibility of the MnO$_2$ dissolution in the subsequent cycles is strongly influenced by Mn$^{3+}$ stabilized on the electrode surface. In the acetate electrolytes, the coordination effect of acetate will gradually be weakened because there is always a surplus of $H^+$ that will combine with acetate to form acetic acid during cycling (Figure 4.8d). Meanwhile, the sacrifice of acetate on the electrode surface increases the likelihood of $Mn^{3+}$ intermediate formation. In this context, liquid flow devices with forced convection characteristics can have the following advantages: this design effectively promotes the reaction of $Mn^{3+}$ on the electrode surface, reduces the risk of electrolyte deterioration, instantaneously refuels the electrolytes on electrodes, and diminishes concentration polarization and provides sufficient charge carriers for high-area-capacity batteries.[69] As a consequence, the energy and voltage efficiency, output voltage, and cycling stability of flow batteries are apparently improved (Figure 4.8c and 4.8d).[56,57,65,68] However, liquid flow devices rely on pumps and additional electrolyte tanks, which makes them suitable only for large-scale static energy storage.

### 4.4.3 Innovation of Battery Configurations

Admittedly, in a strongly acidic reaction system, the zinc anode is subject to corrosion by the acidic electrolyte together with intensive HER. In view of that, decoupled devices, in which the highly acidic electrolyte on the cathode side is separated from the alkaline solution on the Zn anode side by a membrane, have significant ramifications. This decoupled battery not only expands the ESPW ($\approx 3.4$ V) along with the suppression of HER, OER, and excessive dissolution of Zn, it also introduces the anodic reaction with a lower electrode potential ($Zn/Zn(OH)_4^{2-}$, $-1.199$ V vs. SHE) (Figure 4.9a).

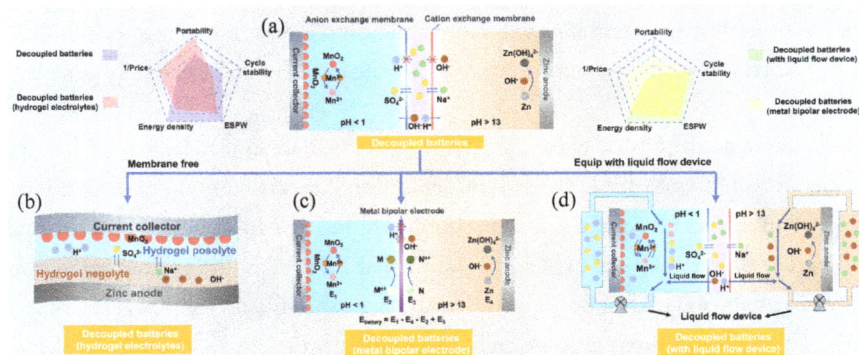

**Figure 4.9.** Decoupled $Zn//MnO_2$ batteries. (a) Schematic of the architecture of a decoupled $Zn//MnO_2$ battery with (b) a hydrogel electrolyte, (c) metal bipolar electrode, and (d) liquid flow device configuration along with various property indices.

The electrochemical performance of decoupled batteries is largely influenced by ion-exchange membranes.[70] For example, bipolar membranes (BPM) composed of proton- and anion-exchange membranes and a middle catalyst layer can not only slow down the pH changes of the anode and cathode electrolytes, but also shorten the distance between the anode and cathode, thereby reducing the electrode overpotential.[13a] However, the energy required for water ionization during discharge and the narrow ion storage space are the drawbacks of BPMs. It is clear that Zn//MnO$_2$ batteries with this two-membrane three-chamber configuration are not suitable for powering consumer electronics. However, they are appropriate for storing electricity generated by wind and photovoltaic hybrid power systems, because the intermediate neutral chamber can accommodate sufficient ions for transporting to provide higher discharge capacity (3000–6000 mAh). Hence, this novel decoupled device is considered as the state-of-the-art zinc battery for large-scale energy storage.[13b]

Considering the expense of ion-exchange membranes, low-cost decoupled hydrogel electrolytes are possible instead of decoupled liquid electrolytes (Figure 4.9b). First, the low mobility of ions caused by the low water content suppresses the ion crossover in decoupled hydrogel electrolytes, which prevents the mixing of cathode and anode electrolytes even without the use of functional separators.[58] Second, the excellent electrochemical stability, mechanical flexibility, and renewability of the hydrogel electrolytes can broaden the configuration scope for this type of Zn//MnO$_2$ battery, including cathode-less batteries and flexible batteries.[54,58,59] Although the design of decoupled devices can effectively enhance the working voltage and energy density of batteries (over 2.4 V and up to 1700 Wh kg$^{-1}$ in Zn//MnO$_2$ batteries), the poor cycling stability and low voltage efficiency caused by the unavoidable ion crossover remain non-trivial issues that need to be addressed. Currently, a metal bipolar electrode (BPE: Cu, Ag) can be used instead of an ion-exchange membrane in decoupled batteries.[60] It can not only participate in the redox reaction to transfer charge, but also realize the separation of electrolyte without ion crossover (Figure 4.9c). However, this BPE requires very strict specific surface area and conductivity, and the potential difference between the reactions on both sides of its surface can affect the overall operating voltage of the batteries.

Although liquid flow batteries were developed decades ago, their integration with decoupled batteries provides new opportunities based on their respective advantages to overcome some issues (Figure 4.9d). Therefore, based on this upgraded battery configuration, various high-capacity and high-voltage zinc batteries have also been proposed, including Zn//MnO$_2$,[65] Zn//Br$_2$,[64] Zn//Fe,[63] Zn//V,[61] and Zn//Ce[62] batteries (Table 4.3). Ion crossover is a common phenomenon in decoupled batteries due to the fact that water will ionize into H$^+$ and OH$^-$; thus OH$^-$ will cross the anion-exchange membrane driven by an electric field to neutralize the acidic electrolyte, and conversely, H$^+$ will cross the cation-exchange membrane to neutralize the alkaline electrolyte. The intervention of the liquid flow device might be an ideal solution to alleviate the unavoidable ion crossover.

## 4.5 Summary and Outlook

Ways of improving the energy density of the various types of batteries have been common topics in the field of energy storage, which is also closely related to human life and social development. Zinc batteries are regarded as a promising alternative technology for next-generation safe batteries. However, the main pitfall is the insufficient energy density. Despite great progress in high-energy-density aqueous zinc batteries in recent years, many issues still need to be addressed. The overall excellent performance of aqueous zinc batteries must come from the holistic improvement of various parts instead of individual parts. Herein, we propose some perspectives and prospects in view of these issues (Figure 4.10):

*Mechanisms that allow both high-voltage and high-capacity energy storage must be further exploited or optimized.* Sulfur in conversion-type materials has a natural capacity advantage, yet the proposed S redox reactions so far are not equipped with the ideal Gibbs free energy change ($\Delta G$), resulting in low electrode potential. Therefore, besides the MnO$_2$/Mn$^{2+}$ mechanism, designing a new high-voltage conversion reaction mechanism for S may be a promising pathway to achieve ultra-high-energy density. Firstly, in theory, the reaction systems with high voltage and high theoretical capacity for cathodes should be designed according to the relationship between $\Delta G$ of the redox reactions and the electrode potential, along with the consideration of the side reactions (HER, OER, etc.). Secondly, the feasibility of this mechanism must be further investigated in zinc batteries through the holistic optimization of the device architecture (coin-type batteries, decoupled batteries, flow batteries, all-solid-state batteries, etc.),

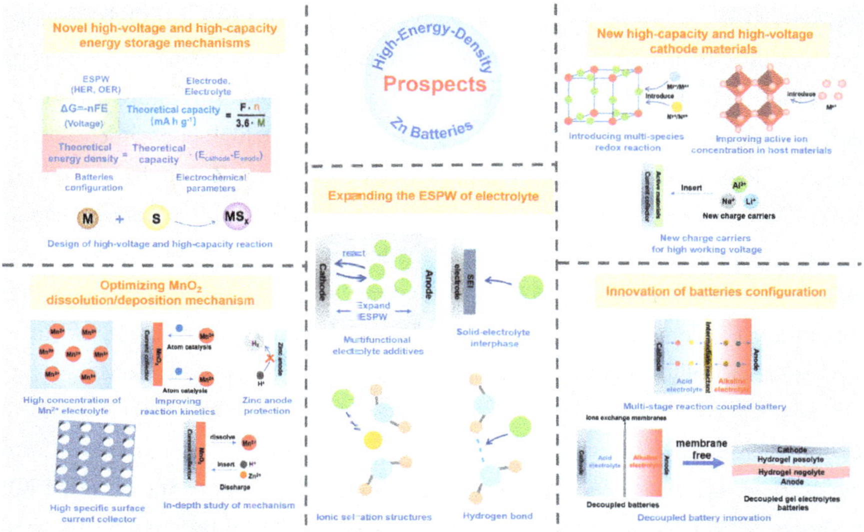

**Figure 4.10.** Future prospects for high-energy-density zinc batteries.

current collector (carbon cloth, carbon felt, Ti foil, stainless steel, etc.), electrolyte (species, concentration, ESPW, etc.), and electrochemical parameters (voltage range, current density, charging/discharging capacity, etc.).

*High-capacity and high-voltage cathode materials must be developed.* Cathode materials such as PBAs, cobalt-based oxides, and polyanionic compounds exhibit unique high-voltage characteristics, while their theoretical capacities are not satisfactory. Therefore, it is imperative to focus on improving the capacity of high-voltage cathode materials. Some strategies, such as introducing multi-species redox reaction and adjusting the concentration of ions in host materials, can be implemented to increase the electron transfer number and the mass proportion of active species, thereby improving the intrinsic capacity of the redox-active materials. On the other hand, for high-capacity cathode materials, the intrinsic low electrode potential is still difficult to break through. Under the premise of ensuring the stable operation of zinc negative electrode, attempting to introduce multifarious charge carriers, such as $Al^{3+}$, $Li^+$, $Na^+$, etc.,[71] might provide a higher discharge platform, and of course, the crystal structure of cathode materials should also match with these charge carriers.

*The MnO$_2$/Mn$^{2+}$ electrolytic battery is promising but also challenging.* MnO$_2$ is the most common cathode material but its Zn storage mechanism in aqueous solutions is by no means simple. The MnO$_2$/Mn$^{2+}$ conversion reaction mechanism featuring low cost, high capacity, and high voltage has been implemented in ultrahigh-energy-density zinc batteries. However, the strong acid solution severely corrodes the Zn anode. In addition, the H$^+$/Zn$^{2+}$ (de)intercalation at low potentials as well as the inevitable accumulation of Mn$^{3+}$ near the electrode surface lead to the limitations of the Zn//MnO$_2$ battery, including decrease in energy efficiency and average voltage, along with poor cycling stability and Coulombic efficiency. Meanwhile, the inherent low conductivity of MnO$_2$ is an issue especially in high-capacity charging, where an excessively thick MnO$_2$ layer will slow the carrier transport across the electrode. The specific surface area and hydrophilicity of the current collector also affect the deposition rate, discharge platform and depth of manganese dioxide. Moreover, the maximum capacity of the Zn//MnO$_2$ battery is determined by the concentration of Mn$^{2+}$ in the electrolyte. According to theoretical calculations, the theoretical maximum energy densities (calculated from the total mass of 1, 2 and 3 mol MnSO$_4$ and 1 L H$_2$O) of the Zn//MnO$_2$ battery are 92.7, 163.9 and 220.3 Wh kg$^{-1}$, respectively. However, a high concentration of Mn$^{2+}$ will inhibit the dissolution of MnO$_2$ and greatly reduce the energy and Coulombic efficiency, and might also induce different storage processes. Hence, the relationship between Mn$^{2+}$ concentration and capacity and energy density should also be controlled carefully to assure both high specific capacity and potential as well as high reversibility.

*The decoupled configuration is an innovation.* Of the many different aqueous battery configurations, we are optimistic that the decoupled battery design holds the greatest potential towards high energy density. This design can greatly expand the ESPW and increase the potential difference between the anode and cathode, leading to high energy densities. However, the use of ion-exchange membranes substantially increases the cost of the batteries, and likely also the mass and volume of the entire battery, causing a decrease in total energy density. Following the idea of the decoupled electrolyte, the application of gel electrolytes avoids the intervention of ion-exchange membranes and also simplifies the structure of batteries. Furthermore, the combined use of liquid and gel electrolytes can compensate for the poor ionic conductivity. Meanwhile, decoupled batteries with multifunctional BPE are also worthy of investigation, especially if these BPEs exhibit anti-HER and -OER effects to suppress side reactions.

*Developing electrolytes with wide ESPW is imperative.* The energy density of aqueous zinc batteries is closely dependent on the ESPW of the electrolyte, which determines the maximum reaction voltage that the battery system can withstand. In most cases, one must consider a tradeoff between mass, cost, and ESPW of an electrolyte. However, the commonly used high-concentration electrolytes introduce too many inactive ions in addition to their high cost. Multifunctional electrolyte additives are being extensively studied for high-voltage reactions (as well as for stable Zn striping/plating on the anode). The additives may introduce extra charge carriers for specific conversion reactions and expand the ESPW while maximizing the concentration of redox-active ions. Moreover, research on the connection between electrolyte additives and ionic solvation structures, hydrogen bonding with water, and SEI is highly relevant. The emergent deep eutectic solvent electrolytes show interesting effects but more comprehensive study is mandatory.

The relatively slow improvement in the energy density of electrochemical energy storage devices is the main hurdle to the fast development of electric vehicles, electronic chips, and other high-tech devices. While the rapid progress of aqueous zinc batteries offers exciting opportunities, their energy density is still far from expectations. There is a long way to go and innovation never stops. We hope this chapter has provided useful insights into the challenges and future development of higher-energy-density zinc batteries.

## Acknowledgements

This work was supported by the National Natural Science Foundation of China (Grant Nos. 51972346, 51932011), and the Hunan Outstanding Youth Talents (2021JJ10064). H.J.F. acknowledges the financial support from the Singapore Ministry of Education by a Tier 1 grant (RG157/19).

## References

[1]  a) D. Larcher, J. M. Tarascon, *Nat. Chem.* **2014**, 7, 19; b) S. Chu, Y. Cui, N. Liu, *Nat. Mater.* **2016**, 16, 16; c) R. Fang, S. Zhao, Z. Sun, D. Wang, H. Cheng, F. Li, *Adv. Mater.* **2017**, 29, 1606823; d) X. Gao, H. Yang, *Energy Environ. Sci.* **2010**, 3, 174.

[2]  a) H. Ao, Y. Zhao, J. Zhou, W. Cai, X. Zhang, Y. Zhu, Y. Qian, *J. Mater. Chem. A* **2019**, 7, 18708; b) A. Zhou, Y. Liu, X. Zhu, X. Li, J. Yue, X. Ma, L. Gu, Y.-S. Hu, H. Li,

X. Huang, L. Chen, L. Suo, *Energy Stor. Mater.* **2021**, 42, 438; c) X. Xu, F. Xiong, J. Meng, X. Wang, C. Niu, Q. An, L. Mai, *Adv. Funct. Mater.* **2020**, 30, 1904398; d) F. Wu, X. Gao, X. Xu, Y. Jiang, X. Gao, R. Yin, W. Shi, W. Liu, G. Lu, X. Cao, *ChemSusChem* **2020**, 13, 1537.

[3] a) F. Wan, Z. Hao, S. Wang, Y. Ni, J. Zhu, Z. Tie, S. Bi, Z. Niu, J. Chen, *Adv. Mater.* **2021**, 33, e2102701; b) C. Li, X. Shi, S. Liang, X. Ma, M. Han, X. Wu, J. Zhou, *Chem. Eng. J.* **2020**, 379, 122248; c) P. He, Y. Quan, X. Xu, M. Yan, W. Yang, Q. An, L. He, L. Mai, *Small* **2017**, 13, 1702551; d) X. Xu, Y. Chen, D. Zheng, P. Ruan, Y. Cai, X. Dai, X. Niu, C. Pei, W. Shi, W. Liu, F. Wu, Z. Pan, H. Li, X. Cao, *Small* **2021**, 17, e2101901; e) B. Li, X. Zhang, T. Wang, Z. He, B. Lu, S. Liang, J. Zhou, *Nano-Micro Lett.* **2021**, 14, 6.

[4] a) D. Chao, S.-Z. Qiao, *Joule* **2020**, 4, 1846; b) X. Guo, J. Zhou, C. Bai, X. Li, G. Fang, S. Liang, *Mater. Today Energy* **2020**, 16, 100396; c) Z. Liu, Y. Huang, Y. Huang, Q. Yang, X. Li, Z. Huang, C. Zhi, *Chem. Soc. Rev.* **2020**, 49, 180; d) V. Verma, S. Kumar, W. Manalastas, M. Srinivasan, *ACS Energy Lett.* **2021**, 6, 1773; e) X. Wang, Z. Zhang, B. Xi, W. Chen, Y. Jia, J. Feng, S. Xiong, *ACS Nano* **2021**, 15, 9244; f) Y. Fang, X. Xie, B. Zhang, Y. Chai, B. Lu, M. Liu, J. Zhou, S. Liang, *Adv. Funct. Mater.* **2021**, 32, 2109671.

[5] a) J. Song, K. Xu, N. Liu, D. Reed, X. Li, *Mater. Today* **2021**, 45, 191; b) J. Yan, E. H. Ang, Y. Yang, Y. Zhang, M. Ye, W. Du, C. C. Li, *Adv. Funct. Mater.* **2021**, 31, 2010213; c) P. Ruan, X. Xu, J. Feng, L. Yu, X. Gao, W. Shi, F. Wu, W. Liu, X. Zang, F. Ma, X. Cao, *Mater. Res. Bull.* **2021**, 133, 111077.

[6] N. Fan, C. Sun, D. Kong, Y. Qian, *J. Power Sources* **2014**, 254, 323.

[7] S. Ovshinsky, M. R. J. Fetcenko, *Science* **1993**, 260, 176.

[8] K. Mizushima, P. C. Jones, P. J. Wiseman, J. B. Goodenough, *Mat. Res. Bull.* **1980**, 15, 783.

[9] a) D. Kundu, B. D. Adams, V. Duffort, S. H. Vajargah, L. F. Nazar, *Nat. Energy* **2016**, 1, 16119; b) C. Li, X. Xie, H. Liu, P. Wang, C. Deng, B. Lu, J. Zhou, S. Liang, *Natl. Sci. Rev.* **2021**, 9, nwab177.

[10] Y. Lu, Y. Lu, Z. Niu, J. Chen, *Adv. Energy Mater.* **2018**, 8, 1702469.

[11] a) Q. Li, Q. Zhang, C. Liu, Z. Zhou, C. Li, B. He, P. Man, X. Wang, Y. Yao, *J. Mater. Chem. A* **2019**, 7, 12997; b) X. Wang, Y. Li, S. Wang, F. Zhou, P. Das, C. Sun, S. Zheng, Z. Wu, *Adv. Energy Mater.* **2020**, 10, 2000081.

[12] a) Y. Xu, P. Cai, K. Chen, Y. Ding, L. Chen, W. Chen, Z. Wen, *Angew. Chem. Int. Ed.* **2020**, 59, 23593; b) Q. Ni, H. Jiang, S. Sandstrom, Y. Bai, H. Ren, X. Wu, Q. Guo, D. Yu, C. Wu, X. Ji, *Adv. Funct. Mater.* **2020**, 30, 2003511.

[13] a) D. Chao, C. Ye, F. Xie, W. Zhou, Q. Zhang, Q. Gu, K. Davey, L. Gu, S. Z. Qiao, *Adv. Mater.* **2020**, 32, e2001894; b) C. Zhong, B. Liu, J. Ding, X. Liu, Y. Zhong, Y. Li, C. Sun, X. Han, Y. Deng, N. Zhao, W. Hu, *Nat. Energy* **2020**, 5, 440.

[14] a) Y. Yang, Y. Tang, G. Fang, L. Shan, J. Guo, W. Zhang, C. Wang, L. Wang, J. Zhou, S. Liang, *Energy Environ. Sci.* **2018**, 11, 3157; b) Y. Wu, Z. Zhu, Y. Li, D. Shen, L. Chen, T. Kang, X. Lin, Z. Tong, H. Wang, C. S. Lee, *Small* **2021**, 17, e2008182; c) X. Liu, G. Xu, Q. Zhang, S. Huang, L. Li, X. Wei, J. Cao, L. Yang, P. K. Chu, *J. Power Sources* **2020**, 463, 228223; d) F. Wu, Y. Wang, P. Ruan, X. Niu, D. Zheng, X. Xu, X. Gao, Y. Cai, W. Liu, W. Shi, X. Cao, *Mater. Today Energy* **2021**, 21, 100842; e) L. Shan, Y. Wang, S. Liang, B. Tang, Y. Yang, Z. Wang, B. Lu, J. Zhou, *InfoMat* **2021**, 3, 1028; f) X. Ma, X. Cao, M. Yao, L. Shan, X. Shi, G. Fang, A. Pan, B. Lu, J. Zhou, S. Liang, *Adv. Mater.* **2021**, 34, 2105452.

[15] P. He, J. Liu, X. Zhao, Z. Ding, P. Gao, L.-Z. Fan, *J. Mater. Chem. A* **2020**, 8, 10370.

[16] a) J. Ding, H. Zheng, H. Gao, Q. Liu, Z. Hu, L. Han, S. Wang, S. Wu, S. Fang, S. Chou, *Adv. Energy Mater.* **2021**, 11, 2100973; b) H. Luo, B. Wang, F. Wang, J. Yang, F. Wu, Y. Ning, Y. Zhou, D. Wang, H. Liu, S. Dou, *ACS Nano* **2020**, 14, 7328; c) K. Zhu, T. Wu, K. Huang, *Energy Stor. Mater.* **2021**, 38, 473; d) S. Wei, S. Chen, X. Su, Z. Qi, C. Wang, B. Ganguli, P. Zhang, K. Zhu, Y. Cao, Q. He, D. Cao, X. Guo, W. Wen, X. Wu, P. M. Ajayan, L. Song, *Energy Environ. Sci.* **2021**, 14, 3954; e) J. Ding, Z. Du, B. Li, L. Wang, S. Wang, Y. Gong, S. Yang, *Adv. Mater.* **2019**, 31, e1904369.

[17] a) X. Wang, L. Chen, F. Lu, J. Lu, X. Chen, G. Shao, *ChemElectroChem* **2019**, 6, 3644; b) J. Cui, Z. Guo, J. Yi, X. Liu, K. Wu, P. Liang, Q. Li, Y. Liu, Y. Wang, Y. Xia, J. Zhang, *ChemSusChem* **2020**, 13, 2160.

[18] a) Z. Tie, Z. Niu, *Angew. Chem. Int. Ed.* **2020**, 59, 21293; b) Q. Zhao, W. Huang, Z. Luo, L. Liu, Y. Lu, Y. Li, L. Li, J. Hu, H. Ma, J. Chen, *Sci. Adv.* **2018**, 4, 1761; c) T. Sun, H. J. Fan, *Curr. Opin. Electrochem.* **2021**, 30, 100799.

[19] W. Wang, V. S. Kale, Z. Cao, Y. Lei, S. Kandambeth, G. Zou, Y. Zhu, E. Abouhamad, O. Shekhah, L. Cavallo, M. Eddaoudi, H. N. Alshareef, *Adv. Mater.* **2021**, 33, e2103617.

[20] Y. Zhao, Y. Huang, F. Wu, R. Chen, L. Li, *Adv. Mater.* **2021**, 33, 2106469.

[21] W. Li, K. Wang, K. Jiang, *Adv. Sci.* **2020**, 7, 2000761.

[22] a) C. Dai, X. Jin, H. Ma, L. Hu, G. Sun, H. Chen, Q. Yang, M. Xu, Q. Liu, Y. Xiao, X. Zhang, H. Yang, Q. Guo, Z. Zhang, L. Qu, *Adv. Energy Mater.* **2021**, 11, 2003982; b) C. Dai, L. Hu, X. Jin, H. Chen, X. Zhang, S. Zhang, L. Song, H. Ma, M. Xu, Y. Zhao, Z. Zhang, H. Cheng, L. Qu, *Adv. Mater.* **2021**, 33, 2105480; c) X. Wu, A. Markir, L. Ma, Y. Xu, H. Jiang, D. P. Leonard, W. Shin, T. Wu, J. Lu, X. Ji, *Angew. Chem. Int. Ed.* **2019**, 58, 12640.

[23] W. Li, Y. Ma, P. Li, X. Jing, K. Jiang, D. Wang, *Adv. Funct. Mater.* **2021**, 31, 2101237.

[24] Z. Chen, F. Mo, T. Wang, Q. Yang, Z. Huang, D. Wang, G. Liang, A. Chen, Q. Li, Y. Guo, X. Li, J. Fan, C. Zhi, *Energy Environ. Sci.* **2021**, 14, 2441.

[25] J. Shin, D. S. Choi, H. J. Lee, Y. Jung, J. W. Choi, *Adv. Energy Mater.* **2019**, 9, 1900083.

[26] a) Li W., Han C., Gu Q., Chou S., Wang J., Liu H., Dou S., *Adv. Energy Mater.* **2020**, 10, 202001852; b) D. Bin, W. Huo, Y. Yuan, J. Huang, Y. Liu, Y. Zhang, F. Dong, Y. Wang, Y. Xia, *Chem* **2020**, 6, 968; c) S. Liu, H. Zhu, B. Zhang, G. Li, H. Zhu, Y. Ren, H. Geng, Y. Yang, Q. Liu, C. C. Li, *Adv. Mater.* **2020**, 32, 2001113; d) T.-T. Lv, Y.-Y. Liu, H. Wang, S.-Y. Yang, C.-S. Liu, H. Pang, *Chem. Eng. J.* **2021**, 411, 128533.

[27] a) D. Bin, Y. Liu, B. Yang, J. Huang, X. Dong, X. Zhang, Y. Wang, Y. Xia, *ACS Appl. Mater. Interfaces* **2019**, 11, 20796; b) T. Sun, Q. Nian, S. Zheng, X. Yuan, Z. Tao, *J. Power Sources* **2020**, 478, 228758.

[28] M. Zhang, W. Wu, J. Luo, H. Zhang, J. Liu, X. Liu, Y. Yang, X. Lu, *J. Mater. Chem. A* **2020**, 8, 11642.

[29] T. Xiong, Z. G. Yu, H. Wu, Y. Du, Q. Xie, J. Chen, Y. W. Zhang, S. J. Pennycook, W. S. V. Lee, J. Xue, *Adv. Energy Mater.* **2019**, 9, 1803815.

[30] a) Y. Wu, M. Wang, Y. Tao, K. Zhang, M. Cai, Y. Ding, X. Liu, T. Hayat, A. Alsaedi, S. Dai, *Adv. Funct. Mater.* **2019**, 30, 1907120; b) X. Zhu, Z. Cao, W. Wang, H. Li, J. Dong, S. Gao, D. Xu, L. Li, J. Shen, M. Ye, *ACS Nano* **2021**, 15, 2971.

[31] a) Y. Zeng, X. Zhang, Y. Meng, M. Yu, J. Yi, Y. Wu, X. Lu, Y. Tong, *Adv. Mater.* **2017**, 29, 1700274; b) Y. Fu, Q. Wei, G. Zhang, X. Wang, J. Zhang, Y. Hu, D. Wang, L. Zuin, T. Zhou, Y. Wu, S. Sun, *Adv. Energy Mater.* **2018**, 8, 1801445; c) L. Chen, Z. Yang, F. Cui, J. Meng, Y. Jiang, J. Long, X. Zeng, *Mater. Chem. Front.* **2020**, 4, 213; d) B. Wu, G. Zhang, M. Yan, T. Xiong, P. He, L. He, X. Xu, L. Mai, *Small* **2018**, 14, 1703850; e) P. Ruan, X. Xu, X. Gao, J. Feng, L. Yu, Y. Cai, X. Gao, W. Shi, F. Wu, W. Liu, X. Zang, F. Ma, X. Cao, *Sustain. Mater. Technol.* **2021**, 28, e00254.

[32] J. Ji, H. Wan, B. Zhang, C. Wang, Y. Gan, Q. Tan, N. Wang, J. Yao, Z. Zheng, P. Liang, J. Zhang, H. Wang, L. Tao, Y. Wang, D. Chao, H. Wang, *Adv. Energy Mater.* **2020**, 11, 2003203.

[33] Z. Luo, S. Zheng, S. Zhao, X. Jiao, Z. Gong, F. Cai, Y. Duan, F. Li, Z. Yuan, *J. Mater. Chem. A* **2021**, 9, 6131.

[34] S. Wang, Z. Yuan, X. Zhang, S. Bi, Z. Zhou, J. Tian, Q. Zhang, Z. Niu, *Angew. Chem. Int. Ed.* **2021**, 60, 7056.

[35] H. Pan, Y. Shao, P. Yan, Y. Cheng, K. S. Han, Z. Nie, C. Wang, J. Yang, X. Li, P. Bhattacharya, K. T. Mueller, J. Liu, *Nat. Energy* **2016**, 1, 16039.

[36] D. Chao, W. Zhou, C. Ye, Q. Zhang, Y. Chen, L. Gu, K. Davey, S. Z. Qiao, *Angew. Chem. Int. Ed.* **2019**, 58, 7823.

[37] Y. Zou, T. Liu, Q. Du, Y. Li, H. Yi, X. Zhou, Z. Li, L. Gao, L. Zhang, X. Liang, *Nat. Commun.* **2021**, 12, 170.

[38] J. Hao, L. Yuan, B. Johannessen, Y. Zhu, Y. Jiao, C. Ye, F. Xie, S. Z. Qiao, *Angew. Chem. Int. Ed.* **2021**, 60, 25114.

[39] S. Deng, Z. Yuan, Z. Tie, C. Wang, L. Song, Z. Niu, *Angew. Chem. Int. Ed.* **2020**, 59, 22002.

[40] D. Chen, M. Lu, B. Wang, R. Chai, L. Li, D. Cai, H. Yang, B. Liu, Y. Zhang, W. Han, *Energy Stor. Mater.* **2021**, 35, 679.

[41] L. Hu, Z. Wu, C. Lu, F. Ye, Q. Liu, Z. Sun, *Energy Environ. Sci.* **2021**, 14, 4095.

[42] L. Wang, Z. Cao, P. Zhuang, J. L., H. Chu, Z. Ye, D. Xu, H. Zhang, J. Shen, M. Ye, *ACS Appl. Mater. Interfaces* **2021**, 13, 13338.

[43] W. Qiu, Y. Li, A. You, Z. Zhang, G. Li, X. Lu, Y. Tong, *J. Mater. Chem. A* **2017**, 5, 14838.

[44] L. Zhang, L. Chen, X. Zhou, Z. Liu, *Adv. Energy Mater.* **2015**, 5, 1400930.

[45] L. Ma, S. Chen, C. Long, X. Li, Y. Zhao, Z. Liu, Z. Huang, B. Dong, J. A. Zapien, C. Zhi, *Adv. Energy Mater.* **2019** 9, 1902446.

[46] L. Ma, S. Chen, H. Li, Z. Ruan, Z. Tang, Z. Liu, Z. Wang, Y. Huang, Z. Pei, J. A. Zapien, C. Zhi, *Energy Environ. Sci.* **2018**, 11, 2521.

[47] Z. Liu, Q. Yang, D. Wang, G. Lang, Y. Zhu, F. Mo, Z. Huang, X. Li, L. Ma, T. Tang, Z. Lu, C. Zhi, *Adv. Energy Mater.* **2019**, 9, 1902473.

[48] Z. Hou, X. Zhang, X. Li, Y. Zhu, J. Liang, Y. Qian, *J. Mater. Chem. A* **2017**, 5, 730.

[49] W. Pan, Y. Wang, X. Zhao, Y. Zhao, X. Liu, J. Xuan, H. Wang, D. Y. C. Leung, *Adv. Funct. Mater.* **2021**, 31, 2008782.

[50] F. Wan, Y. Zhang, L. Zhang, D. Liu, C. Wang, L. Song, Z. Niu, J. Chen, *Angew. Chem. Int. Ed.* **2019**, 58, 7062.

[51] L. Ma, N. Li, C. Long, B. Dong, D. Fang, Z. Liu, Y. Zhao, X. Li, J. Fan, S. Chen, S. Zhang, C. Zhi, *Adv. Funct. Mater.* **2019**, 29, 1906142.

[52] N. Li, G. Li, C. Li, H. Yang, G. Qin, X. Sun, F. Li, H. M. Cheng, *ACS Appl. Mater. Interfaces* **2020**, 12, 13790.

[53] X. Zeng, J. Liu, J. Mao, J. Hao, Z. Wang, S. Zhou, C. D. Ling, Z. Guo, *Adv. Energy Mater.* **2020**, 10, 1904163.

[54] L. Dai, Y. Wang, L. Sun, Y. Ding, Y. Yao, L. Yao, N. E. Drewett, W. Zhang, J. Tang, W. Zheng, *Adv. Sci.* **2021**, 8, 2004995.

[55] X. Shen, X. Wang, Y. Zhou, Y. Shi, L. Zhao, H. Jin, J. Di, Q. Li, *Adv. Funct. Mater.* **2021**, 31, 2101579.

[56] C. Xie, T. Li, C. Deng, Y. Song, H. Zhang, X. Li, *Energy Environ. Sci.* **2020**, 13, 135.

[57]  G. Li, W. Chen, H. Zhang, Y. Gong, F. Shi, J. Wang, R. Zhang, G. Chen, Y. Jin, T. Wu, Z. Tang, Y. Cui, *Adv. Energy Mater.* **2020**, 10, 1902085.

[58]  H. Tang, Y. Yin, Y. Huang, J. Wang, L. Liu, Z. Qu, H. Zhang, Y. Li, M. Zhu, O. G. Schmidt, *ACS Energy Lett.* **2021**, 6, 1859.

[59]  G. G. Yadav, D. Turney, J. Huang, X. Wei, S. Banerjee, *ACS Energy Lett.* **2019**, 4, 2144.

[60]  C. Liu, X. Chi, C. Yang, Y. Liu, *Energy Environ. Mater.* **2023**, 6, e12300.

[61]  M. Ulaganathan, S. Suresh, K. Mariyappan, P. Periasamy, R. Pitchai, *ACS Sustain. Chem. Eng.* **2019**, 7, 6053.

[62]  S. Gu, K. Gong, E. Z. Yan, Y. Yan, *Energy Environ. Sci.* **2014**, 7, 2986.

[63]  K. Gong, X. Ma, K. M. Conforti, K. J. Kuttler, J. B. Grunewald, K. L. Yeager, M. Z. Bazant, S. Gu, Y. Yan, *Energy Environ. Sci.* **2015**, 8, 2941.

[64]  F. Yu, L. Pang, X. Wang, E. R. Waclawik, F. Wang, K. Ostrikov, H. Wang, *Energy Stor. Mater.* **2019**, 19, 56.

[65]  C. Liu, X. Chi, Q. Han, Y. Liu, *Adv. Energy Mater.* **2020**, 10, 1903589.

[66]  X. Xie, H. Fu, Y. Fang, B. Lu, J. Zhou, S. Liang, *Adv. Energy Mater.* **2021**, 12, 2102393.

[67]  Z. Liu, Y. Yang, S. Liang, B. Lu, J. Zhou, *Small Struct.* **2021**, 2, 2100119.

[68]  J. Lei, Y. Yao, Z. Wang, Y.-C. Lu, *Energy Environ. Sci.* **2021**, 14, 4418.

[69]  T. Xue, H. J. Fan, *J. Energy Chem.* **2021**, 54, 194.

[70]  J. Huang, Y. Xie, L. Yan, B. Wang, T. Kong, X. Dong, Y. Wang, Y. Xia, *Energy Environ. Sci.* **2021**, 14, 883.

[71]  C. Wang, L. Sun, M. Li, L. Zhou, Y. Cheng, X. Ao, X. Zhang, L. Wang, B. Tian, H. J. Fan, *Sci. China Chem.* **2021**, 65, 399.

© 2024 World Scientific Publishing Company
https://doi.org/10.1142/9789811278327_0005

Chapter 5

# Anode Design for Suppressing Zn Metal Dendrites

Yizhou Wang, Wenxi Wang, Husam N. Alshareef*

*Materials Science and Engineering, King Abdullah University of Science and Technology (KAUST), Thuwal 23955-6900, Saudi Arabia*

Aqueous Zn-ion batteries attract widespread attention toward future stationary energy storage systems due to their high safety and low cost. However, Zn metal anodes suffer from the dendrite issue, which could easily cause battery internal short circuit and consequent immature battery failure. In this regard, suppressing the Zn dendrite growth is of great importance for the future development of aqueous Zn-ion batteries. In this chapter, we will introduce three anode design strategies for suppressing Zn metal dendrites, *i.e.*, substrate design, protective layer design, and porous structure design, where exemplary work from our group is presented for each strategy.

## 5.1 Introduction

With the global public goal of carbon neutrality, promoting the large-scale application of renewable energy (*e.g.*, solar energy, wind energy) is essential. In this regard, developing supporting stationary energy storage systems to ensure renewable energy's stable and continuous output becomes an important research direction for scientists and enterprises.[1] Lithium-ion battery, as the dominant technology in the current battery market, is widely used in various applications, ranging from electric cars to portable intelligent electronics. However, it is not an ideal candidate for stationary energy storage systems due to the limited lithium resources and high cost of raw materials.[2]

---

* Corresponding author: husam.alshareef@kaust.edu.sa

Tremendous attention has been focused on developing energy storage systems with low-cost characteristics, including sodium-ion batteries,[3] potassium-ion batteries,[4] and Zn-ion batteries (ZIB).[5] Among these battery systems, ZIB is highly promising thanks to several advantages: (1) Zn metal possesses high specific capacities (820 mAh g$^{-1}$ and 5855 mAh cm$^{-3}$);[6] (2) Zn metal has high compatibility in water and a reasonably low electrochemical potential (−0.76 V *vs.* standard hydrogen electrode), which enables its application in aqueous battery systems with extremely high safety;[7] (3) Zn has higher abundance than lithium in the earth crust, and the mature production technology makes the price of Zn extremely cost-effective.[8] However, Zn metal's performance in aqueous ZIBs suffers from several problems, particularly dendrite formation.[9] Many factors can contribute to Zn dendrite formation, especially the uneven distribution of surface charge density at the anode and the inhomogeneous ion flux in the electrolyte, causing Zn to deposit unevenly and form Zn tips on the surface of the anode. Such as-formed Zn metal tips tend to exhibit locally concentrated surface charge density and promote the growth of Zn over other areas, which is often referred to as the notorious "tip effect".[10] These Zn tips eventually grow into prominent Zn dendrites during cycling and finally cause an internal battery short circuit. Thus, a stable Zn metal anode with dendrite-free deposition behavior is a key requirement for the broad application of aqueous ZIBs.

In this chapter, we will present three anode design strategies for suppressing Zn metal dendrites toward substrate design, protective layer design, and porous structure design. In each strategy, we will introduce it using typical exemplary work from our group:

(1) MoS$_2$-Mediated Epitaxial Plating of Zn Metal Anodes.[11] (Scheme 5.1a, published in *Adv. Mater.* **2023**, 35, 2208171, https://doi.org/10.1002/adma.202208171)

(2) Controlled Deposition of Zinc-Metal Anodes via Selectively Polarized Ferroelectric Polymers.[12] (Scheme 5.1b, published in *Adv. Mater.* **2022**, 34, 2106937, https://doi.org/10.1002/adma.202106937)

(3) Organic Acid Etching Strategy for Dendrite Suppression in Aqueous Zinc-Ion Batteries.[13] (Scheme 5.1c, published in *Adv. Energy Mater.* **2022**, 12, 2102797, https://doi.org/10.1002/aenm.202102797)

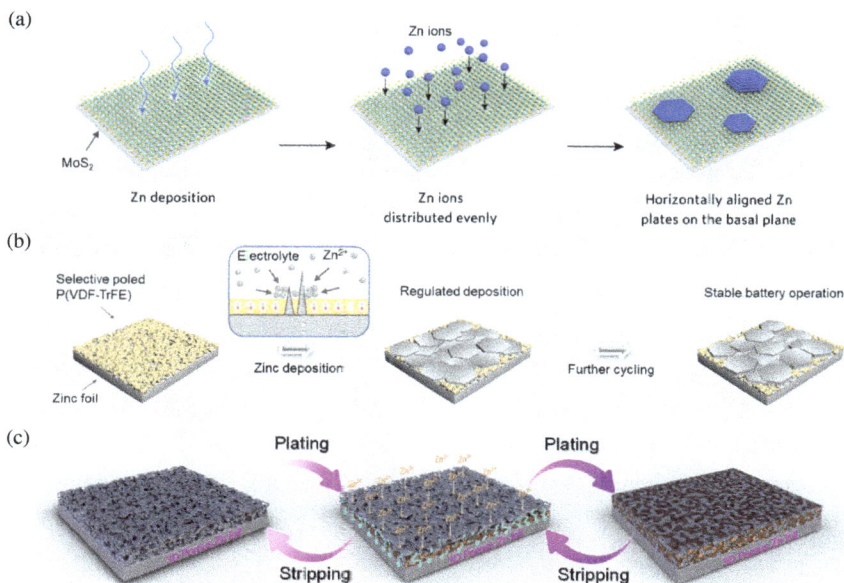

**Scheme 5.1.** Anode design strategies for suppressing Zn metal dendrites. (a) Schematic illustration showing the possible reaction pathways of Zn deposition on the basal plane of $MoS_2$ substrate. Reproduced with permission.[11] Copyright 2023, Wiley-VCH. (b) Schematic illustration showing the Zn deposition behaviors on selectively poled P(VDF-TrFE)-coated Zn. P(VDF-TrFE): poly(vinylidene fluoride-trifluoroethylene). Reproduced with permission.[14] Copyright 2022, Wiley-VCH. (c) Schematic illustration of the proposed plating/stripping process using the TFA-AN@Zn electrode. TFA: trifluoromethanesulfonic acid; AN: acetonitrile. Reproduced with permission.[15] Copyright 2022, Wiley-VCH.

## 5.2 MoS₂-Mediated Epitaxial Plating of Zn Metal Anodes

Herein, large-area quasi-single-crystalline two-dimensional (2D) material ($MoS_2$) is used as the substrate to study the Zn metal deposition behavior. Via density functional theory (DFT) calculation and various experimental characterizations (optical microscopy, scanning electron microscopy (SEM), X-ray diffraction (XRD), *etc.*), we confirmed that the edges of $MoS_2$ would induce the preferential deposition of Zn and thereby affect the crystalline quality of the deposited Zn film. Furthermore, we obtained a Zn film with almost all (002) crystal plane orientation, as demonstrated by the hexagonal reflections in the XRD pole figure analysis after wrapping the edges of the mono-oriented $MoS_2$ substrate. Confirmation of anode epitaxy by pole figure is done here for the first time in Zn metal anode research. These results validated the epitaxial growth of Zn on the $MoS_2$ substrate and confirmed the effectiveness of using large-area mono-oriented substrates with similar lattice structures to suppress metal dendrites (Scheme 5.1a).

To realize the epitaxial deposition of Zn on the substrate, one prerequisite is that the substrate needs to have a single-crystalline (or quasi-single-crystalline) structure with a well-defined orientation over the whole area, and the substrate should also have a similar lattice structure to Zn. Most 2D materials with hexagonal structure can meet these needs well. However, one thing should be noted that, while applying 2D materials in various application scenarios, their edges generally possess huge impact on the overall performance. For example, more exposed edges in 2D materials could enhance reaction rates, especially in the fields of energy storage and conversion.[14] Based on these considerations, the thought that the edges of the 2D material might be an important factor during the Zn electrodeposition process is worth exploring. Therefore, we prepared large-area mono-oriented $MoS_2$ film because of its similar lattice structure to Zn, which can enable the epitaxial deposition of Zn; in addition, the large area can facilitate a controlled study of the influence of the $MoS_2$ layer edges on the Zn deposition process.

The $MoS_2$ film was prepared via an epitaxial phase conversion method developed by our group.[15] The as-prepared $MoS_2$ film, grown on a wafer with two-inch diameter, is shown in Figure 5.1a. Compared to the pristine $Al_2O_3$ substrate, the $MoS_2$ film presented a uniform light green color. To confirm the successful preparation and thickness uniformity of the $MoS_2$ film, Raman measurements on different regions across the film were conducted (Figure 5.1b). The frequency peaks at 383.9 $cm^{-1}$ and 408 $cm^{-1}$ correspond well to the typical $E_{2g}^1$ and $A_{2g}$ vibration modes of 2H-phase $MoS_2$, respectively. The Raman frequency peaks and peak differences retrieved in different regions showed a high consistency, indicating the uniform thickness distribution across the $MoS_2$ film.[16]

XRD analysis was then conducted to characterize the crystalline structure of the $MoS_2$ film. As shown in the symmetric two-theta scan XRD curve (Figure 5.1c), the $MoS_2$ film exhibits only a series of peaks related to the $MoS_2$ basal plane, including the (002) plane at 14.6°, (004) plane at 29.1°, (006) plane at 44.2°, and (008) plane at 60.2°. In the 2D XRD pattern, the signals of these basal planes showed a concentrated distribution at the middle of the 2D XRD frame, corresponding to their high intensity

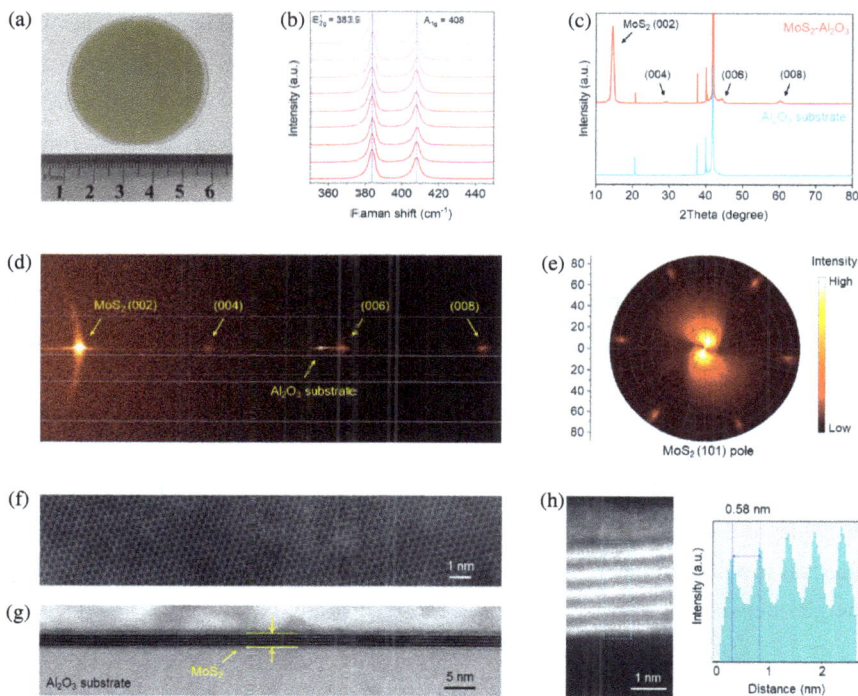

**Figure 5.1.** Characterization of the wafer-scale mono-oriented $MoS_2$ film. (a) Digital photo of the wafer-scale $MoS_2$. (b) Raman spectra of the wafer-scale $MoS_2$ retrieved from different regions. (c) Theta-2theta XRD pattern, (d) 2D XRD pattern, and (e) XRD pole figure of $MoS_2$ on $Al_2O_3$ substrate. STEM images of the wafer-scale $MoS_2$ in (f) top surface, (g) cross-section, and (h) the corresponding interlamellar space profile. Reproduced with permission.[11] Copyright 2023, Wiley-VCH.

along the out-of-plane axis (Figure 5.1d).[17] These results indicate the highly ordered basal-plane orientation of the wafer-scale $MoS_2$ film. XRD pole figure test was carried out to evaluate the in-plane orientation of the $MoS_2$ film. As shown in Figure 5.1e, signals were found at a tilt angle of 77.5° in the $MoS_2$ (101) pole figure, which corresponds to the angle between the (101) plane and the (002) plane of $MoS_2$. These signals showed a narrow distribution and an azimuthal separation of 60°, indicating the in-plane mono-oriented structure of $MoS_2$ across the wafer-scale film. To further visualize the crystal structures of the $MoS_2$, the $MoS_2$ sample was transformed onto a Cu grid for scanning transmission electron microscopy (STEM) characterization. As shown in Figure 5.1f, the ordered honeycomb-like hexagonal structure on the top surface indicated a high crystal quality of $MoS_2$. The STEM image of the sample's cross-section exhibited a clear layered structure of the MoS$_2$ film (Figure 5.1g), in which the interplanar spacing of MoS$_2$ was measured to be 0.58 nm (Figure 5.1h), corresponding well to the MoS$_2$ (002) lattice planes. Such uniform layered structure and their horizontal alignment against the substrate further indicate the highly ordered lattice characteristic and the van der Waals epitaxial growth of the wafer-scale MoS$_2$ film. All these characterizations proved that the prepared MoS$_2$ film possessed a high uniformity across a large area. Since the MoS$_2$ film is highly smooth over a large area, the surface roughness/nanostructure-induced uneven electrical field distribution does not occur during metal deposition on the MoS$_2$ substrate.[18] Moreover, the wafer-scale MoS$_2$ film can be readily transferred onto other substrates (e.g., silicon wafer, metal), enabling the subsequent Zn electrodeposition studies on it.

Subsequently, we applied theoretical calculations to study the influence of MoS$_2$ substrate on Zn deposition. To probe the influence of different sites (i.e., edge and basal plane) of MoS$_2$ on metallic Zn plating, DFT calculations were performed to study the binding energies between the Zn atom and the MoS$_2$ substrate (Figure 5.2). As shown in Figure 5.2a-c, the binding energies of Zn atom on MoS$_2$ at S edge, Mo edge site 1, and Mo edge site 2 are –1.98 eV, –2.23 eV, and –2.24 eV, respectively. In sharp contrast, the binding energy of Zn atoms on the MoS$_2$ basal plane is only –0.2~–0.21 eV (Figure 5.2d-f). Therefore, if the edge of the MoS$_2$ substrate is exposed with either sulfur or molybdenum atoms, there is a strong interaction between the Zn atom and the MoS$_2$ edge; on the other hand, the basal plane of MoS$_2$ is not so conducive to the adsorption of Zn atoms. The considerable difference in binding energies for Zn atoms on the MoS$_2$ edges and the basal plane (Figure 5.2g) also indicates that, if the edge and the basal plane of MoS$_2$ both exist, the edges would act as the seeding points of Zn, resulting in preferred Zn nucleation at the edge rather than the basal plane.[19] Furthermore, the interface energy between the Zn (002) plane and the MoS$_2$ (002) plane is calculated by DFT (Figure 5.2h). The calculation results show a linear relationship between the interface energy and the number of Zn atoms in Zn (002)/MoS$_2$ (002) models (Figure 5.2i). As reported in previous work,[20] the intercept of the fitted line represents the interface energy. The interface energy herein is estimated to be 1.4 meV Å$^{-2}$, and such positive value indicates a stable interface between the MoS$_2$ (002) plane and the Zn (002) plane.

Based on theoretical calculations, we then designed a transparent electrochemical cell to observe the electrodeposition of Zn on MoS$_2$. The cell configuration is shown in

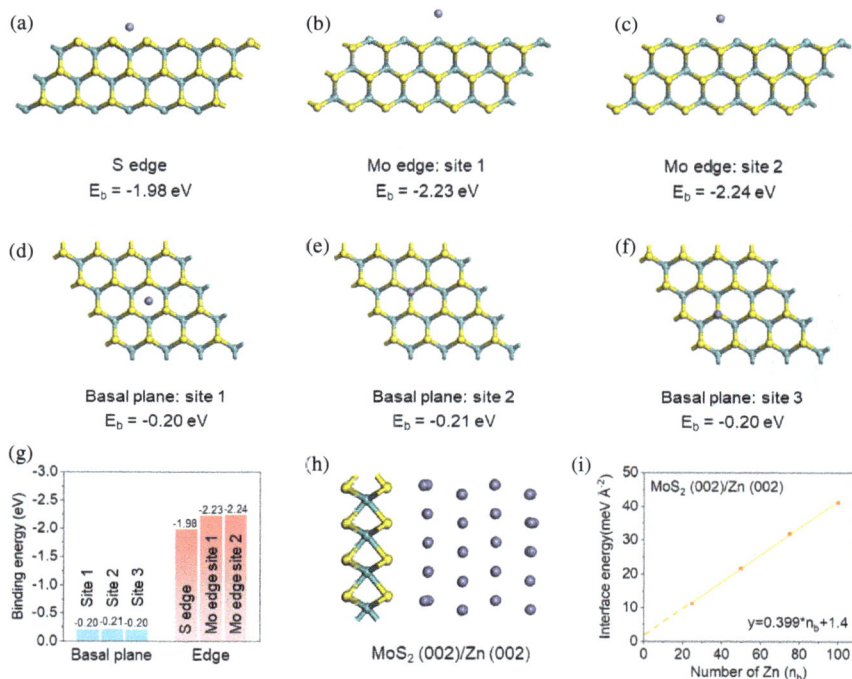

**Figure 5.2.** DFT calculations. The models of Zn atoms adsorbed on (a) S edge, (b) Mo edge site 1, (c) Mo edge site 2, (d) basal plane site 1, (e) basal plane site 2, and (f) basal plane site 3. (g) The calculated binding energies of Zn atoms at different sites of the $MoS_2$ substrate. (h) Atomic structure of the $MoS_2$ (002)-Zn (002) interface and (i) the calculated interface energy $\gamma$ of the $MoS_2$ (002) plane versus the Zn (002) plane with different Zn atoms. Zn, S and Mo atoms are symbolized as gray, yellow, and cyan spheres, respectively. Reproduced with permission.[11] Copyright 2023, Wiley-VCH.

Figure 5.3a, where the Au pad coated by $MoS_2$ was employed as the working electrode, and commercial Zn foil acted as the counter electrode. Given the fact that a 2 M $ZnSO_4$ aqueous solution was used as the electrolyte in most reported aqueous Zn ion batteries,[21] this work used the same solution as the electrolyte. All the components were placed in the middle of a hollow-square gel tape, and a transparent glass coverslip was placed on the top to facilitate the *in situ* observation. It should be noted that although 2H-phase $MoS_2$ is a typical semiconducting material, it can exhibit an enhanced electrical conductivity via a "self-gating" mechanism while applying a negative potential.[22] Thus, at the working electrode, once a negative voltage is applied, Zn ions in the electrolyte could be deposited as metallic Zn on $MoS_2$. We first fully covered the Au pad by $MoS_2$, and generated some edges on $MoS_2$ outside the Au pad (Figure 5.3b and 5.3c).

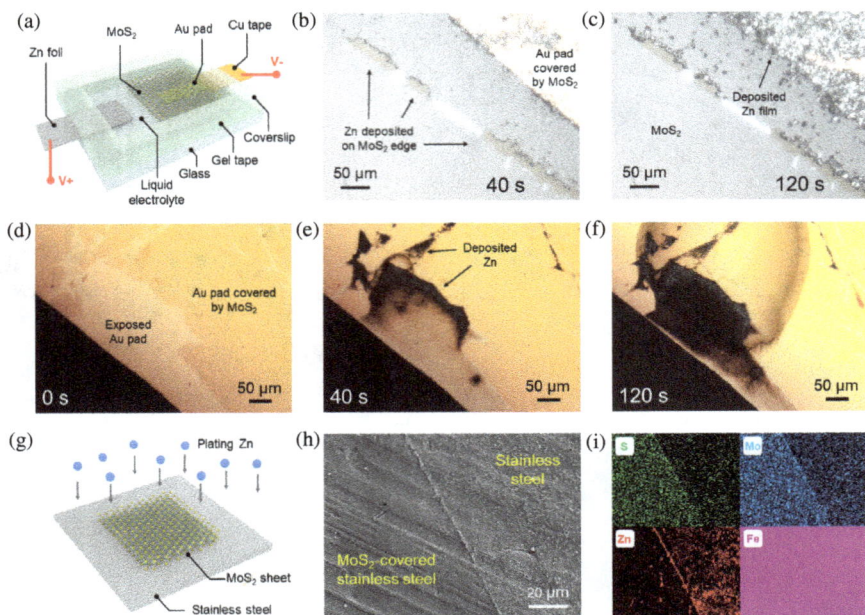

**Figure 5.3.** Investigation of the influence of MoS$_2$ edges on Zn deposition. (a) Schematic illustration showing the configuration of the optical cell used for *in situ* observation of Zn deposition. (b-f) Optical microscopy images of depositing Zn on MoS$_2$-covered Au at –0.2 V for different times. In (b-c), the Au pad was fully covered by MoS$_2$, and edges of MoS$_2$ were generated by carefully scratching the outside of the Au pad. In (d-f), randomly distributed MoS$_2$ breakages were generated via an ultrasonic treatment. (g) Schematic illustration showing the procedure of electrodepositing Zn onto SS plate that was partially covered by MoS$_2$. (h) SEM image and (i) the corresponding element mappings of the MoS$_2$ edge region after depositing a small amount of Zn (10 mA cm$^{-2}$ for 1 s). Reproduced with permission.[11] Copyright 2023, Wiley-VCH.

According to the "self-gating" mechanism, the region of MoS$_2$ closer to the Au pad should exhibit better conductivity, and thus be more conducive to Zn deposition. However, it is interesting that, after we deposited Zn for a certain time, significantly more Zn was deposited at the MoS$_2$ edges than in the surrounding areas. Next, we exposed part of the Au pad in the electrolyte by randomly generating a crack in the MoS$_2$ area. Under the same electrochemical protocols, Zn was first deposited on the edge of MoS$_2$ and then spread along the exposed Au as time proceeded (Figure 5.3d–5.3f). These optical microscopy results indicate that the edge of MoS$_2$ significantly enhances Zn deposition, and thus affects the uniformity of the deposited Zn film.

To explore more details of the deposited Zn on the $MoS_2$ edge, we transferred a piece of $MoS_2$ film onto the stainless steel (SS) plate, and characterized its morphology after depositing a tiny amount of Zn (Figure 5.3g). Interestingly, after deposition, a higher density of Zn nanoplates could be found at the $MoS_2$ edge rather than the surrounding, and these densely packed Zn nanoplates even made the edge of $MoS_2$ more obvious (Figure 5.3h). To learn the elemental distribution at this region, energy dispersive spectroscopy (EDS) mapping was conducted (Figure 5.3i). We can easily find a distinct dividing line between the exposed SS and $MoS_2$-covered SS from the EDS result. Meanwhile, on this dividing line, the Zn element signal was much stronger than the surrounding while the Fe element exhibited the opposite, further indicating the gathering of the deposited Zn at the $MoS_2$ edge. All these experimental results correspond to the DFT theoretical calculations, confirming that edges of $MoS_2$ are favored sites for the deposition of Zn.

Since we determined the colossal influence of $MoS_2$ edges on Zn deposition behavior, we decided to study Zn deposition on $MoS_2$ films free of any edges. We used the epitaxial conversion process described earlier to prepare large-area, mono-oriented, and continuous $MoS_2$ films. The wafer-scale process has been developed in our lab and can give edge-free $MoS_2$ film for Zn deposition. We transferred the wafer-scale $MoS_2$ film onto the SS plate to fully cover it (denoted as $MoS_2$-SS) for subsequent Zn deposition tests. For comparison, bare SS plate and one plate partially covered with wafer-scale $MoS_2$ (where $MoS_2$ edges were exposed, denoted as $MoS_2$-SS-E) were prepared. To avoid any possible cracks of $MoS_2$, we employed a customized template to conduct the Zn electrodeposition process. The same amount of Zn (10 mA cm$^{-2}$ for 10 s) was deposited on different substrates for subsequent analysis. To study the Zn crystal quality, both in-plane and out-of-plane XRD techniques were performed to study the crystallographic character of as-deposited Zn films. In XRD symmetric out-of-plane two-theta scans, Zn (002) plane peak showed a dominant strength in the $MoS_2$-SS sample, especially compared to that of the bare SS sample. By contrast, in the case of $MoS_2$-SS-E sample, although Zn (002) plane peak exhibited a high intensity, a certain intensity of the Zn (101) plane peak could also be observed. Then 2D XRD characterization was conducted to show the out-of-plane orientation of Zn. As shown in Figure 5.4a, the Zn (002) plane in the $MoS_2$-SS sample delivered a concentrated intensity along the out-of-plane direction, revealing a highly oriented

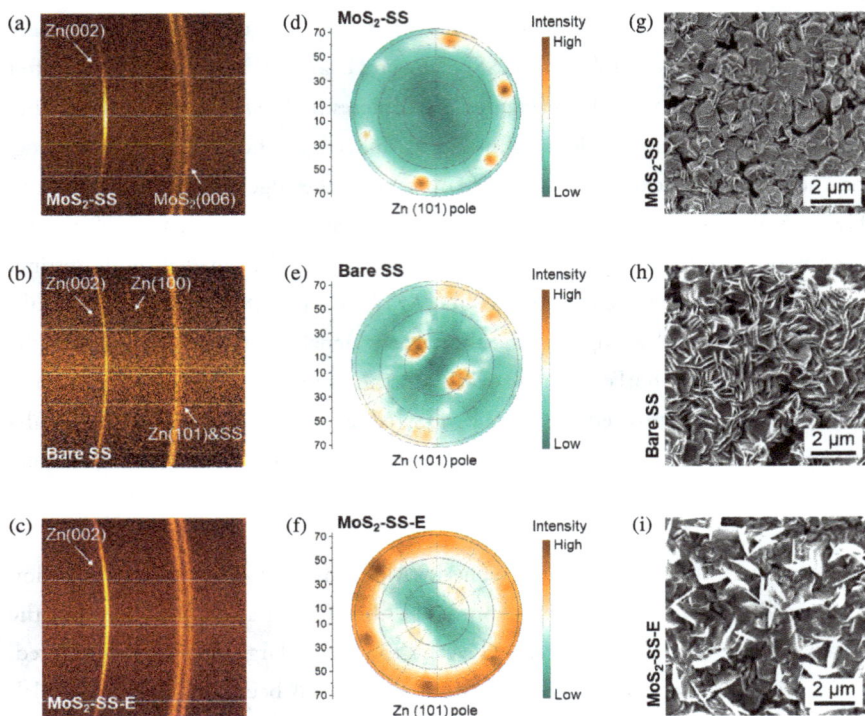

**Figure 5.4.** Characterizations of the electrodeposited Zn films. (a) 2D XRD patterns of the (a) MoS$_2$-SS, (b) bare SS, and (c) MoS$_2$-SS-E deposited with a certain amount of Zn. XRD pole figures of the (d) MoS$_2$-SS, (e) bare SS, and (f) MoS$_2$-SS-E. SEM images showing the typical morphology of the above three samples: (g) MoS$_2$-SS, (h) bare SS, and (i) MoS$_2$-SS-E. Reproduced with permission.[11] Copyright 2023, Wiley-VCH.

Zn (002) plane crystallites.[23] On the other hand, in the bare SS (Figure 5.4b) sample, the Zn (002) plane signal showed isotropic orientation, indicating a random Zn lattice plane distribution. In the MoS$_2$-SS-E sample (Figure 5.4c), it can be seen that the Zn (002) plane signal showed a stronger intensity (or brighter light point) in the middle of the frame, but it was not as strong as that in the MoS$_2$-SS sample, which revealed that Zn deposited on MoS$_2$-SS-E has a certain orientation, but such orientation was weaker than that of MoS$_2$-SS. XRD pole figures were characterized to evaluate the in-plane orientation of Zn. As shown in Figure 5.4d, in the MoS$_2$-SS sample, the Zn (101) pole figure showed six spots separated azimuthally by 60° at a tilt angle (*psi* angle) of 65.7°, which was consistent with the (101) lattice plane projection in the

Zn (002) plane.[24] The hexagonal-distributed signals of Zn correspond well to the pole figure of the $MoS_2$ substrate as shown in Figure 5.4e, suggesting the effective epitaxial growth of Zn on $MoS_2$. On the contrary, Zn deposited on bare SS showed random orientation (Figure 5.4e). Interestingly, while depositing Zn on $MoS_2$-SS-E, the Zn (101) pole figure showed a weak signal of 60°-azimuthal-distributed spots (Figure 5.4f). Such a phenomenon further indicates that $MoS_2$ can enable the epitaxial electrodeposition of Zn, but edges may reduce the Zn crystalline quality. Then we employed SEM to observe the morphology of Zn deposited on different substrates. Significant differences were found among these three substrates. The Zn deposited on $MoS_2$-SS showed a film structure consisting of horizontally aligned nanoplates (Figure 5.4g). Oppositely, the Zn plates on bare SS showed a random distribution of vertical standing Zn flakes (Figure 5.4h). In $MoS_2$-SS-E, most Zn plates were horizontally aligned, but some vertically interspersed Zn flakes could also be observed (Figure 5.4i). Based on these results, we infer that the mono-oriented substrate with similar lattice structure can enable the epitaxial electrodeposition of metal, but the edges of the substrate, as highly active reaction sites, would weaken such epitaxial growth relationship.

We must note that the crystal quality of the electrodeposited Zn film on $MoS_2$ herein is not perfect, as some extra signals in addition to the hexagonal-distributed spots could be observed in the XRD pole figure (Figure 5.4d). But still, considering the limit of the methodology (*i.e.*, electrochemical deposition, which is inherently a much more complicated process than those high-quality single-crystalline film preparation methods such as high-vacuum chemical vapor deposition), the results obtained in this work effectively proved the feasibility of introducing this epitaxial strategy in suppressing dendrites in metal anodes. Moreover, as the high reaction activity of 2D material edges plays an unfavorable role in the epitaxial deposition of metal, this work clearly indicates the following optimization directions regarding the epitaxy strategy. For example, passivating the edges of the epitaxial substrate (*e.g.*, pre-absorbing or grafting electrochemically inert groups), or introducing strong metal-philic sites on the substrate's basal plane (*e.g.*, building vacancies or doping with heteroatoms), can be powerful approaches to enhance the effectiveness of the epitaxial substrates. Thus, we believe our discovery herein can inspire new possibilities for future research on metal anode protection and advanced high-energy batteries.

## 5.3 Controlled Deposition of Zn-Metal Anodes via Selectively Polarized Ferroelectric Polymers

Herein, we suppress Zn-dendrite growth by introducing a selectively polarized ferroelectric polymer material P(VDF-TrFE) as the protective layer on Zn anodes. Since a direct-current polarization process can alter the dipole direction of the polymer molecule, we selectively regulated the polarization direction, and endowed the negative dipole center of the ferroelectric polymer at the electrolyte-side surface. Such a protective layer could provide an inner electrostatic field between polymer coating and Zn metal, inducing the locally concentrated Zn ions at the polymer coating surface, thus suppressing Zn dendrite growth and enabling the Zn to grow horizontally along the coating surface (Scheme 5.1b).

Since P(VDF-TrFE) is easy to crystallize into the ferroelectric β-phase (Figure 5.5a), it was selected as the ferroelectric polymer protective layer for Zn metal anodes. A facile spin coating process was used to ensure the uniform coating of such ferroelectric polymer material. As shown in the XRD patterns (Figure 5.5b), after the heat treatment, the β-phase peak (2 Theta = ~19.9°) of the polymer coating appeared. SEM images also confirmed the crystallization of this polymer (Figure 5.5c). After heat treatment, the morphology of P(VDF-TrFE) changed from a smooth surface to numerous needle-like grains (Figure 5.5c), which corresponds to the reported literature.[25] After the crystallization into β-phase, the thickness of the P(VDF-TrFE) on Zn foil was measured to be ~0.5 μm, and such a thin coating would ensure that the impact on the battery energy density is minimal. Here we should note that a small amount of Zn trifluoromethanesulfonate ($Zn(OTf)_2$) was added to the spinning polymer solution. These salt nanoparticles can be readily dissolved in an aqueous electrolyte to form a porous morphology of the P(VDF-TrFE) layer with randomly distributed ~0.1 μm-diameter pores (Figure 5.5c). These pores inside the coating layer can function as channels for Zn ion diffusion and Zn metal growth. To confirm the pore-forming effect of the $Zn(OTf)_2$ salt, we prepared a bare P(VDF-TrFE) coating layer with the same protocols (only without $Zn(OTf)_2$ addition), and no pores can be observed. The ferroelectricity of the polymer coating was validated by piezoresponse force microscopy (PFM). By applying positive and negative bias voltages on the different regions of the P(VDF-TrFE) coating, an obvious phase contrast (dark and bright colors) was observed (Figure 5.5d), indicating two different poling states of the P(VDF-TrFE) coating. Thus, the ferroelectricity of the polymer coating provides the feasibility for us to adjust its surface

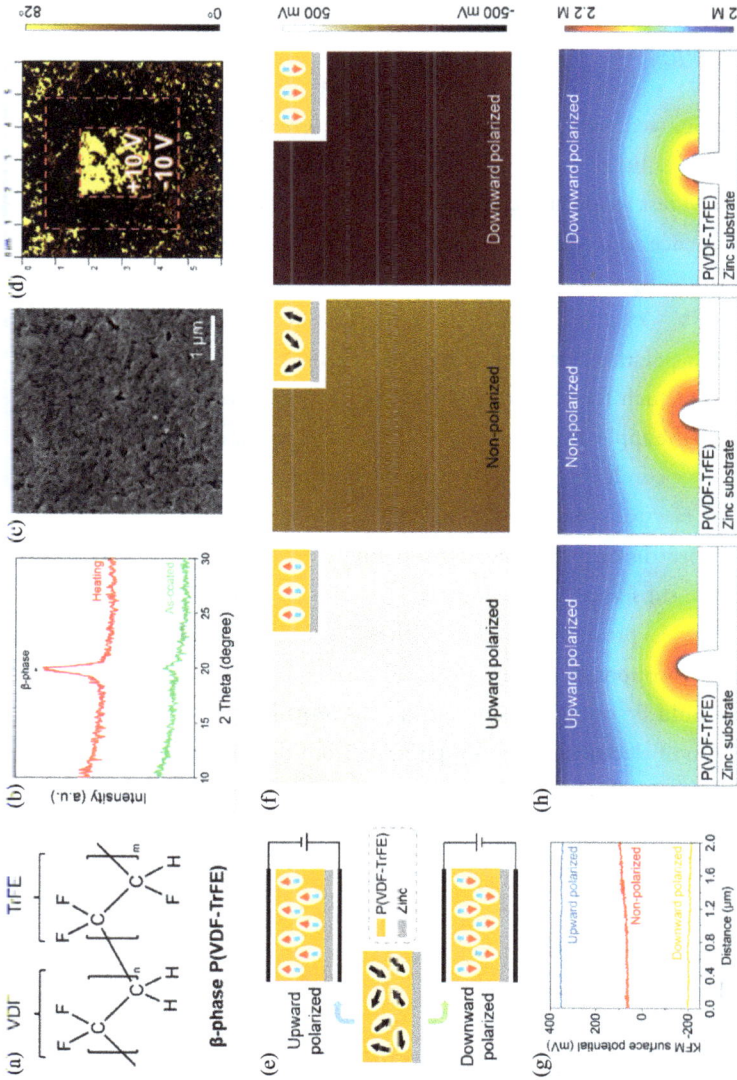

**Figure 5.5.** Characterizations of the ferroelectric polymer coating layer. (a) Molecular structure of β-phase P(VDF-TrFE). (b) XRD patterns of the P(VDF-TrFE) before and after heat treatment. (c) SEM images of the P(VDF-TrFE) coating after sinking in 2 M ZnSO₄ aqueous solution for 30 min. (d) PFM image of the P(VDF-TrFE)-coated Zn. The P(VDF-TrFE)-coated Zn was previously treated with a potential bias pattern as labeled in the figure. (e) Schematics showing the polarization process of P(VDF-TrFE) coating. (f) KPFM images of P(VDF-TrFE)-coated Zn anodes with varying dipole directions over an area of 4 μm² and (g) the corresponding KPFM surface potential profiles of different samples retrieved from the middle of each image. (h) Zn ion concentration simulations on P(VDF-TrFE)-coated Zn foil with varying states of polarization. Reproduced with permission.[12] Copyright 2022, Wiley-VCH.

electrical properties. In order to selectively polarize the ferroelectric polymer, the polymer-coated Zn films were subjected to an external electric field, as schematically shown in Figure 5.5e. Herein, the polarized P(VDF-TrFE) with negative dipole end on the surface of Zn is denoted as upward polarized, and that with positive dipole end on the surface of Zn is denoted as downward polarized.

Subsequently, we employed Kelvin Probe Force Microscopy (KPFM) technique to evaluate the surface potential of the ferroelectric polymers with different polarization directions. The KPFM test results indicated that the selective polarization procedures had a noticeable influence on the surface potentials of different samples. The measured surface potentials of the upward polarized, non-polarized, and downward polarized samples were ~350 mV, ~75 mV, and ~-200 mV, respectively (Figure 5.5f and 5.5g). Compared to the non-polarized state, the polarization process can effectively change the dipole direction of such ferroelectric polymer and increase/decrease the surface potential according to the applied external electric field. Based on the modulated surface potential situations, we conducted finite element simulations on the distributions of the electrostatic field, electrochemical potential, and consequent Zn ion concentration in the case of one Zn protrusion grown across the coating. Regarding the P(VDF-TrFE) coating as a dielectric substance, electrostatic field could be generated between the coating surface and the Zn protrusion once the dielectric substance is polarized. When there are coatings with opposite surface electrical properties, the direction of the electrostatic field would be opposite between the Zn protrusion and the polymer coating. The direction of electric field lines is pointing from the Zn protrusion to the coating surface in the downward polarized case, and the direction of electric field lines is opposite in the upward polarized case. The existing electrostatic fields can affect the electrochemical potential distribution during Zn deposition. The upward polarized coating is expected to concentrate the electrochemical potential locally at the tip of the Zn protrusion. The non-polarized coating is expected to lead to a relatively even potential distribution around the Zn protrusion. In contrast, the downward polarized coating is expected to produce a potential concentrated at the corner of polymer coating and Zn protrusion. Consequently, the modulated electrochemical potential is expected to result in Zn ion concentration distributed with the same trends (Figure 5.5h). Such interesting electric field and ion concentration distributions mean that the polarized ferroelectric polymer coating could play a significant role during Zn deposition, as we show next.

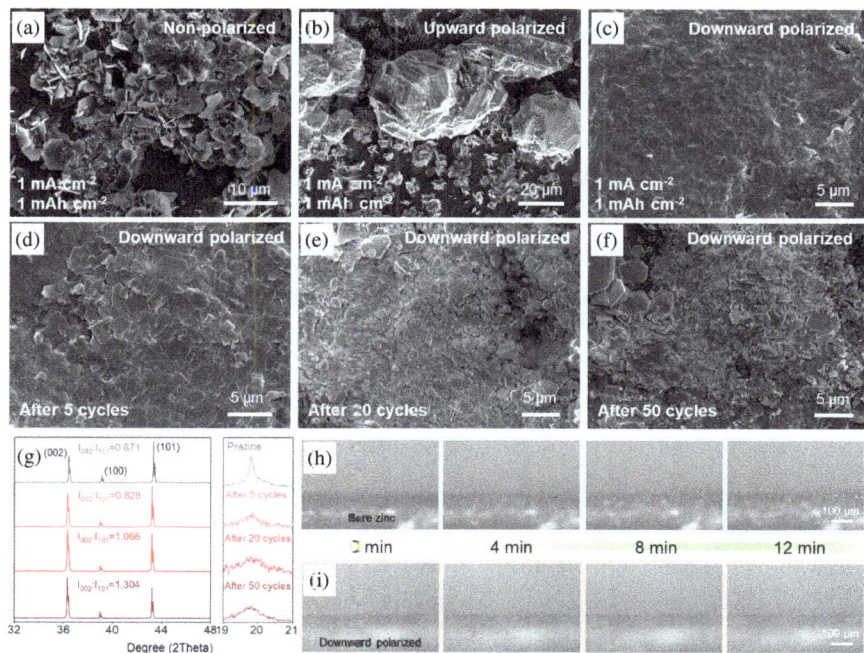

**Figure 5.6.** Morphology evolution of Zn metal anodes coated with downward polarized P(VDF-TrFE) film. SEM images of (a) non-polarized, (b) upward polarized, and (c) downward polarized P(VDF-TrFE)-coated Zn metal anodes after depositing a certain amount of Zn (1 mA cm$^{-2}$ for 1 hour). SEM images of downward polarized P(VDF-TrFE)-coated Zn metal anode deposited/stripped at 1 mA cm$^{-2}$, 1 mAh cm$^{-2}$ after (d) 5 cycles, (e) 20 cycles, and (f) 50 cycles, and (g) the corresponding XRD patterns. *In situ* optical microscopy images of plating Zn on (h) bare Zn foil and (i) downward polarized P(VDF-TrFE)-coated Zn foil. Reproduced with permission.[12] Copyright 2022, Wiley-VCH.

To further validate the effect of the polarized P(VDF-TrFE) layer on Zn deposition behavior, the morphology of different anodes after depositing a certain amount of Zn (1 mA cm$^{-2}$ for 1 hour) was observed by *ex situ* SEM characterization. As shown in Figure 5.6a-c, after the Zn deposition, Zn foil coated with differently polarized P(VDF-TrFE) showed different surface morphologies. For the non-polarized P(VDF-TrFE)-coated Zn, the deposited Zn showed the morphology of randomly distributed Zn plates (Figure 5.6a). With upward polarized P(VDF-TrFE) coating, the Zn grew locally into polyhedral Zn dendrites (Figure 5.6b), which was even worse and might easily cause battery internal short circuit upon cycling. While for the downward polarized P(VDF-TrFE) coating (Figure 5.6c), the Zn could grow along the surface of the polymer coating into horizontally aligned Zn plates.

Considering downward polarized P(VDF-TrFE) could render a favorable Zn deposition behavior, we assembled downward polarized P(VDF-TrFE)-coated Zn into symmetric batteries and studied its morphology after cycling different times under 1 mA cm$^{-2}$ for 1 h (Figure 5.6d-f). Interestingly, after five plating/stripping cycles, there were still some horizontally aligned Zn plates on the surface of the P(VDF-TrFE) coating. The horizontally aligned Zn plate morphology was maintained and the thickness of the Zn plate was increased after more cycles, indicating that these remaining Zn plates could play the role of the seeding layer and guide subsequently deposited Zn to grow horizontally. Furthermore, XRD patterns showed that, with such polymer coating, the basal plane ((002) plane) of Zn was strengthened as the cycling proceeds (Figure 5.6g). For the pristine downward polarized P(VDF-TrFE)-coated Zn, the intensity ratio between the (002)-plane peak and the (101)-plane peak was 0.671. In comparison, this value increased to 0.828, 1.066, and 1.304 after 5, 20, and 50 cycles, respectively. Interestingly, the shape of the β-P(VDF-TrFE) XRD peak turned from a sharp peak to a small bump, indicating that the polymer surface was covered by Zn after cycling. The XRD results are consistent with the morphology of parallel-arranged layered Zn observed by SEM, suggesting that the selectively polarized ferroelectric polymer surface could modulate the Zn deposition behavior and lead to the (002) plane preferential deposition of Zn on the polymer surface.

Furthermore, to visualize the effect of the ferroelectric polymer coating on the Zn deposition behavior, a transparent symmetric cell was constructed with different Zn anodes as working electrodes, where the morphology evolution during deposition can be *in situ* observed via an optical microscope. Herein, Zn was deposited on anodes with a relatively large current density (10 mA cm$^{-2}$) for 12 mins (corresponding to a total capacity of 2 mAh cm$^{-2}$). As shown in Figure 5.6h, while Zn was deposited on bare Zn, the Zn nuclei were distributed randomly, and Zn was deposited uncontrollably into dendrites that could be readily observed. On the contrary, with the downward polarized P(VDF-TrFE) coating, the Zn was deposited homogeneously, and the height of the deposited Zn increased evenly as the deposition time increased (Figure 5.6i). All these results confirmed that the selectively polarized P(VDF-TrFE) layer could effectively suppress the growth of Zn dendrites.

Next, symmetrical Zn cells with galvanostatic cycling measurements were conducted to test the electrochemical performance of the ferroelectric polymer-coated Zn anodes.

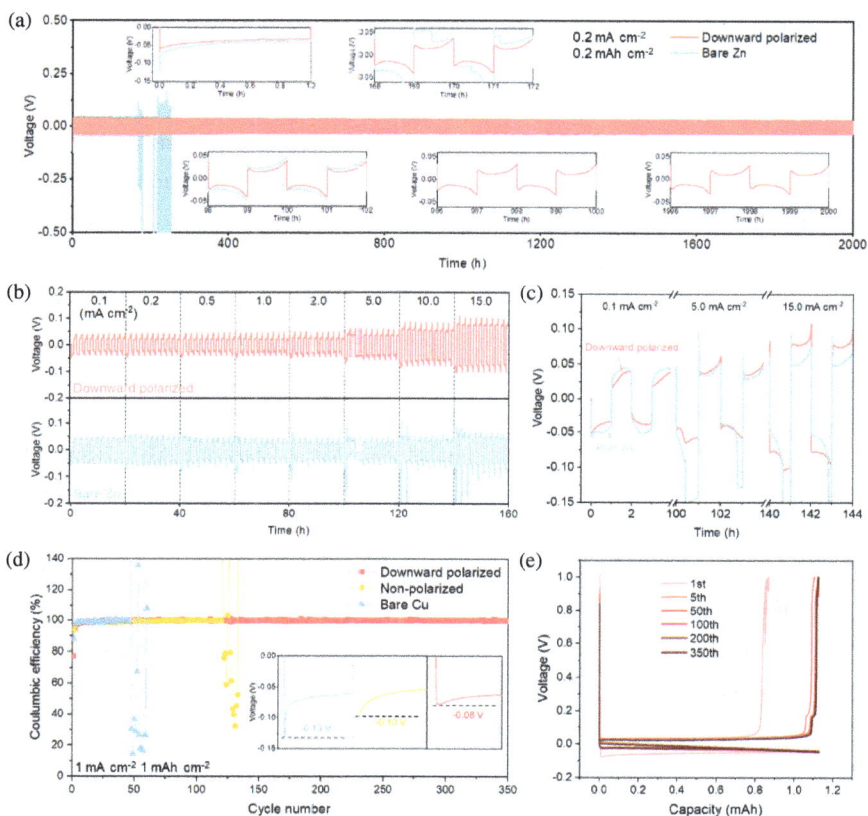

**Figure 5.7.** Electrochemical performance of symmetric Zn∥Zn batteries and asymmetric Zn∥Cu batteries that employed downward polarized P(VDF-TrFE) coating. (a) Cycling performances of symmetric batteries using different anodes galvanostatically charging/discharging at 0.2 mA cm$^{-2}$ for 1 hour. Insets are the enlarged details at typical cycling hours. (b) Rate performances of symmetric batteries using different anodes galvanostatically charging/discharging at various current densities for 1 hour and (c) the corresponding magnified curves at certain cycling hours. (d) Coulombic efficiency evaluations of asymmetric batteries using different copper foils at 1 mA cm$^{-2}$ and 1 mAh cm$^{-2}$. Insets are the enlarged voltage profiles at the initial cycle. (e) Voltage profiles of the asymmetric batteries using downward polarized P(VDF-TrFE)-coated copper foil at different cycles. Reproduced with permission.[12] Copyright 2022, Wiley-VCH.

As shown in Figure 5.7a, charging/discharging the symmetrical cell at 0.2 mA cm$^{-2}$ for 1 hour, the cell with downward polarized P(VDF-TrFE) coating on Zn anodes showed a stable operation for 2,000 hours. In contrast, the bare Zn-based cell showed apparent fluctuation in the voltage profile and died after around 200 hours. Furthermore, the downward polarized P(VDF-TrFE) coating on Zn anodes showed obvious lower nucleation overpotential and deposition overpotential compared to the bare Zn anodes (Figure 5.7a, insets). These improved overpotentials could be ascribed to the locally concentrated Zn ion distribution induced by the polarized polymer coating and indicate improved Zn deposition kinetics. Same cycling tests were also conducted on non-polarized P(VDF-TrFE) coating on Zn anodes. Subsequently, continuously charging/discharging the symmetrical cells at various galvanostatic current densities from 0.1 mA cm$^{-2}$ to 15 mA cm$^{-2}$ was conducted to evaluate the rate performances of the Zn anodes. The downward polarized P(VDF-TrFE) coating on Zn showed stable deposition overpotentials as the deposition current density increased (Figure 5.7b and 5.7c). On the contrary, for the bare Zn and non-polarized P(VDF-TrFE)-coated Zn, when the plating/stripping current increased, the voltage profiles became unstable and significantly higher deposition overpotentials could be observed, which could be attributed to the formation of severe dendrites and might be a sign that the battery is about to die (Figure 5.7b and 5.7c).

Furthermore, to study the excellent performance of the downward polarized P(VDF-TrFE) coating, we used the same procedure to prepare such coating on copper foils for asymmetric batteries. With the downward polarized P(VDF-TrFE) coating, the Zn‖Cu asymmetric battery exhibited a long cycling life of 350 cycles (700 hours) plating/striping at 1 mA cm$^{-2}$ or 0.2 mA cm$^{-2}$ for 1 hour, much better than bare copper and the non-polarized P(VDF-TrFE)-coated copper (Figure 5.7d and 5.7e). Similarly, as shown in the inset of Figure 5.7d, the downward polarized P(VDF-TrFE) coating exhibited the lowest nucleation overpotential (–0.08 V) compared to the non-polarized P(VDF-TrFE) coated copper (–0.10 V) and bare copper (–0.13 V), indicating improved Zn nucleation kinetics. Although high coulombic efficiency is generally considered a prerequisite for good cycle performance for anode protection research,[26] the copper foil with downward polarized P(VDF-TrFE) coating showed lower coulombic efficiencies in the initial several cycles compared to other copper anodes. Such phenomenon was observed using SEM images. Some horizontally aligned Zn plates would be maintained after stripping, and they could be used as the template to guide further deposited Zn to grow horizontally, thus providing a longer lifespan for the cells (Figure 5.6c-f). These results indicated that downward polarized P(VDF-TrFE) coating had a pronounced mediation effect on the Zn deposition behavior and could significantly enhance the Zn deposition stability.

## 5.4 Organic Acid Etching Strategy for Dendrite Suppression in Aqueous Zn-Ion Batteries

Herein, we propose to corrode Zn foil using an organic acid strategy, in which the etched Zn foil generates a well-developed 3D hierarchical pore architecture and a clean appearance without a passivation layer after chemical corrosion. As a result, the interfacial wettability and carrier pathways of 3D porous Zn foil are significantly improved, which increases zincophilicity and the number of nucleation sites while reducing the local current density. Benefiting from the 3D porous structure, the electrodeposition of Zn preferentially occurs in the cavities accompanying the visible suppression of dendrite growth due to more homogeneous ion flux accessible to the liquid-metal interface (Scheme 5.1c).

The schematic diagram portraying the preparation of the 3D porous Zn foil using the organic acid etching process is shown in Figure 5.8a. According to Eq. (5.1), the dissociated proton in the trifluoromethanesulfonic acid (TFA) in acetonitrile (AN) solution can be displaced by metallic Zn and subsequently generate hydrogen, leading to the formation of the porous structure after soaking Zn foil in TFA-AN solution (Figure 5.8a).

$$Zn + 2CF_3SO_3H \rightarrow Zn(CF_3SO_3)_2 + H_2\uparrow \qquad (5.1)$$

In this study, the experimental parameters (*e.g.*, acid volume, acid concentration, and etching duration) were systematically controlled to optimize the pore structure and distribution. After optimization, the Zn foil etched with 1.5 ml 0.5 M TFA-AN acid for 24 h shows a well-developed porous structure and the lowest polarization potential, which is labeled as TFA-AN@Zn.

The surface and cross-sectional microarchitectures of the 3D porous Zn foil etched with TFA-AN solution (*i.e.*, TFA-AN@Zn) were characterized with SEM technique. As shown in Figure 5.8b, the TFA-AN@Zn surface distributes massive pores with a diameter ranging from nanoscale to microscale after systematic optimization. Further, the etching depth of Zn foil reaches 10 μm, and the porous structure is still visible from the cross-sectional SEM image in Figure 5.8c. In contrast, the pristine Zn foil shows an uneven surface and non-porous cross-section (Figure 5.8d-e). Thus, the organic acid undertakes surface cleaning and pore-foaming function. Additionally,

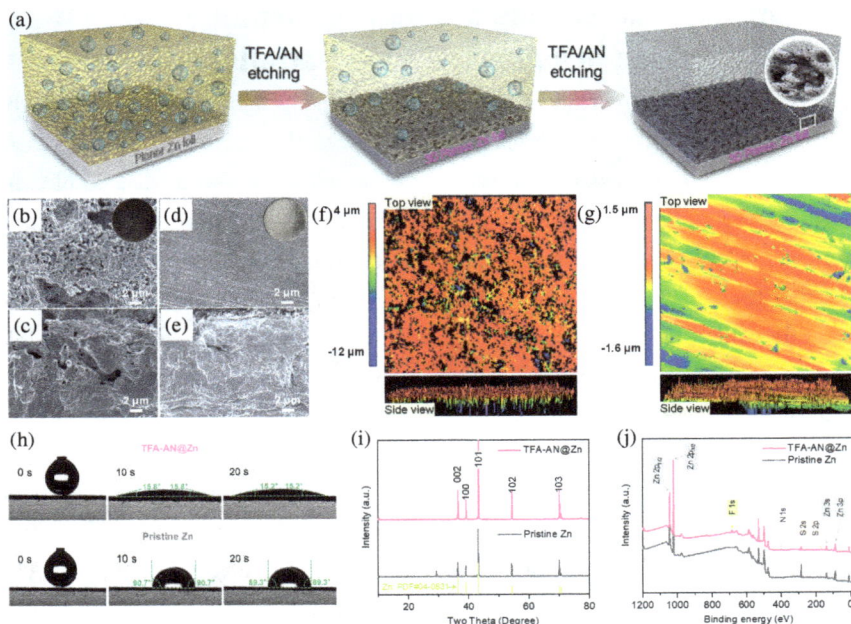

**Figure 5.8.** (a) Schematic diagram showing the fabrication of 3D porous Zn foil in non-aqueous organic TFA-AN acid. Top-view and cross-sectional SEM images of (b, c) TFA-AN@Zn and (d, e) pristine Zn (inset is the digital photo of TFA-AN@Zn and pristine Zn). Top and side view of optical surface profiles for (f) TFA-AN@Zn and (g) pristine Zn. (h) Contact angles of 1.0 M $ZnSO_4$ aqueous electrolyte on the surface of TFA-AN@Zn and pristine Zn. (i) XRD patterns and (j) overall XPS spectra of TFA-AN@Zn and pristine Zn. Reproduced with permission.[13] Copyright 2022, Wiley-VCH.

unlike the silver color of metallic Zn foil in Figure 5.8d, TFA-AN@Zn displays an ash-black surface, suggesting that the Zn foil is successfully corroded by the organic acid (Figure 5.8b).

The 3D optical surface measurement also illustrates a porous architecture and a smoother surface for TFA-AN@Zn (Figure 5.8f), whereas the pristine Zn shows a ridge-like surface with a height gap of several micrometers (Figure 5.8g). These results are perfectly consistent with SEM images in Figure 5.8b and 5.8d. Several reasons probably trigger these differences in the surface and inner structure of Zn foil: (i) The acid etching reaction of Zn foil in HCl-AN and HCl-$H_2O$ is found to be more vigorous, resulting in a rapid depletion of $H^+$ at the solution-Zn foil interface, and the formation of inhomogeneous corrosive surface. However, the etching reaction of Zn foil in TFA-AN solution is relatively slow and mild, allowing sufficient H$^+$ to continually etch the same places, resulting in the formation of a porous structure. (ii) The gas bubbles generated during the etching reaction in the TFA-AN solution are almost invisible, while the gas bubbles in HCl-AN are dense and large. In principle, the intimate contact of H$^+$ with Zn foil is significantly impeded due to the presence of gas bubbles on the surface of Zn foil. Accordingly, the following etching reaction can only occur until the gas bubbles break or separate from the Zn foil surface. (iii) The Zn foil surface develops by-products (e.g., $Zn_5(OH)_8Cl_2 \cdot H_2O$) during the etching reaction in HCl-AN and HCl-H$_2$C solution due to the existence of H$_2$O, which can affect the accessibility of H$^+$ to Zn foil and then the formation of the porous structure.

A hydrophilic surface of Zn foil is highly desirable to homogenize Zn plating in aqueous electrolytes,[27] which was evaluated by contact angle measurement (Figure 5.8h). Benefiting from 3D porous architecture and chemically modified surface, the intrinsic wettability of TFA-AN@Zn in 1.0 M ZnSO$_4$ aqueous electrolyte is superior to the pristine Zn foil, where the contact angle on TFA-AN@Zn is only 15.2° after 20 s resting. By contrast, the contact angle of the pristine Zn foil almost stabilizes at 89.3° even after 20 s. The superior hydrophilic properties of TFA-AN@Zn enable improved surface accessibility to the electrolyte and hence facilitate a homogeneous Zn deposition under more even ion transport.[28] XRD verifies that the main peaks of the TFA-AN@Zn and pristine Zn foil (e.g., (002), (100), and (101)) in Figure 5.8i fit well with the standard hexagonal-phased Zn (JCPDS No. 04-0831). Surface analysis was characterized with XPS to further confirm the products on the surface of Zn foil. Comparing the pristine Zn foil, the XPS spectra of TFA-AN@Zn show additional fluorine (F), nitrogen (N), and sulfur (S) in Figure 5.8j. The chemical components on the surface of TFA-AN@Zn are different from pristine Zn foil and hence affect the surface wettability of Zn foil.

The electrochemical plating/stripping performance was investigated by long-term galvanostatic cycling of a typical Zn||Zn symmetric cell illustrated in Figure 5.9a. As shown in Figure 5.9b, the symmetric TFA-AN@Zn cell demonstrates extended longevity and superior stability at the current density of 4.0 mA cm$^{-2}$. The cycling performance of the TFA-AN@Zn electrode reaches 930 h at 4.0 mA cm$^{-2}$ with the cut-off capacity of 2.0 mAh cm$^{-2}$.

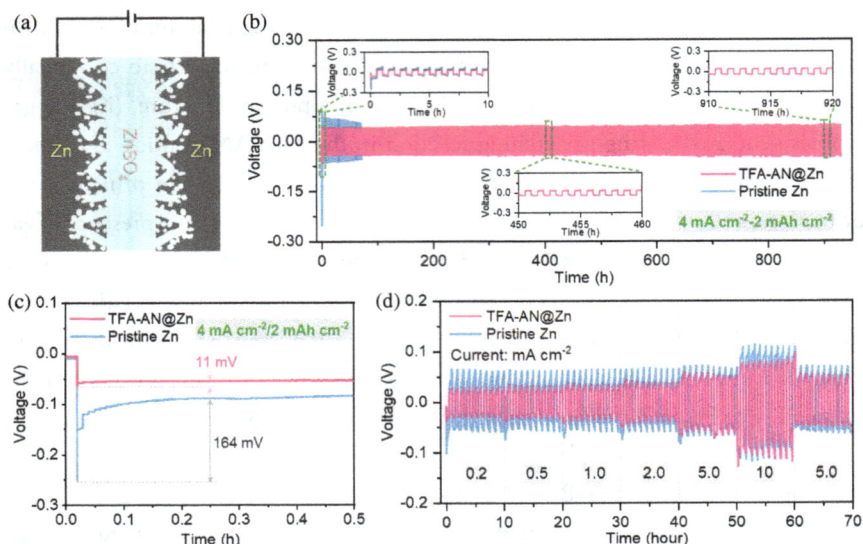

**Figure 5.9.** Electrochemical performance of symmetric cell in 1.0 M ZnSO$_4$ aqueous electrolyte. (a) Symmetric cell prototype assembled with TFA-AN@Zn. (b) Long-term plating/stripping performance of TFA-AN@Zn and pristine Zn (insets are their magnified voltage profiles). (c) Voltage profiles of TFA-AN@Zn and pristine Zn at the 1st plating under the current density of 4.0 mA cm$^{-2}$ with the capacity of 2.0 mAh cm$^{-2}$. (d) Rate performance of TFA-AN@Zn and pristine Zn at various current densities from 0.2 to 10 mA cm$^{-2}$. Reproduced with permission.[13] Copyright 2022, Wiley-VCH.

The difference in electrochemical performance was also reflected in the 1st plating voltage profiles of the TFA-AN@Zn and pristine Zn foil at the current density of 4.0 and 1.0 mA cm$^{-2}$ (Figure 5.9c). The TFA-AN@Zn electrode displays a nucleation overpotential of 11 mV in the voltage profiles, much smaller than the pristine Zn foil (164 mV) at 4.0 mA cm$^{-2}$. More notably, the pristine Zn foil retains a nucleation overpotential of 102 mV. Such low overpotential is probably attributed to the unique hierarchical 3D porous structure and clean surface of TFA-AN@Zn, leading to the enhanced affinity with Zn and lower resistance for Zn nucleation.[29] Thus, the superior cycling performance, low overpotential of TFA-AN@Zn foil at the current density of 4.0 and 1.0 mA cm$^{-2}$, and higher active surface area demonstrate the advantage of 3D Zn anode and the effective etching strategy of non-aqueous organic acid.

Figure 5.9d presents the rate performance of TFA-AN@Zn and pristine Zn at various current densities. The symmetric TFA-AN@Zn cell consistently exhibits a lower voltage hysteresis than the pristine Zn when increasing the current densities from 0.2 to 10 mA cm$^{-2}$. The voltage profiles of TFA-AN@Zn are still lower than the pristine Zn when the current density returns to 5.0 mA cm$^{-2}$. This stable rate performance suggests that the TFA-AN@Zn foil with 3D hierarchical porous structure provides a lower polarization of metallic anode and reduces local current density, leading to a highly efficient Zn plating/stripping process.

To visualize the Zn plating/stripping behavior in the aqueous electrolyte, the morphology evolution of the TFA-AN@Zn and pristine Zn electrode at different duration were characterized with *ex situ* SEM measurements and *in situ* optical microscopy. Figure 5.10a shows that the pristine Zn presents massive vertical flake-like

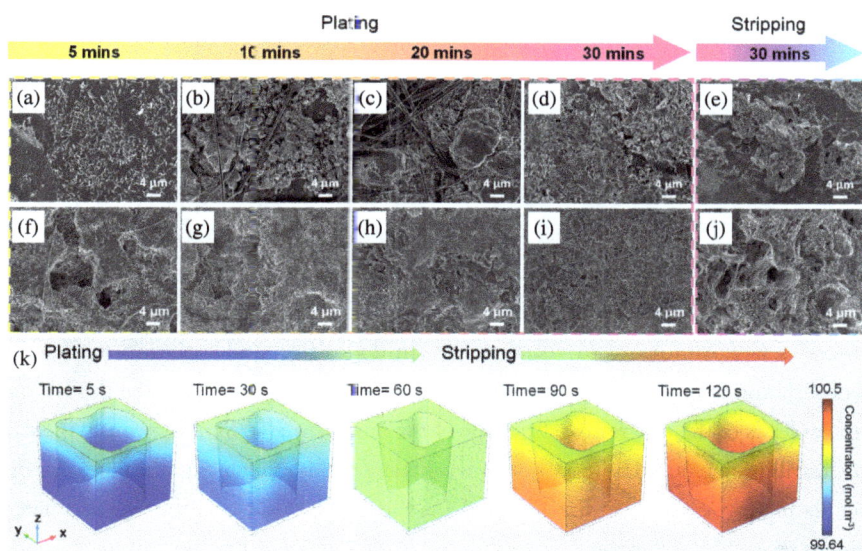

**Figure 5.10.** Electrochemical Zn plating/stripping behavior. Morphology evolution of (a-e) pristine Zn and (f-j) TFA-AN@Zn electrode during the plating/stripping process at the current density of 4.0 mA cm$^{-2}$ with different duration. (k) 3D COMSOL simulations for the variation of Zn ion concentration and single pore geometry during the Zn plating/stripping process. Reproduced with permission.[13] Copyright 2022, Wiley-VCH.

Zn protrusions at the initial plating stage. After that, the persistent accumulation of Zn around the protuberant tip can induce dendrite growth and inhomogeneous deposition morphologies, leading to the failure of the battery.[30] As expected, the dendrite morphologies composed of the stacking hexagonal flakes with different orientations and uneven particles on the surface of pristine Zn are developed with the deposition duration from 10 to 30 minutes (Figure 5.10b-d). Expectedly, the protruding Zn plates still stand on the surface of Zn foil even after stripping for 30 minutes (Figure 5.10e). By contrast, the TFA-AN@Zn electrode illustrates an entirely different deposition process (Figure 5.10f-i). The plating process of the TFA-AN@Zn electrode starts from the bottom of the pores until those pores fill up to form a flat and dendrite-free Zn surface after 30 minutes. This process is similar to the reported 3D lattice electrode.[31] In subsequent, the porous structure nearly reappears after stripping for 30 minutes (Figure 5.10j). Thus, the difference in the morphology evolution between the pristine Zn and TFA-AN@Zn electrode during the plating/stripping process verifies the spatially confined Zn growth and the superior Zn dendrite suppression in the TFA-AN@Zn electrode.

Based on the experimental observations, we further performed 3D finite element analysis using COMSOL multiphysics to simulate the electrochemical plating/stripping dynamic process on the TFA-AN@Zn foil. The color map shows that the overpotential near the bottom of the porous Zn is higher than the top surface, revealing the high priority of Zn deposition at the bottom region along the pore walls. The plating/stripping dynamic processes were simulated to visualize the Zn deposition and variation of Zn ion concentration along with the pore wall. As illustrated in Figure 5.10k, the deposition process leads to the progressive shrinkage of the pore diameter. Simultaneously the Zn ion distribution tends to be more homogeneous as Zn ions migrate into the pores during the plating process (*i.e.*, from 5 to 60 s). At the dissolution stage, the pore diameter widens reversibly, accompanying the increase of Zn ion concentration around the pore bottom. Accordingly, the simulated deposition nearly resembles the *ex situ* SEM results, where the deposition starts from the bottom region and grows along the pore wall, and then the pore structure is restored in the dissolution process (Figure 5.10f-k). This combined information offers decisive evidence that the 3D porous Zn electrode can control the reaction kinetics of Zn ions and diminish the dendrite growth. This improvement is mainly ascribed to the well-developed hierarchical pores and voids, which provide abundant nucleation sites and change the electric distribution.[28]

## 5.5 Conclusions

In summary, we introduced three anode design strategies for suppressing Zn metal dendrites via exemplary work from our group. These three works inhibit the formation of dendrites in the Zn anode by substrate design, protective layer design, and porous structure design. More specifically,

(1) In the work "MoS$_2$-Mediated Epitaxial Plating of Zn Metal Anodes", we have developed a strategy for deposition of epitaxial Zn anodes over a large area using explored mono-oriented MoS$_2$ as the substrate. Through a series of experiments and theoretical calculations, we confirmed that the edge of MoS$_2$ exhibited a much stronger Zn deposition-inducing effect than the basal plane, and would therefore affect the deposition behavior of Zn. In contrast, when deposited on MoS$_2$ free of edges, the Zn film grows epitaxially with high degree of orientation in both in-plane and out-of-plane. This work elucidates the effect of edges in the epitaxial substrate on the anodic metal electrodeposition process, and indicates the superiority of large-area mono-oriented substrates in studying the mechanisms of electrochemical processes.

(2) In the work "Controlled Deposition of Zinc-Metal Anodes via Selectively Polarized Ferroelectric Polymers", a polarized ferroelectric polymer (P(VDF-TrFE)) coating layer was demonstrated to improve Zn deposition behaviors and to achieve long-lifespan Zn metal anodes. Via polarizing the polymer coating with an external electric field in a downward direction, an inner electrostatic field could be generated, thus inducing locally concentrated Zn ion at the polymer coating surface. The horizontally arranged Zn nanosheet morphology and enhanced Zn basal plane ((002) plane) peak intensity in XRD curves revealed that such ferroelectric polymer coating could effectively modulate the Zn deposition behavior and limit the growth of Zn dendrites. Using such Zn anodes, the symmetric/asymmetric Zn batteries delivered improved electrochemical performances and high cycling stability. The design of such a ferroelectric polymer coating layer provides a new perspective to protect the Zn anode and could play an enlightening role in protecting other metal anodes.

(3) In the work "Organic Acid Etching Strategy for Dendrite Suppression in Aqueous Zinc-Ion Batteries", we have demonstrated an effective chemical etching strategy for constructing a hierarchical 3D porous Zn anode using the non-aqueous organic acid. Unlike the conventionally vigorous etching reaction in the aqueous

inorganic acid (*i.e.*, HCl), the chemical corrosion of TFA to Zn is relatively sluggish and controllable. This approach efficiently removes the surface passivation layer of Zn foil and generates abundant porous architectures with the diameter from nanoscale to microscale on the Zn surface, enabling higher accessibility to electrolyte and increasing the number of nucleation sites. As a result, the 3D porous TFA-AN@Zn electrode induces a bottom-top Zn deposition behavior with low overpotential on the plating/stripping, demonstrating the novelty of this study. More importantly, the TFA-AN@Zn electrode inhibits the growth of dendrite morphology, leading to the stable cycling performance of over 900 h at the current density of 4.0 mA cm$^{-2}$ with a cut-off capacity of 2.0 mAh cm$^{-2}$ in symmetric cells.

We believe that these strategies can inspire future research on designing non-dendrite Zn metal anodes and contribute to developing high-performance Zn ion batteries.

## Acknowledgment
The research reported in this chapter is supported by King Abdullah University of Science & Technology (KAUST), Saudi Arabia.

## References
[1] J. Yin, W. Zhang, N. A. Alhebshi, N. Salah, H. N. Alshareef, *Adv. Energy Mater.* **2021**, 11, 2100201.

[2] Y. Wang, D. Zhou, V. Palomares, D. Shanmukaraj, B. Sun, X. Tang, C. Wang, M. Armand, T. Rojo, G. Wang, *Energy Environ. Sci.* **2020**, 13, 3848.

[3] C. Yang, S. Xin, L. Mai, Y. You, *Adv. Energy Mater.* **2021**, 11, 2000974.

[4] R. Rajagopalan, Y. Tang, X. Ji, C. Jia, H. Wang, *Adv. Funct. Mater.* **2020**, 30, 1909486.

[5] J. Ming, J. Guo, C. Xia, W. Wang, H. N. Alshareef, *Mater. Sci. Eng. R Rep.* **2019**, 135, 58.

[6] a) Y. Zhu, J. Yin, X. Zheng, A.-H. Emwas, Y. Lei, O. F. Mohammed, Y. Cui, H. N. Alshareef, *Energy Environ. Sci.* **2021**, 14, 4463; b) L. E. Blanc, D. Kundu, L. F. Nazar, *Joule* **2020**, 4, 771.

[7] a) F. Wang, O. Borodin, T. Gao, X. Fan, W. Sun, F. Han, A. Faraone, J. A. Dura, K. Xu, C. Wang, *Nat. Mater.* **2018**, 17, 543; b) D. Kundu, B. D. Adams, V. Duffort, S. H. Vajargah, L. F. Nazar, *Nat. Energy* **2016**, 1, 16119.

[8] Y.-P. Deng, R. Liang, G. Jiang, Y. Jiang, A. Yu, Z. Chen, *ACS Energy Lett.* **2020**, 5, 1665.

[9] a) Q. Yang, Q. Li, Z. Liu, D. Wang, Y. Guo, X. Li, Y. Tang, H. Li, B. Dong, C. Zhi, *Adv. Mater.* **2020**, 32, 2001854; b) L. Ma, Q. Li, Y. Ying, F. Ma, S. Chen, Y. Li, H. Huang, C. Zhi, *Adv. Mater.* **2021**, 33, 2007406; c) Du Yuan, J. Zhao, H. Ren, Y. Chen, R. Chua, E. T. J. Jie, Y. Cai, E. Edison, W. M. Jr., M. W. Wong, M. Srinivasan, *Angew. Chem. Int. Ed.* **2021**, 60, 7213; d) J. F. Parker, C. N. Chervin, I. R. Pala, M. Machler, M. F. Burz, J. W. Long, D. R. Rolison, *Science* **2017**, 356, 415; e) B. J. Hopkins, C. N. Chervin, J. F. Parker, J. W. Long, D. R. Rolison, *Adv. Energy Mater.* **2020**, 10, 2001287.

[10] F. Xie, H. Li, X. Wang, X. Zhi, D. Chao, K. Davey, S.-Z. Qiao, *Adv. Energy Mater.* **2021**, 11, 2003419.

[11] Y. Wang, X. Xu, J. Yin, G. Huang, T. Guo, Z. Tian, R. Alsaadi, Y. Zhu, H. N. Alshareef, *Adv. Mater.* **2023**, 35, 2208171.

[12] Y. Wang, T. Guo, J. Yin, Z. Tian, Y. Ma, Z. Liu, Y. Zhu, H. N. Alshareef, *Adv. Mater.* **2022**, 34, 2106937.

[13] W. Wang, G. Huang, Y. Wang, Z. Cao, L. Cavallo, M. N. Hedhili, H. N. Alshareef, *Adv. Energy Mater.* **2022**, 12, 2102797.

[14] a) Y. Zhu, L. Peng, Z. Fang, C. Yan, X. Zhang, G. Yu, *Adv. Mater.* **2018**, 30, 1706347; b) Z. Liu, L. Zhang, L. Sheng, Q. Zhou, T. Wei, J. Feng, Z. Fan, *Adv. Energy Mater.* **2018**, 8, 1802042; c) Y. Luo, G.-F. Chen, L. Ding, X. Chen, L.-X. Ding, H. Wang, *Joule* **2019**, 3, 279; d) A. Bruix, H. G. Füchtbauer, A. K. Tuxen, A. S. Walton, M. Andersen, S. Porsgaard, F. Besenbacher, B. Hammer, J. V. Lauritsen, *ACS Nano* **2015**, 9, 9322.

[15] X. Xu, Z. Wang, S. Lopatin, M. A. Quevedo-Lopez, H. N. Alshareef, *2D Mater.* **2019**, 6, 015030.

[16] a) X. Xu, G. Das, X. He, M. N. Hedhili, E. D. Fabrizio, X. Zhang, H. N. Alshareef, *Adv. Funct. Mater.* **2019**, 29, 1901070; b) X. Xu, C. Zhang, M. K. Hota, Z. Liu, X. Zhang, H. N. Alshareef, *Adv. Funct. Mater.* **2020**, 30, 1908040.

[17] T. Takahashi, C. Ando, M. Saito, Y. Miyata, Y. Nakanishi, J. Pu, T. Takenobu, *NPJ 2D Mater. Appl.* **2021**, 5, 31.

[18] a) Q. Zhang, J. Luan, X. Huang, L. Zhu, Y. Tang, X. Ji, H. Wang, *Small* **2020**, 16, 2000929; b) X. Xu, Y. Chen, D. Zheng, P. Ruan, Y. Cai, X. Dai, X. Niu, C. Pei, W. Shi, W. Liu, F. Wu, Z. Pan, H. Li, X. Cao, *Small* **2021**, 17, 2101901.

[19] X. Tang, D. Zhou, P. Li, X. Guo, B. Sun, H. Liu, K. Yan, Y. Gogotsi, G. Wang, *Adv. Mater.* **2020**, 32, 1906739.

[20] X. Fan, X. Ji, F. Han, J. Yue, J. Chen, L. Chen, T. Deng, J. Jiang, C. Wang, *Sci. Adv.* **2018**, 4, eaau9245.

[21] a) T. Zhang, Y. Tang, S. Guo, X. Cao, A. Pan, G. Fang, J. Zhou, S. Liang, *Energy Environ. Sci.* **2020**, 13, 4625; b) C. Liu, X. Xie, B. Lu, J. Zhou, S. Liang, *ACS Energy Lett.* **2021**, 6, 1015; c) S. Guo, L. Qin, T. Zhang, M. Zhou, J. Zhou, G. Fang, S. Liang, *Energy Stor. Mater.* **2021**, 34, 545.

[22] Y. He, Q. He, L. Wang, C. Zhu, P. Golani, A. D. Handoko, X. Yu, C. Gao, M. Ding, X. Wang, F. Liu, Q. Zeng, P. Yu, S. Guo, B. I. Yakobson, L. Wang, Z. W. Seh, Z. Zhang, M. Wu, Q. J. Wang, H. Zhang, Z. Liu, *Nat. Mater.* **2019**, 18, 1098.

[23] a) J. Zheng, J. Yin, D. Zhang, G. Li, D. C. Bock, T. Tang, Q. Zhao, X. Liu, A. Warren, Y. Deng, S. Jin, A. C. Marschilok, E. S. Takeuchi, K. J. Takeuchi, C. D. Rahn, L. A. Archer, *Sci. Adv.* **2020**, 6, eabb1122; b) J. Wan, J. Xie, X. Kong, Z. Liu, K. Liu, F. Shi, A. Pei, H. Chen, W. Chen, J. Chen, X. Zhang, L. Zong, J. Wang, L.-Q. Chen, J. Qin, Y. Cui, *Nat. Nanotech.* **2019**, 14, 705.

[24] a) M. V. Kelso, N. K. Mahenderkar, Q. Chen, J. Z. Tubbesing, J. A. Switzer, *Science* **2019**, 364, 166; b) N. K. Mahenderkar, Q. Chen, Y.-C. Liu, A. R. Duchild, S. Hofheins, E. Chason, J. A. Switzer, *Science* **2017**, 355, 1203.

[25] D. Guo, N. Setter, *Macromolecules* **2013**, 46, 1883.

[26] J. Chen, J. Zhao, L. Lei, P. Li, J. Chen, Y. Zhang, Y. Wang, Y. Ma, D. Wang, *Nano Lett.* **2020**, 20, 3403.

[27] R. Yuksel, O. Buyukcakir, W. K. Seong, R. S. Ruoff, *Adv. Energy Mater.* **2020**, 10, 1904215.

[28] H. Tian, Z. Li, G. Feng, Z. Yang, D. Fox, M. Wang, H. Zhou, L. Zhai, A. Kushima, Y. Du, Z. Feng, X. Shan, Y. Yang, *Nat. Commun.* **2021**, 12, 237.

[29] Y. Zeng, X. Zhang, R. Qin, X. Liu, P. Fang, D. Zheng, Y. Tong, X. Lu, *Adv. Mater.* **2019**, 31, e1903675.

[30] S. Li, J. Fu, G. Miao, S. Wang, W. Zhao, Z. Wu, Y. Zhang, X. Yang, *Adv. Mater.* **2021**, 33, e2008424.

[31] G. Zhang, X. Zhang, H. Liu, J. Li, Y. Chen, H. Duan, *Adv. Energy Mater.* **2021**, 11, 2003927.

https://doi.org/10.1142/9789811278327_0006

Chapter 6

# Aqueous Electrolytic Zinc-Manganese Dioxide Batteries[1]

Mingming Wang,[a] Xinhua Zheng,[a] Yi Cui,[b,c] Wei Chen[a,*]

[a]Department of Applied Chemistry, School of Chemistry and Materials
Science, Hefei National Research Center for Physical Sciences
at the Microscale, University of Science and Technology of China,
Hefei, Anhui 230026, China
[b]Department of Materials Science and Engineering, Stanford University,
Stanford, California 94305, United States
[c]Stanford Institute for Materials and Energy Sciences, SLAC National
Accelerator Laboratory, Menlo Park, California 94025, United States

Rechargeable aqueous zinc-manganese dioxide ($Zn-MnO_2$) batteries have been attracting significant attention owing to their advantages of low cost, high safety and ease of manufacturing, which are promising attributes for grid-scale energy storage applications. However, most traditional Mn-based batteries with solid-state conversion and intercalation reactions suffer from low capacity and poor long-term cycling stability. The recent novel storage mechanism based on cathode $Mn^{2+}/MnO_2$ deposition/stripping chemistry has fundamentally tackled these issues, enabling a new generation of electrolytic $Zn-MnO_2$ batteries with superior electrochemical performance. Here we review the recent advances in aqueous Mn-based batteries with the $Mn^{2+}/MnO_2$ deposition/stripping chemistry represented by electrolytic $Zn-MnO_2$ batteries. We provide a summary of the development of $Zn-MnO_2$ batteries with different storage mechanisms and highlight new opportunities of the emerging $Mn^{2+}/MnO_2$ chemistry in the latest generation. We then present the current understanding of the $Mn^{2+}/MnO_2$ charge storage mechanism and its potential in manganese-based batteries for large-scale energy storage applications. Moreover, insights into opportunities and future directions for electrolytic $Zn-MnO_2$ batteries with the $Mn^{2+}/MnO_2$ chemistry are proposed.

---

* Corresponding author: weichen1@ustc.edu.cn
[1] Adapted with permission from M. Wang, X. Zheng, Y. Cui, W. Chen, *Adv. Energy Mater.* **2021**, 11, 2002904.

## 6.1 Introduction

In order to alleviate the drastic impact of fossil fuels on climate change, environmental pollution and quality of life, there has never been a greater and more urgent demand of sustainable energy.[1–3] The intermittency of the renewable energy resources such as solar and wind calls for the development of low-cost and scalable energy storage technologies.[4–8] Rechargeable batteries are regarded as the most efficient energy storage technologies that have been widely applied to portable electronics, electric vehicles and grid-scale energy storage.[9–11] Although lithium ion batteries are dominating the current market of electric vehicles and portable electronic devices,[12–14] their application in grid-scale energy storage is just beginning due to relatively high cost, limited service life and safety concerns.[15–17] Other existing rechargeable batteries such as sodium-sulfur (Na-S), lead-acid, and redox-flow batteries have been gradually applied to grid storage, but they have encountered different obstacles that need to be overcome, as summarized in Figure 6.1. For example, the Na-S batteries have potentially severe safety issues due to their operation at high temperature (~350°C). The lead-acid batteries exhibited poor cycling stability (typically less than 1,000 cycles). The redox-flow batteries demonstrated relatively low energy density and high cost at system level. By contrast, aqueous rechargeable batteries offer an alternative energy storage technology for grid storage due to their ease of fabrication, fast operation rates and good safety.[18–20] Among them, aqueous $Zn\text{-}MnO_2$ batteries have been attracting much research and industry interests owing to their additional advantages of low cost,[21,22] environmental friendliness,[23] high theoretical capacity and energy density (Figure 6.1).[24,25] Moreover, they offer rich redox chemistry thanks to the existence of various valence states such as $Mn^0$, $Mn^{2+}$, $Mn^{3+}$, $Mn^{4+}$, $Mn^{6+}$ and $Mn^{7+}$.[26–29] Due to the relatively high electrochemical potentials between $Mn^{2+}$, $Mn^{3+}$ and $Mn^{4+}$, which are the most commonly available valences of the Mn element,[30–32] the Mn-based electrodes in different forms of oxides and hydroxides (e.g., $Mn(OH)_2$, $MnOOH$, $Mn_2O_3$, and $MnO_2$) are typically utilized as cathodes in the rechargeable aqueous batteries.[33–35]

Historically, the $MnO_2$ cathodes have been studied extensively for Mn-based aqueous batteries.[36,37] Dating back to the early 1860s, Georges Leclanché invented the first primary $Zn\text{-}MnO_2$ cell on the basis of $MnO_2$-carbon black cathode, Zn foil anode, and a mixture of $ZnCl_2$ and $NH_4Cl$ electrolyte.[38] Thereafter, the studies of

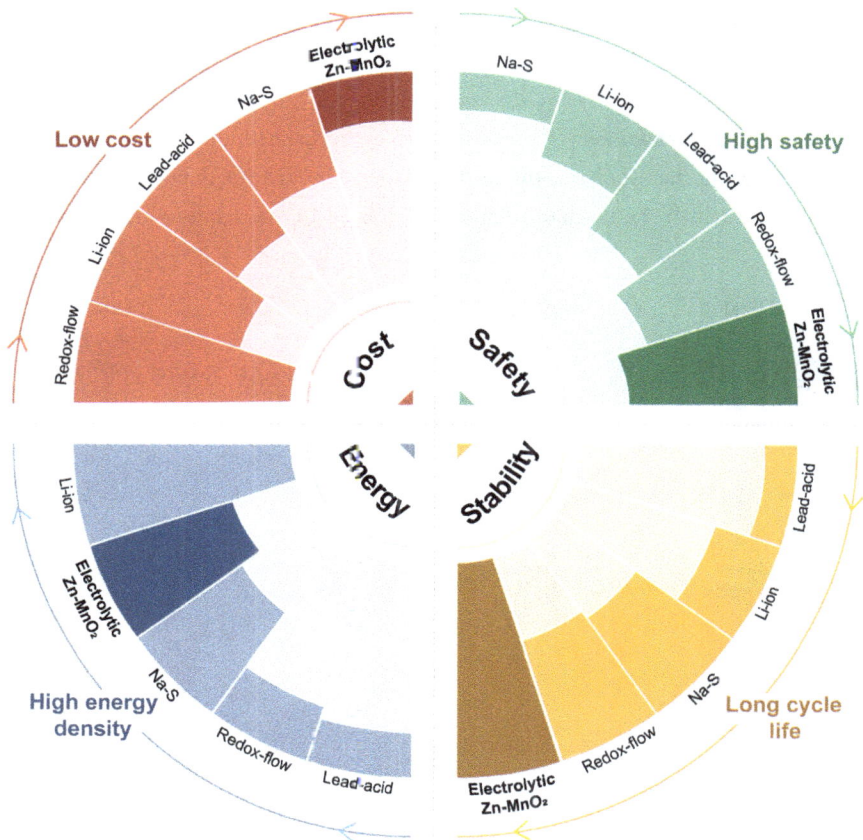

**Figure 6.1.** A comparison of aqueous electrolytic Zn-MnO$_2$ batteries with other battery technologies of Li-ion, Na-S, lead-acid and redox-flow batteries in terms of cost, safety, stability and energy density for grid-scale energy storage. All data are comparative but not absolute values based on literature.

MnO$_2$ and its composite materials in different rechargeable batteries have increased exponentially.[39–42] Much effort has been devoted to the improvement of electrochemical performance of the aqueous Mn-based batteries. Despite significant progress achieved over the past decades, the Mn-based batteries still encountered fundamental issues that severely hindered their practical applications, such as low capacity, slow charge/discharge rate and poor electrochemical stability.[43–46] The past attempts on the exploration of Mn-based batteries were based on storage mechanisms of solid-state conversion and cation (i.e., Zn$^{2+}$, Na$^+$, H$^+$) intercalation into the Mn-based cathodes, in which complex solid-phase behaviors cause limited capacity and poor reaction reversibility (Figure 6.2).

Recently, Cui and coworkers[47] proposed and demonstrated a completely different charge storage mechanism based on the liquid/solid $Mn^{2+}/MnO_2$ deposition/stripping chemistry, in which the discharged state is soluble $Mn^{2+}$ aqueous solution and the

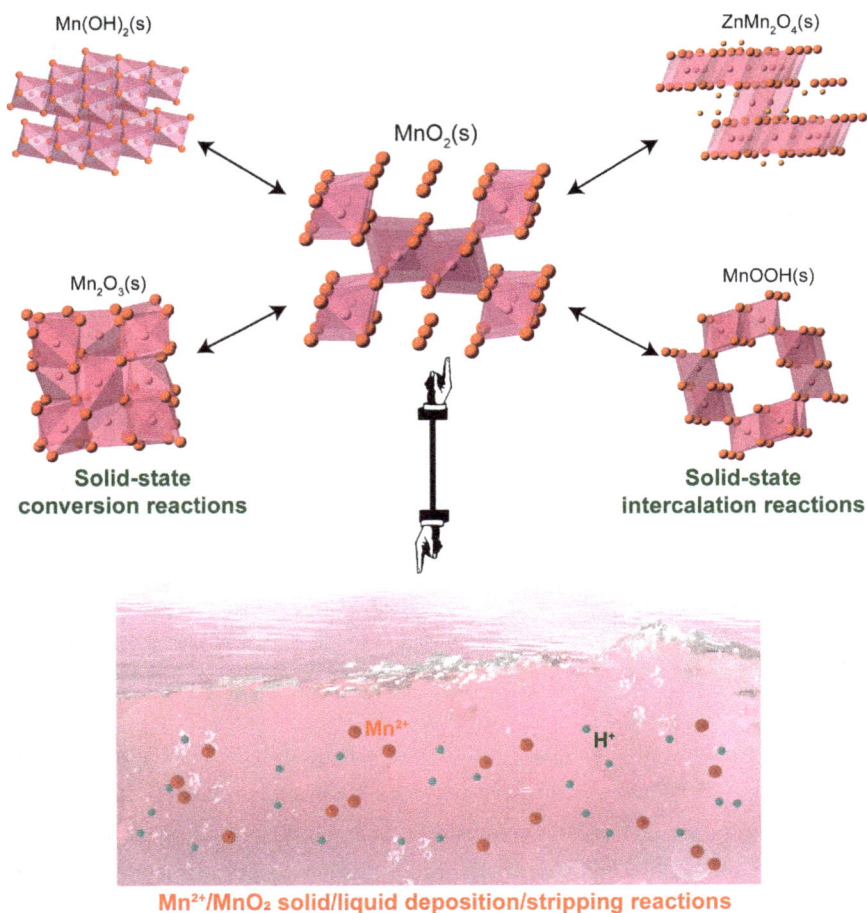

$Mn(OH)_2(s)$

$ZnMn_2O_4(s)$

$MnO_2(s)$

$Mn_2O_3(s)$

$MnOOH(s)$

**Solid-state conversion reactions**

**Solid-state intercalation reactions**

$Mn^{2+}$     $H^+$

**$Mn^{2+}/MnO_2$ solid/liquid deposition/stripping reactions**

**Figure 6.2.** Charge storage mechanisms in aqueous $Zn$-$MnO_2$ batteries. The development of $Zn$-$MnO_2$ batteries can be classified by the storage mechanisms, which include solid-state conversion, intercalation reactions and solid/liquid deposition/stripping reactions, the latter being the storage mechanism of the new generation $Zn$-$MnO_2$ batteries.

charged state is solid-state $MnO_2$ (Figure 6.2). Using a new full cell chemistry of manganese dioxide-hydrogen gas ($MnO_2$-$H_2$) battery as an example, it achieved high cathode specific capacity and ultrastable cycling performance (10,000 cycles). This study evoked significant research interest in the exploration of the cathode $Mn^{2+}$/$MnO_2$ deposition/stripping chemistry to combine with a variety of anode chemistries, creating different Mn-based batteries such as electrolytic $Zn$-$MnO_2$,[48–50] $MnO_2$-$Cu$,[51,52] $MnO_2$-carbon,[53] and $MnO_2$-$Pb$.[54] to name a few, with superb electrochemical performance. Although the development of the $Mn^{2+}$/$MnO_2$ deposition/stripping chemistry is at its early stage, it presents great opportunities for the design of high-performance Mn-based batteries for practical large-scale energy storage applications.

This chapter aims to discuss recent advances in the current state-of-the-art rechargeable electrolytic $Zn$-$MnO_2$ batteries with the new $Mn^{2+}$/$MnO_2$ chemistry. The development of $Zn$-$MnO_2$ batteries with different storage mechanisms is illustrated, with a focus on the latest generation electrolytic $Zn$-$MnO_2$ batteries in the $Mn^{2+}$/$MnO_2$ chemistry. In addition, we present the current understanding of the emerging $Mn^{2+}$/$MnO_2$ storage mechanism by a combination of advanced characterization techniques with simulation and theoretical calculations. The potential large-scale energy storage applications of the electrolytic $Zn$-$MnO_2$ batteries are also discussed. Furthermore, opportunities and future directions on the development of aqueous Mn-based batteries with the $Mn^{2+}$/$MnO_2$ chemistry are proposed.

## 6.2 The Development of Zn-MnO$_2$ Batteries with Different Mechanisms

As depicted in Figure 6.2, the $Zn$-$MnO_2$ batteries have gone through two major generations that are characterized by their distinct charge storage mechanisms.[55–57] The first generation $Zn$-$MnO_2$ batteries are based on the solid-state conversion and intercalation reactions of $MnO_2$ cathode, where alkaline and mild aqueous electrolytes are typically utilized.[58,59] The new generation $Zn$-$MnO_2$ batteries batteries are based on the cathode $Mn^{2+}$/$MnO_2$ deposition/stripping chemistry, where acidic aqueous solutions are generally required to realize the efficient deposition/stripping reactions.[47]

Since the discovery of the primary $Zn$-$MnO_2$ battery,[60] the very first generation Mn-based batteries (Generation 1A) were typically operated in alkaline electrolytes.[61]

However, they typically showed limited capacities and poor cycling performance.[62,63] The general failure mechanism of $MnO_2$ cathode in the alkaline electrolyte is ascribed to the partial irreversible conversion reactions between $MnO_2$ and the reduced $Mn_2O_3$ and $Mn(OH)_2$, or the gradual dissolution of $MnO_2$ into the strong alkaline electrolytes (Figure 6.2).[64,65] In order to improve the $MnO_2$ cathode instability issue, researchers have developed various strategies such as utilizing mixed KOH and LiOH instead of pure KOH as the electrolyte,[66,67] or different metals doping of $MnO_2$.[68,69] It was found that the battery capacity retention rate and the number of cycles can be largely increased. In the case of mixed KOH and LiOH electrolyte, the enhancement of cycle performance is attributed to the formation of reversible $Li_xMnO_2$ (0 < x < 1) by lithium ion insertion into the $MnO_2$. Hertzberg and coworkers[62] used the alkaline electrolyte mixture of KOH and LiOH with a molar ratio of 1:3 to study the performance of Zn-$MnO_2$ battery. Results showed that the battery can run over 60 cycles with a capacity of ~150 mAh g$^{-1}$, better than the cell in pure KOH electrolyte. In addition, some researchers have confirmed that the alkaline Zn-$MnO_2$ battery is rechargeable for extended cycles by using a shallow cycle protocol which is typically less than 0.2–0.5 electron transfer per $MnO_2$.[70,71] However, the delivered capacity was very limited at the shallow depth of discharge such as 10%.[72] Attempt to achieve high-capacity $MnO_2$ cathodes in alkaline electrolytes have been demonstrated to be effective by different metal doping into the $MnO_2$ structures. Wroblowa, Conway and coworkers[73,74] found that the alkaline Zn-$MnO_2$ batteries with a small amount of bismuth-doped $MnO_2$ cathode showed a relatively stable lifetime of up to 1,000 cycles with about 80% of two-electron capacity of 616 mAh g$^{-1}$. However, these alkaline Bi/$MnO_2$-Zn batteries need to be operated by deploying ion-selective membranes to block the $ZnO_2^{2-}$ ion transfer from the Zn anode to the $MnO_2$ cathode, and their operational voltages are ~300 mV lower than the typical alkaline Zn-$MnO_2$ batteries. In 2016, Banerjee and coworkers[69] synthesized a Cu ion-intercalated Bi/$MnO_2$ that showed nearly two-electron charge capacity, high areal capacity of more than 20 mAh cm$^{-2}$, excellent rate performance and a stable lifetime of more than 6,000 cycles, which demonstrated promising advancement towards the widespread applications of the alkaline Zn-$MnO_2$ batteries.

The solid-state cation intercalation storage mechanisms in mild aqueous electrolytes account for the advanced first-generation Zn-$MnO_2$ batteries (Generation 1B), which is dominated by the Zn-MnO$_2$ battery chemistry.[75-77] In the 1980s, Yamamoto and coworkers[78,79] introduced the concept of a rechargeable Zn-MnO$_2$ battery in a mild aqueous electrolyte of ZnSO$_4$, which showed operational voltage of 1.4 V and relatively good reversibility of 30 cycles. Later, Kang and coworkers[59] expanded this Zn-MnO$_2$ battery chemistry with intercalation mechanism by using mild ZnSO$_4$ or Zn(NO$_3$)$_2$ solutions as the electrolyte. At the cathode Zn$^{2+}$ can be reversibly intercalated into the tunnels of $\alpha$-MnO$_2$, and the anode zinc can be electrochemically stripped to Zn$^{2+}$ and deposited back reversibly.[80,81] The battery showed a relatively large capacity of 210 mAh g$^{-1}$ and decent capacity retention at low charge/discharge rates.[59] However, the Zn-MnO$_2$ batteries with the solid-state intercalation mechanism in mild aqueous electrolytes exhibited low capacities at high charge/discharge rates and significant capacity decay for long-term cycling. To elucidate the failure mechanism, a variety of *ex situ* and *in situ* techniques were conducted but controversial mechanisms were proposed by different groups.[50,82,83] For example, Kang and coworkers[59] proposed the Zn$^{2+}$ insertion/extraction mechanism in the Zn-MnO$_2$ batteries in mild ZnSO$_4$ or Zn(NO$_3$)$_2$ electrolyte by the demonstration of reversible Zn$^{2+}$ insertion and desertion into/out of $\alpha$-MnO$_2$ with the assistance of *ex situ* X-ray diffraction (XRD). However, Liu and coworkers[84] proposed the mechanism of reversible H$^+$ insertion/extraction reactions that were accompanied by the deposition of Zn$_4$(SO$_4$)(OH)$_6$·nH$_2$O in the electrolyte of ZnSO$_4$ with a MnSO$_4$ additive *via* transmission electron microscopy. Though complicated due to different phases and structures of MnO$_2$ materials and complex battery systems, it is generally accepted that the proton insertion into the MnO$_2$ cathode promotes the reversibility of the Mn-based batteries in mild aqueous electrolytes (Figure 6.2).[84,85] Wang and coworkers[86] reported a Zn-MnO$_2$ battery in a mild electrolyte, where the MnSO$_4$ was added into ZnSO$_4$ solution to inhibit the dissolution of MnO$_2$ during the cycling.[84,87] The electrochemical and structural analysis identified that the MnO$_2$ cathode underwent a consequent H$^+$ and Zn$^{2+}$ insertion/extraction process with high reversibility and cycling stability.[88] In order to further improve the electrochemical performance of the Mn-based batteries in mild aqueous electrolytes, Chen and coworkers[85] reported a rechargeable aqueous Zn-MnO$_2$ battery with 3 M Zn(CF$_3$SO$_3$)$_2$ and 0.1 M Mn(CF$_3$SO$_3$)$_2$ additive as the electrolyte, which delivered a highly reversible capacity of 225 mAh g$^{-1}$ and 94% capacity retention after 2,000 cycles. It is subjected to the bulky anion CF$_3$SO$_3^-$ that is beneficial to the reactivity and stability of the Zn anode and the spinel ZnMn$_2$O$_4$ cathode,[90] where the pre-added

$Mn(CF_3SO_3)_2$ can effectively hamper the $MnO_2$ dissolution, and the electrode integrity can be maintained because of the *in situ* generated amorphous $MnO_x$ layer in the charge process. Although the capacity and long-term cycling performance of the Mn-based batteries have been greatly improved with the intercalation storage mechanism in mild aqueous electrolytes, it is still desirable to develop novel Mn-based battery chemistry with high output voltage and energy density to fulfill the extensive promotion of the large-scale energy storage market.[91–93]

The development of the new generation electrolytic $Zn$-$MnO_2$ batteries stems from the cathode $Mn^{2+}/MnO_2$ deposition/stripping chemistry (Generation 2) proposed by Cui and coworkers in 2017 (Figure 6.3a).[47] Although the $Mn^{2+}/MnO_2$ chemistry has been studied previously,[94–96] it is more related to the mechanism of $MnO_2$ electrodeposition, and it has not yet been deployed in the battery field due to unsatisfactory electrochemical performance. Since then researchers from all over the world have applied this emerging $Mn^{2+}/MnO_2$ storage mechanism to a variety of Mn-based batteries including electrolytic $Zn$-$MnO_2$,[48,50] $MnO_2$-$Cu$,[51,52] $MnO_2$-$Bi$,[97] $MnO_2$-$Pb$,[54] $MnO_2$-$Cd$,[98] $MnO_2$-$Cr$,[99] and $MnO_2$-carbon[53] and achieved unprecedented electrochemical performance, as shown in the chronological development of $Zn$-$MnO_2$ batteries (Figure 6.3). Table 6.1 shows a summary of the two generations of $Zn$-$MnO_2$ batteries with their cell parameters and electrochemical performance in different electrolytes. With the gradual development of the storage mechanism, the $Zn$-$MnO_2$ battery performance in terms of output voltage, capacity and energy density has been greatly improved, reaching the highest values in the new generation thanks to the cathode $Mn^{2+}/MnO_2$ chemistry. The emerging $Mn^{2+}/MnO_2$ energy storage mechanism is gaining revived attention as promising redox chemistry for the development of the state-of-the-art aqueous electrolytic $Zn$-$MnO_2$ batteries.

# 6.3 Aqueous Mn-based Batteries with $Mn^{2+}/MnO_2$ Chemistry

## 6.3.1 *$MnO_2$-hydrogen Gas Battery*

The cathode $Mn^{2+}/MnO_2$ deposition/stripping reactions were initially demonstrated in the $MnO_2$-$H_2$ battery by coupling it with a catalytic hydrogen gas anode as reported

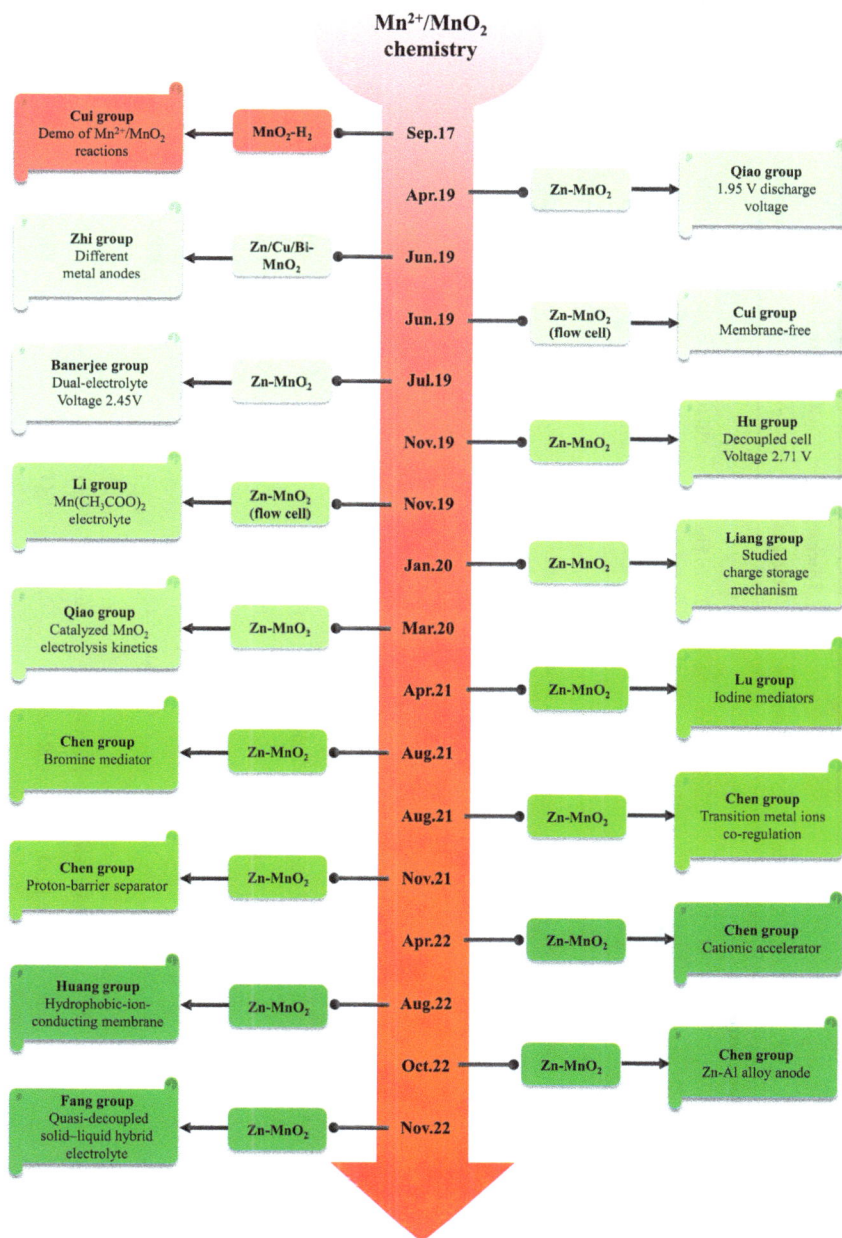

**Mn²⁺/MnO₂ chemistry**

| | | |
|---|---|---|
| **Cui group** Demo of Mn²⁺/MnO₂ reactions | MnO₂-H₂ ← Sep.17 | |
| | Apr.19 → Zn-MnO₂ → | **Qiao group** 1.95 V discharge voltage |
| **Zhi group** Different metal anodes ← Zn/Cu/Bi-MnO₂ ← Jun.19 | | |
| | Jun.19 → Zn-MnO₂ (flow cell) → | **Cui group** Membrane-free |
| **Banerjee group** Dual-electrolyte Voltage 2.45V ← Zn-MnO₂ ← Jul.19 | | |
| | Nov.19 → Zn-MnO₂ → | **Hu group** Decoupled cell Voltage 2.71 V |
| **Li group** Mn(CH₃COO)₂ electrolyte ← Zn-MnO₂ (flow cell) ← Nov.19 | | |
| | Jan.20 → Zn-MnO₂ → | **Liang group** Studied charge storage mechanism |
| **Qiao group** Catalyzed MnO₂ electrolysis kinetics ← Zn-MnO₂ ← Mar.20 | | |
| | Apr.21 → Zn-MnO₂ → | **Lu group** Iodine mediators |
| **Chen group** Bromine mediator ← Zn-MnO₂ ← Aug.21 | | |
| | Aug.21 → Zn-MnO₂ → | **Chen group** Transition metal ions co-regulation |
| **Chen group** Proton-barrier separator ← Zn-MnO₂ ← Nov.21 | | |
| | Apr.22 → Zn-MnO₂ → | **Chen group** Cationic accelerator |
| **Huang group** Hydrophobic-ion-conducting membrane ← Zn-MnO₂ ← Aug.22 | | |
| | Oct.22 → Zn-MnO₂ → | **Chen group** Zn-Al alloy anode |
| **Fang group** Quasi-decoupled solid–liquid hybrid electrolyte ← Zn-MnO₂ ← Nov.22 | | |

**Figure 6.3.** Historical timeline for the development of new generation electrolytic Zn-MnO₂ batteries with Mn²⁺/MnO₂ chemistry.

Table 6.1. Summary of the parameters and electrochemical performance of Zn-MnO$_2$ batteries with different storage mechanisms.

| Electrolyte | pH | Cathode | Anode | Discharge Voltage (V) | Capacity (mAh g$^{-1}$) | Retention/Cycles (rate) | Ref. |
|---|---|---|---|---|---|---|---|
| **Generation 1A: Alkaline electrolytes** | | | | | | | |
| 4 M KOH | ~13.5 | MnO$_2$/Bi$_2$O$_3$ | Zn plate | -0.3 vs. Hg/HgO | 230 (C/3) | 0/15 (C/3) | [62] |
| 1 M KOH + 3 M LiOH | ~13.2 | MnO$_2$/Bi$_2$O$_3$ | Zn plate | -0.3 vs. Hg/HgO | 300 (C/3) | 43%/60 (C/3) | [62] |
| Saturated LiOH + 1 M ZnSO$_4$ | ~8.5 | MnO$_2$ | Zn plate | 1.60 | 148 | 20%/50 | [66] |
| 1M NaOH or 9M KOH | ~14 | MnO$_2$ | CMC-gell Zn powders | -0.4 vs. Hg/HgO | N/A | N/A | [100] |
| 7 M KOH | ~14 | MnO$_2$ | Zn foil | 1.40 | 330 | 80%/2 | [101] |
| Saturated LiOH + 1M ZnSO$_4$ | ~8.5 | MnO$_2$-BGM | Zn plate | 1.60 | 165 | 65%/2 | [102] |
| 1 M KOH + 3 M LiOH | ~14.6 | β-MnO$_2$/Bi$_2$O$_3$ | Zn plate | 1.12 | 316 | 100%/100 (C/10) | [103] |
| 40 wt% KOH | ~14 | γ-MnO$_2$ nanowires | Zn powders | 1.2 | 280 | 31 h (100 mA) | [104] |
| 9 M KOH | ~14.6 | Ag$_4$Bi$_2$O$_5$-MnO$_2$ | N/A | 1.2 | 481 | ~34%/115 (1000 mA g$^{-1}$) | [105] |
| 15–45 wt% KOH | 12–14 | Cu-Bi-birnessite | Zn/ZnO | 1.1 | 616 | 81%/900 | [106] |
| 9 M KOH | ~14.6 | CM-MnO$_2$ | Zn | 1.00 | 492 (1 C) | 40%/1000 | [73] |
| 45 wt% KOH | ~14.5 | MnO$_2$ | Zn/ZnO nickel mesh | 1.43 | 31 | 80%/3000 (C/2) | [72] |
| 25 wt% KOH | ~14.3 | Cu-Bi-birnessite | NiOOH | -0.4 vs. Hg/HgO | 616 | 6000 (40 C) | [69] |
| 25 wt% KOH | ~14.3 | MnO/Bi$_2$O$_3$ | CNT-Zn | -0.5 vs. Hg/HgO | 750 | 80%/350 (C/4) | [107] |

**Generation 1B: Mild aqueous electrolytes**

| | | | | | | | |
|---|---|---|---|---|---|---|---|
| 1 M ZnSO$_4$ | ~4.0 | α-MnO$_2$ | Zn-birnessite | 1.25 | 195 (C/20) | 70%/30 | [108] |
| 0.1 M Zn(NO$_3$)$_2$ | ~5.2 | α-MnO$_2$ | Zn foil | 1.4 | 210 (0.5 C) | 100%/100 (6 C) | [59] |
| 1 M ZnSO$_4$ | ~4.0 | α-MnO$_2$ | Zn foil | 1.3 | 115 (83 mA g$^{-1}$) | 36%/75 (83 mA g$^{-1}$) | [109] |
| 1 M ZnSO$_4$ | ~4.0 | δ-MnO$_2$ | Zn foil | 1.38 | 112 (83 mA g$^{-1}$) | 44%/100 (83 mA g$^{-1}$) | [76] |
| 1 M ZnSO$_4$ | ~4.0 | α-MnO$_2$ | Zn foil | 1.3 | 233 (83 mA g$^{-1}$) | 63%/50 (83 mA g$^{-1}$) | [110] |
| 1 M ZnSO$_4$ | ~4.0 | α-MnO$_2$ | Zn foil | 1.4 | 108 (C/2) | 43/50 (C/2) | [75] |
| 2 M ZnSO$_4$ + 0.1 M MnSO$_4$ | ~3.8 | α-MnO$_2$ | Zn foil | 1.44 | 160 (5C) | 92%/5000 (5 C) | [84] |
| 2 M ZnSO$_4$ | ~3.6 | γ-MnO$_2$ | Zn plate | 1.5 | 119 (2 mA) | 83%/50 (2 mA) | [79] |
| 1 M ZnSO$_4$ + 0.1 M MnSO$_4$ | ~3.9 | MnO$_2$-birnessite | Zn foil | 1.39 | 505.9 (200 mA g$^{-1}$) | 100%/400 (500 mA g$^{-1}$) | [88] |
| 2 M ZnSO$_4$ + 0.1 M MnSO$_4$ | ~3.8 | MnO$_2$ | Zn foil | 1.35 | 220 (60 mA g$^{-1}$) | 95%/100 (60 mA g$^{-1}$) | [87] |
| 2 M ZnSO$_4$ + 0.2 M MnSO$_4$ | ~3.6 | MnO$_2$/graphene | Zn foil | 1.4 | 382.2 (3A g$^{-1}$) | 94%/3000 (3 A g$^{-1}$) | [111] |
| 3 M Zn(CF$_3$SO$_3$)$_2$ + 0.1 M Mn(CF$_3$SO$_3$)$_2$ | ~3.3 | β-MnO$_2$ | Zn foil | 1.4 | 225 (6.5 C) | 94%/2000 (6.5 C) | [89] |
| 3 M Zn(CF$_3$SO$_3$)$_2$ | ~3.6 | ZnMn$_2$O$_4$/C | Zn foil | 1.4 | 150 (50 mA g$^{-1}$) | 94%/500 (500 mA g$^{-1}$) | [90] |
| 2 M ZnSO$_4$ + 0.1 M MnSO$_4$ | ~3.8 | PANI- MnO$_2$ | Zn foil | 1.36 | 290 | 96%/200 (200 mA g$^{-1}$) | [112] |

(*Continued*)

**Table 6.1.** (*Continued*)

| Electrolyte | pH | Cathode | Anode | Discharge Voltage (V) | Capacity (mAh g⁻¹) | Retention/Cycles (rate) | Ref. |
|---|---|---|---|---|---|---|---|
| **Generation 2: Acidic electrolytes with Mn²⁺/MnO₂ reactions** | | | | | | | |
| 1 M MnSO₄ + 0.05 M H₂SO₄ | ~1.0 | Carbon felt | Pt/C- carbon felt | 1.3 | 226.2 (10 mA cm⁻²) | 100%/10000 (10 mA cm⁻²) | [47] |
| 1 M ZnSO₄ + 1 M MnSO₄ + 0.1 M H₂SO₄ | ~1.1 | Carbon fiber cloth | Zn foam | 1.95 | 570 (2 mA cm⁻²) | 92%/1800 (30 mA cm⁻²) | [50] |
| 1 M MnSO₄ + 1 M ZnSO₄ | ~3.8 | Carbon felt | Zn foil | 1.78 | 0.5 mA cm⁻² (6 C) | 100%/1000 (4 C) | [49] |
| 0.5 M MnSO₄ + 2.8 M H₂SO₄ | ~0.1 | Carbon felt | Carbon felt | 1.63 | 3.7 Ah L⁻¹ (20 mA cm⁻²) | 91.5%/100 (10 mAh cm⁻²) | [52] |
| 1 M Na₂SO₄ + 1 M MnSO₄ + 0.1 M H₂SO₄ | ~1.1 | Graphite felt | Activated carbon | 1.2 | 42.3 | 100%/7000 (1 mAh cm⁻²) | [53] |
| 2 M ZnSO₄ + 1 M MnSO₄ | ~3.3 | MnO₂ | CNT/Zn | 1.5 | 0.96 mAh cm⁻² (1 mA cm⁻²) | 100%/11000 (5 mA cm⁻²) | [91] |
| 2 M MnSO₄ + 2 M H₂SO₄ | ~0.15 | MnO₂@graphite felt | PTO | 0.82 | 150 (0.4 mA cm⁻²) | 80%/5000 (2.5 C) | [113] |
| 1 M MnSO₄ + 0.5 M H₂SO₄ | ~0.5 | Carbon cloth | Zn foil | 2.42 | 621 (2 mA cm⁻²) | 97.5%/6000 (2 mA cm⁻²) | [48] |
| 0.8 M MnSO₄ + 0.1 M H₂SO₄ | ~0.7 | MnO₂@carbon felt | Zn foil | 2.45 | 308 | 100%/35 (C/2) | [114] |
| 0.1 M MnSO₄ + 3 M H₂SO₄ | ~0.1 | MnO₂@carbon felt | Zn foil | 2.71 | 616 (100 mAh g⁻¹) | 98%/200 h (500 mAh g⁻¹) | [115] |
| 3 M MnSO₄ + 0.3 M H₂SO₄ + 0.06 M NiSO₄ | ~0.5 | Ni-MnO₂@carbon felt | Zn foil | 2.44 | 270 (2 C) | 99%/450 (2 C) | [116] |

BGM: battery-grade manganese dioxide; CM: chemically modified; CMC: carboxymethyl cellulose; CNT: carbon nanotube; PANI: polyaniline; PTO: pyrene-4,5,9,10-tetraone.

by Cui and coworkers.[47,117] The cathode reactions were conducted between soluble $Mn^{2+}$ and solid $MnO_2$ by reversible deposition/stripping, which is fundamentally different from the traditional solid-state cathode reactions in the first-generation Mn-based batteries. The electrochemical reactions of the $MnO_2$-$H_2$ battery can be described in the following:

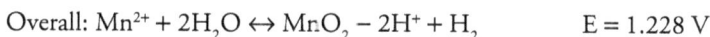

Cathode: $Mn^{2+} + 2H_2O \leftrightarrow MnO_2 + 4H^+ + 2e^-$    $E_0 = 1.228$ V vs. SHE

Anode: $2H^+ + 2e^- \leftrightarrow H_2$    $E_0 = 0$ V vs. SHE

Overall: $Mn^{2+} + 2H_2O \leftrightarrow MnO_2 - 2H^+ + H_2$    $E = 1.228$ V

The $MnO_2$-$H_2$ battery consists of a cathode-less porous carbon felt, a Pt/C catalyst-coated carbon felt anode, a glass fiber separator placed between the cathode and anode, and the aqueous $MnSO_4$ electrolyte. During charge, the soluble $Mn^{2+}$ in the electrolyte was oxidized and coated on the cathode in the form of $MnO_2$, and hydrogen gas was generated from the anode under the active electrocatalysts. During discharge, the deposited $MnO_2$ was stripped back into the electrolyte, and hydrogen gas was oxidized on the anode. When using the electrolyte of 1 M $MnSO_4$ with 0.05 M $H_2SO_4$, the $MnO_2$-$H_2$ battery exhibited a discharge potential of ~1.3 V (Figure 6.4b), fast rates up to 100 C, and a lifetime of over 10,000 cycles without decay at a current density of 10 mA cm$^{-2}$. However, when cycled in pure $MnSO_4$ electrolyte without $H_2SO_4$, the $MnO_2$-$H_2$ battery showed gradual capacity decay over cycling. This is due to the synergetic effects of the cathode and anode reactions in the $H_2SO_4$ additive electrolyte, where higher electrolyte conductivity contributes to faster $Mn^{2+}$/$MnO_2$ reactions on the cathode and more favorable hydrogen evolution and oxidation reactions on the anode. In addition, it was confirmed that protons help promote the dissolution of $MnO_2$ in the discharge process and reduce the cell overpotential.[49] This work demonstrated exciting features of the cathode $Mn^{2+}$/$MnO_2$ deposition/stripping charge storage mechanism and the importance of electrolyte management for the development of the Mn-based batteries.

## 6.3.2 Electrolytic Zn-MnO₂ Batteries

The excellent electrochemical performance of the $MnO_2$-$H_2$ battery with the cathode $Mn^{2+}$/$MnO_2$ chemistry has attracted much attention in the community of the Mn-based batteries. Therefore, researchers have applied the emerging cathode chemistry to many different Mn-based batteries. Owing to the abundance, high theoretical gravimetric capacity of 820 mAh g$^{-1}$ and volumetric capacity of 5855 mAh cm$^{-3}$ of the Zn metal anode, electrolytic Zn-MnO$_2$ batteries have been regarded as one of the potential competitors for grid-scale energy storage.[118,119] We discuss in this section the recent progress in electrolytic Zn-MnO$_2$ batteries using the cathode Mn$^{2+}$/MnO$_2$ chemistry.

Recently, Qiao and coworkers[50,116] reported an aqueous electrolytic Zn-MnO$_2$ battery that was comprised of a Zn-foam anode, a glass fiber separator, a carbon fiber cloth as the cathode substance and a mixture of ZnSO$_4$ with MnSO$_4$ aqueous electrolyte (Figure 6.4a). When charged at 2.2 V to a capacity of 2 mAh cm$^{-2}$, the Zn-MnO$_2$ battery is capable of delivering an initial discharge capacity of 1.3 mAh cm$^{-2}$, which gradually increased to 1.92 and 1.97 mAh cm$^{-2}$ in the 10$^{th}$ and 100$^{th}$ cycle, respectively.

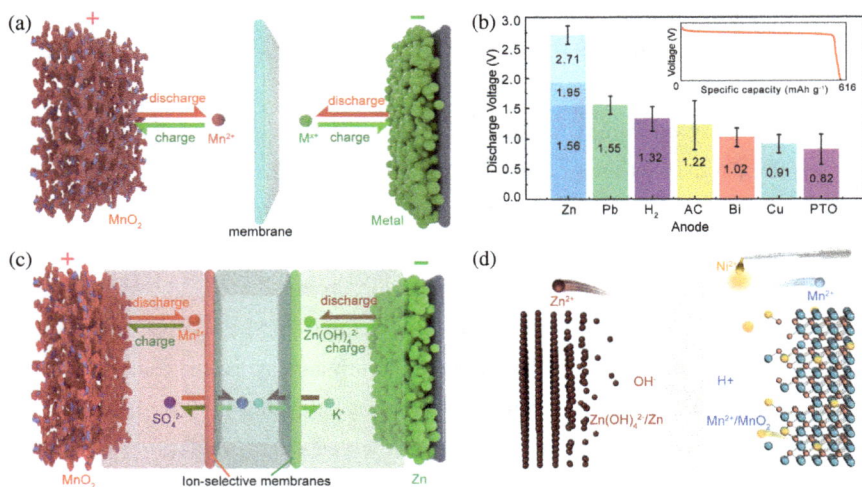

**Figure 6.4.** The aqueous Mn-based batteries with Mn$^{2+}$/MnO$_2$ chemistry. (a) Schematic of the MnO$_2$-M (M: H$_2$, Zn, Cu, Pb, Sn, Bi, carbon, polymers, etc.) batteries with the Mn$^{2+}$/MnO$_2$ chemistry and their working mechanisms. (b) Statistics of the discharge voltages of different MnO$_2$-M batteries that were reported in literature. The inset shows a typical discharge curve of the MnO$_2$-M batteries with a pronounced discharge plateau and the discharge capacity of ~616 mAh g$^{-1}$ is calculated based on the deposited MnO$_2$ with two-electron transfer of the Mn$^{2+}$/MnO$_2$ reactions. (c) Schematic of the electrolyte-decoupled Zn-MnO$_2$ battery with the cathode Mn$^{2+}$/MnO$_2$ chemistry and its working mechanism. (d) Schematic of the electrolyte-decoupled Zn-MnO$_2$ battery with catalyzed electrolysis kinetics via facile introduction of Ni$^{2+}$ into electrolyte. Reproduced with permission from Ref. [116], John Wiley and Sons.

This result agrees well with the electrochemical behavior of the $MnO_2$-$H_2$ battery.[47] Three distinct regions with different electrochemical windows in the discharge curves were used to explain the energy storage mechanism of the electrolytic Zn-$MnO_2$ battery in the $ZnSO_4$ and $MnSO_4$ electrolyte. The first region (2–1.7 V) is responsible for the $MnO_2$/$Mn^{2+}$ reaction, the second region (1.7–1.4 V) is dominated by the proton insertion into $MnO_2$, and the third region (1.4–0.8 V) is a result of the $Zn^{2+}$ insertion into $MnO_2$. The storage mechanism was also confirmed by Liang and coworkers in their electrolytic Zn-$MnO_2$ batteries.[120] In order to maximize the first high-voltage region and minimize the inferior second and third regions, 0.1 M $H_2SO_4$ was added to the electrolyte of $ZnSO_4$ and $MnSO_4$ in the electrolytic Zn-$MnO_2$ battery to facilitate the $Mn^{2+}$/$MnO_2$ deposition/stripping reactions. Therefore, the redox reactions of the electrolytic Zn-$MnO_2$ batteries can be expressed in the following:

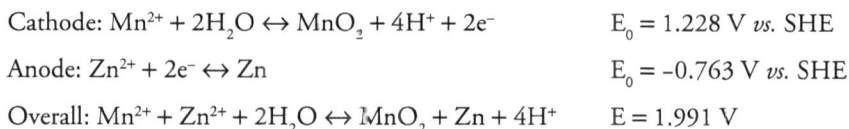

Cathode: $Mn^{2+} + 2H_2O \leftrightarrow MnO_2 + 4H^+ + 2e^-$      $E_0 = 1.228$ V $vs.$ SHE

Anode: $Zn^{2+} + 2e^- \leftrightarrow Zn$      $E_0 = -0.763$ V $vs.$ SHE

Overall: $Mn^{2+} + Zn^{2+} + 2H_2O \leftrightarrow MnO_2 + Zn + 4H^+$      $E = 1.991$ V

It exhibited much improved electrochemical performance such as a high discharge voltage of 1.95 V (Figure 6.4b), a high current density of 60 mA cm$^{-2}$ and good cycling stability for over 1,800 cycles. More importantly, the cost of this battery was estimated to be less than US$10 per kWh, far below that of the lithium ion and lead-acid batteries, showing its great potential for large-scale energy storage applications.[121] The electrolytic Zn-$MnO_2$ battery technology is being industrialized by startups founded by the designers.

After that, the electrolytic Zn-$MnO_2$ batteries have been developed progressively. Lu $et$ $al.$ and Chen $et$ $al.$ successively proposed the strategy of halogen redox mediators[122–124] by introducing trace amounts of iodine or bromine into the electrolyte, and realized good cycling stability of the $MnO_2$ cathodes with high areal capacities. Aiming at the improvement of reaction kinetics of $MnO_2$, Chen $et$ $al.$ introduced transition metal ions represented by Co and Ni into the electrolyte to achieve $in$ $situ$ co-doping of $MnO_2$ during the electrodeposition process and attained a high-rate $MnO_2$ cathode.[125] Later, Chen $et$ $al.$ designed a general strategy of cationic accelerators to regulate the solvation structures of metal ions in solution to improve the reaction kinetics of both $MnO_2$ cathode and Zn anode.[126] However, electrolytic Zn-$MnO_2$ batteries still have many problems, such as serious hydrogen evolution reaction and chemical corrosion at the Zn anodes.[127,128] In order to solve these problems, Chen *et al.* applied different strategies on electrodes, electrolyte, and separator of the $Zn\text{-}MnO_2$ batteries, such as the design of Zn-Al alloys for Zn anode, proton-trapping agents for electrolyte, and novel proton-barrier membrane as separator.[129–131] However, more approaches to improve the overall performance of electrolytic $Zn\text{-}MnO_2$ batteries are still highly desirable in future studies.

### 6.3.3 *Electrolyte-decoupled Zn-MnO$_2$ Battery*

With the improved electrolyte optimization and management, the advantages of the cathode $Mn^{2+}/MnO_2$ chemistry have been further amplified. The reduction potential of Zn anode in alkaline electrolyte is –1.199 V (*vs.* SHE), lower than that of the Zn anode in acidic electrolyte (–0.763 V *vs.* SHE). In order to achieve high overall potential for the $Zn\text{-}MnO_2$ battery, it is in principal feasible to build a battery by separating the cathode and anode in the acidic and alkaline electrolytes, respectively (Figure 6.4c). In 2019, Banerjee and coworkers[114] reported for the first time the development of electrolyte-decoupled aqueous $Zn\text{-}MnO_2$ batteries with high voltages of 2.45 V and 2.8 V without the use of costly ion-selective membranes. The cathode was driven by the $Mn^{2+}/MnO_2$ deposition/stripping or $Mn^{2+}/MnO_4^-$ chemistries in aqueous acidic electrolytes and the anode was driven by the $Zn/Zn(OH)_4^{2-}$ reactions in alkaline electrolyte. The cell was achieved by gelling the anode alkaline electrolyte and using low-cost cellophane separator to separate from the cathode aqueous acidic electrolyte. It is considered as an important advance towards the fabrication of electrolyte-decoupled high-voltage aqueous $Zn\text{-}MnO_2$ batteries. Liu and coworkers[48] built such a high-voltage aqueous $Zn\text{-}MnO_2$ battery by using a bipolar membrane to separate the cathode and anode electrolytes in different pH environments. The $Zn\text{-}MnO_2$ battery with the dual electrolytes showed a high discharge voltage of 2.42 V, high Coulombic efficiency of 98.4% and stable cycling for 1,500 cycles. Very recently, Hu and coworkers[115] developed a similar $Zn\text{-}MnO_2$ battery with a two-membrane three-chamber design as shown in Figure 6.4c. The cathode $MnO_2$ in strong acidic chamber with an electrolyte of $MnSO_4$ and $H_2SO_4$ and the anode Zn in strong alkaline chamber with an electrolyte of KOH, ZnO and vanillin were separated by a neutral chamber with an electrolyte of $K_2SO_4$, where cation and anion exchange membranes were placed in between. The central chamber in neutral electrolyte provides adequate intermediate space for the ion transportation to avoid the neutralization effect caused by the excessive pH difference of the acidic and alkaline electrolytes. The Zn-MnO$_2$ battery with the decoupled stronger acidic and alkaline electrolytes showed a higher discharge voltage of 2.71 V than the previous design (Figure 6.4b). In addition, it showed 200 hours of high-current deep charge-discharge behavior, and stable cycling life of 1,000 times with a 10% state of charge. Enabled by connecting cells in parallel and series, the stacked batteries showed capability to deliver high voltages and capacities. However, the usage of strong acid and alkaline electrolytes and the complex two-membrane three-chamber design will increase the difficulty on battery assembling, housing and welding. Optimization of the battery system complexity in terms of electrolytes and cell structures are highly desirable to the mass production of the electrolyte-decoupled Zn-MnO$_2$ batteries. Furthermore, the usage of two ion-selective membranes will worsen the kinetics and corresponding power density of the device. Further development of high-performance and cost-effective membranes will help improve the overall battery performance and lower the capital cost for practical applications of the electrolyte-decoupled Zn-MnO$_2$ batteries. Recently, Qiao and coworkers reported an electrolyte-decoupled Zn-MnO$_2$ battery with catalyzed Mn$^{2+}$/MnO$_2$ electrolysis kinetics (discharge in 60 s, at 50 C) via facile introduction of trace amounts of Ni$^{2+}$ into the electrolyte (Figure 6.4d).[116] The hybrid aqueous battery demonstrated an electrochemical stability window of over 3.4 V and flat discharge voltage of 2.44 V.

## 6.3.4 *Electrolytic Zn-MnO$_2$ Flow Batteries*

The exploration of Mn-based batteries with the cathode Mn$^{2+}$/MnO$_2$ chemistry has stimulated the development of Mn-based redox flow batteries. Redox flow battery is a rechargeable battery technology in which active substances exist in liquid electrolytes and are stored externally, enabling the separation of energy and power.[2] In addition, redox flow batteries have the characteristics of high safety, high efficiency and long life, showing promises for large-scale energy storage.[132–134] The feasibility of redox flow electrolytic Zn-MnO$_2$ battery (Figure 6.5a) was demonstrated by Qiao and coworkers.[50] A redox-flow battery stack with a capacity of 2 Ah was built, which demonstrated a practical strategy towards large-scale energy storage using the new electrolytic Zn-MnO$_2$ system. The output voltage, energy efficiency, and cost of the flow-battery design were comprehensively analyzed, which performed better than conventional aqueous flow battery systems, such as Zn-Fe, Zn-Br$_2$, Zn-Ce, Zn-air, and all-vanadium flow batteries.[50] In a recent work, Li and coworkers[30] demonstrated an electrolytic Zn-MnO$_2$ flow

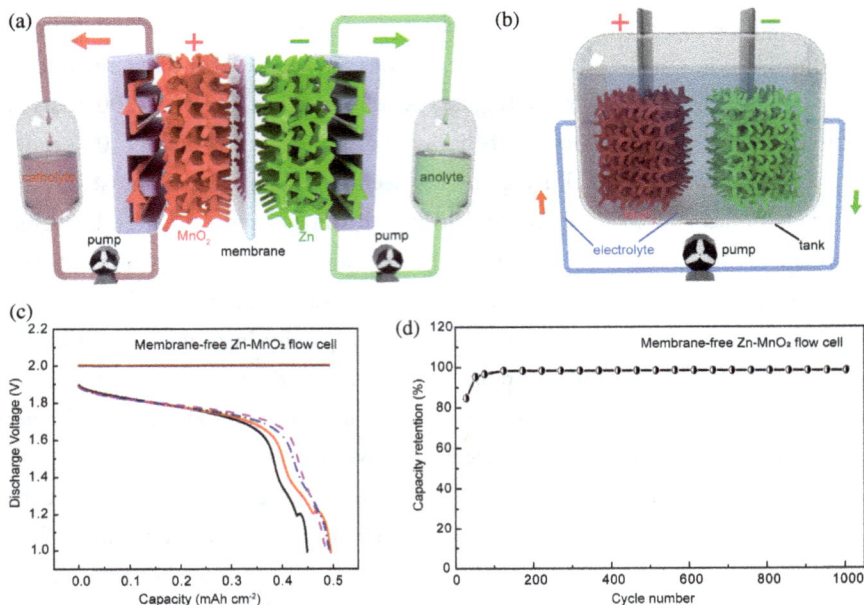

**Figure 6.5.** Electrolytic Zn-MnO$_2$ flow batteries with cathode Mn$^{2+}$/MnO$_2$ chemistry. (a) Schematic illustration of the conventional Zn-MnO$_2$ flow battery. The cathode and anode are separated by a membrane. (b) Schematic of the membrane-free Zn-MnO$_2$ flow battery, which consists of a tank containing cathode, anode and electrolyte, and the pump-drive electrolyte circulation system. (c) The charge and discharge curves of the membrane-free Zn-MnO$_2$ flow battery. The initial Coulombic efficiency of the battery is ~90%, but the capacity ramps up with the cycling and can reach theoretical value after the initial activation. (d) The cycling stability behavior of the membrane-free Zn-MnO$_2$ flow battery with capacity retention over 1,000 cycles. The charge-discharge and cycling tests were performed at a constant charge potential of 2 V to 0.5 mAh cm$^{-2}$ and a constant discharge rate of 4 C. Panels (c) and (d) are reproduced with permission from Ref. [49], John Wiley and Sons.

battery by taking the advantages of the cathode Mn$^{2+}$/MnO$_2$ deposition/stripping reactions and the anode Zn$^{2+}$/Zn reactions. Mn(CH$_3$COO)$_2$ was selected as the active material for the catholyte. Due to the coordination effect of the CH$_3$COO$^-$, Mn$^{2+}$ can be directly converted to MnO$_2$ and reversibly stripped back to Mn$^{2+}$ even in the acid-free neutral electrolyte. The electrolytic Zn-MnO$_2$ flow battery exhibited a discharge platform of ~1.56 V (Figure 6.4b), good rate performance and no significant capacity decay for over 400 cycles. However, the utilization of Mn(CH$_3$COO)$_2$ to replace MnSO$_4$ in the electrolyte has reduced the cell voltage by ~530 mV. Therefore, further research is needed to improve the overall electrochemical performance and lower the cost of the conventional Zn-MnO$_2$ flow battery. In this regard, Cui and coworkers[49] developed a membrane-free electrolytic Zn-MnO$_2$ flow battery by circulating the single electrolyte, instead of the separated catholyte and anolyte in the conventional flow battery, through one tank without using the relatively expensive ion-selective membrane (Figure 6.5b). The membrane-free Zn-MnO$_2$ flow battery was conducted by the cathode Mn$^{2+}$/MnO$_2$ reactions on carbon felt and the anode Zn$^{2+}$/Zn reactions on Zn foil in the electrolyte of 1 M ZnSO$_4$ and 1 M MnSO$_4$. By virtue of the electrolyte flow, the newly generated proton during the charge process can be evenly distributed, facilitating the dissolution of MnO$_2$ during the discharge process. The battery exhibited a high discharge voltage of ~1.8 V (Figure 6.5c), good discharge rates from 0.5 C to 10 C, and stable 1,000 cycles with nearly 100% of Coulombic efficiency (Figure 6.5d). The development of membrane-free electrolytic Zn-MnO$_2$ flow battery provides an economic strategy towards large-scale energy storage applications.

## 6.4 Understanding the Mn$^{2+}$/MnO$_2$ Storage Mechanism

Researchers have attempted to understand the Mn$^{2+}$/MnO$_2$ deposition/stripping mechanism from different aspects including *ex situ* and *in situ* characterizations, dynamic simulation and theoretical calculation. In some early work, Donne and coworkers[135–137] studied the electrodeposition mechanism of MnO$_2$ in acidic and neutral media by using the techniques of rotating disk electrode and rotating ring-disk electrode,[96,138] indicating that the electrodeposition of MnO$_2$ is closely related to the pH value and dependent on the substrate. In addition, the electrochemical kinetic behavior of the MnO$_2$ cathode in aqueous solution was investigated to reveal the reduction process of MnO$_2$ *via* a series of constant current discharge experiments,[137] which helps the exploration of the Mn$^{2+}$/MnO$_2$ deposition/stripping mechanism. In terms of the battery characterization, a combination of different characterization tools is often involved in revealing the dynamic processes of the Mn$^{2+}$/MnO$_2$ reactions, which require high spatial and temporal resolutions. Scanning electron microscopy (SEM) provides a powerful opportunity for researchers to directly observe the evolution of the Mn$^{2+}$/MnO$_2$ reactions on the Mn-based battery cathode. Hu and coworkers[115] analyzed the morphology of the MnO$_2$ cathode by *ex situ* SEM at different depth of charge (DoC) and discharge (DoD) states in a complete charge/discharge cycle (Figure 6.6a). The cathode showed porous carbon nanoparticulate morphology before charge

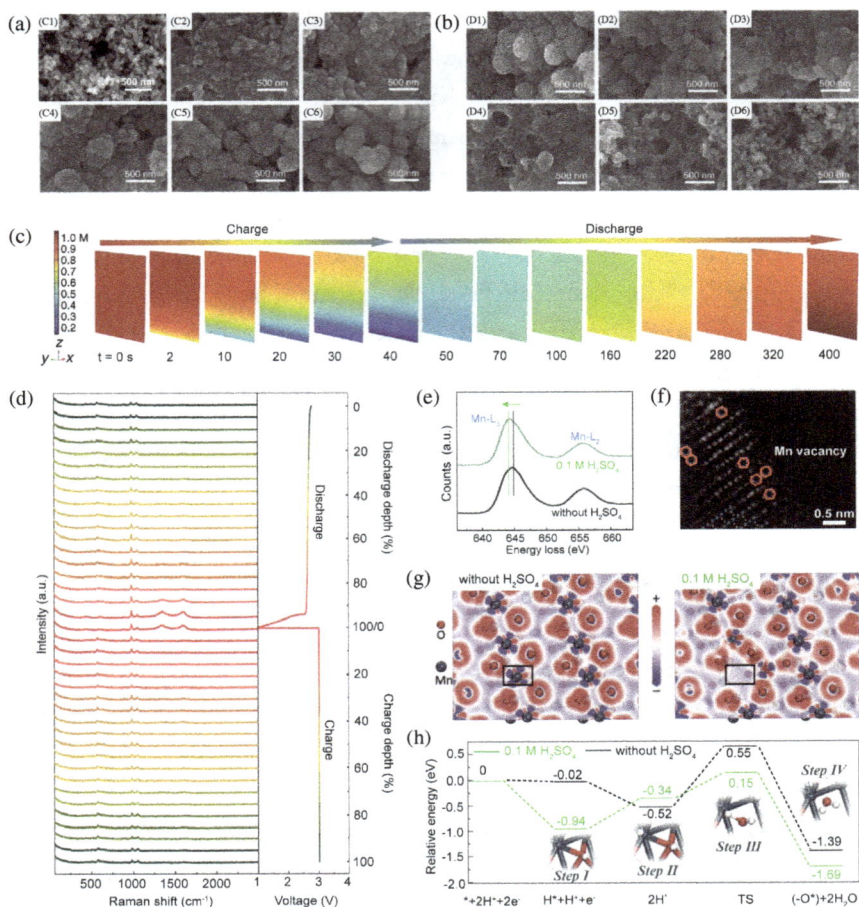

**Figure 6.6.** Energy storage mechanism of the $Mn^{2+}/MnO_2$ chemistry revealed by characterization, simulation and theoretical calculation. (a) SEM images of the $MnO_2$ electrode after charge at DoCs of 0 (C1), ~20% (C2), ~40% (C3), ~60% (C4), ~80% (C5) and ~100% (C6) at a charge voltage of 3 V. (b) SEM images of the $MnO_2$ electrode after discharge at DoDs of 0 (D1), ~20% (D2), ~40% (D3), ~60% (D4), ~80% (D5) and ~100% (D6) at a current density of 200 mA g$^{-1}$. (c) Finite element analysis of the cathode $Mn^{2+}/MnO_2$ deposition/stripping reactions by the simulation of dynamic electrolyte concentration variation over an entire charge-discharge process. The model was built in COMSOL software based on a cell with a cathode and an anode in a square shape of 250 μm × 250 μm, and the electrolyte with thickness of 350 μm which is equivalent to the separator was used to set them apart. The resulting representative spectra correspond to different charge-discharge durations were obtained under a constant charge potential of 1.6 V to a capacity of 1 mAh cm$^{-2}$ and a discharge current density of 10 mA cm$^{-2}$ in 1 M $MnSO_4$ electrolyte. Detailed information about the modelling is available in Ref. [47]. (d) *In situ* Raman spectra of the $MnO_2$ cathode in the charge/discharge processes. The measurements were performed

(Figure 6.6a, C1) and was gradually coated with nanosheet-like $MnO_2$ deposits during charge until the pores and spaces between the carbon particles completely filled up, forming the heavily deposited $MnO_2$ cathode (Figure 6.6a, C2–C6). During discharge, the interconnected $MnO_2$ particles on the cathode were gradually dissolved and reverted to the initial porous carbon nanoparticulate morphology (Figure 6.6b). Energy-dispersive X-ray spectroscopy, ex situ XRD and X-ray photoelectron spectroscopy quantitatively confirmed the highly reversible $Mn^{2+}/MnO_2$ reactions in each charge/discharge stage. Moreover, the electrolyte evolution over the entire charge/discharge processes was monitored through a finite element simulation by modeling the dynamic $MnSO_4$ electrolyte concentration variation.[47] As shown in Figure 6.6c, the representative spectra of the $Mn^{2+}$ concentration distribution demonstrated the fully recoverable spatial variation of the electrolyte concentration after the complete charge/discharge cycle, suggesting the highly reversible $Mn^{2+}/MnO_2$ reactions. This agrees well with the evolution of the $Mn^{2+}/MnO_2$ reactions on the Mn-based battery cathode. Experimentally, the concentrations of the $Mn^{2+}$ in the electrolyte during charge/discharge processes were measured by plasma optical emission spectroscopy,[115] which were consistent with the calculated theoretical concentrations from the amount of deposited $MnO_2$, confirming the highly reversible $Mn^{2+}/MnO_2$ reactions.

In addition, in situ Raman spectroscopy was applied to probe the deposited $MnO_2$ on the cathode carbon substrate of the Zn-$MnO_2$ battery in operation (Figure 6.6d).[115]

---

(Continued on Facing page)   on the decoupled Zn-$MnO_2$ battery at a charge potential of 3 V and a discharge current density of 200 mA $g^{-1}$. (e) Mn-$L_{2,3}$ EELS spectra of the deposited $MnO_2$ with and without 0.1 M $H_2SO_4$ addition in the electrolytes. The shift of Mn-$L_3$ main peak of $MnO_2$ from 644.9 eV (without 0.1 M $H_2SO_4$) to 644.3 eV (with 0.1 M $H_2SO_4$) and the higher $L_3/L_2$ integral peak intensity ratio of $MnO_2$ in 0.1 M $H_2SO_4$ imply the reduction of Mn valence state and the existence of $Mn^{3+}$ in $MnO_2$ that was deposited in the electrolyte with 0.1 M $H_2SO_4$ addition. (f) Atomic-resolution HAADF-STEM image of the deposited $MnO_2$ with Mn vacancies (marked in red hexagons). (g) Electron density variation of the $MnO_2$ with (in 0.1 M $H_2SO_4$) and without Mn vacancies (without $H_2SO_4$). The Mn vacancy is marked with a black rectangle in $MnO_2$ with 0.1 M $H_2SO_4$ on the right image. (h) Relative energy profiles of the $MnO_2$ dissolution reaction pathway with four subsequent steps. Insets are the schematics of the $MnO_2$ crystal structure changes in the four reaction steps from I to IV. Mn, O, and H atoms are labeled as blue, red, and white balls, respectively. Detailed explanation of the four reaction steps is available in Ref. [50]. Panels (a), (b) and (d) are reproduced with permission from Ref. [115], Springer Nature. Panel (c) is reproduced with permission from Ref. [47], Springer Nature. Panels (e)–(h) are reproduced with permission from Ref. [50], John Wiley and Sons.

The characteristic peak of $MnO_2$ at 572 cm$^{-1}$ appears immediately at the beginning of the charge process, which indicates that $MnO_2$ is being deposited, until it suddenly disappeared at the end of the discharge process where the discharge potential dropped rapidly, which is accompanied by the appearance of the characteristic carbon peaks at 1354 cm$^{-1}$ and 1598 cm$^{-1}$, indicating complete $MnO_2$ dissolution from the cathode and exposure of the underlying carbon substrate. This further confirmed the highly reversible $Mn^{2+}/MnO_2$ reactions.

Density functional theory (DFT) calculations were performed to understand the energy storage mechanism of the $Mn^{2+}/MnO_2$ redox reactions in acidic electrolytes.[50] It was suggested by electron energy loss spectroscopy (EELS, Figure 6.6e) that the deposited $MnO_2$ in the acidic electrolyte of $MnSO_4$ with 0.1 M $H_2SO_4$ contains Mn vacancies as opposed to the deposited $MnO_2$ without $H_2SO_4$. The Mn vacancies were directly observed by a high-angle annular dark-field scanning transmission electron microscope image (HAADF-STEM, Figure 6.6f), which was performed on a high-resolution aberration-corrected transmission electron microscope. It was further analyzed that the deposited $MnO_2$ with Mn vacancies has a higher diffusion coefficient and a lower overpotential than that of the pristine $MnO_2$. Such Mn vacancies were believed to facilitate the generation of unsaturated oxygen species and help offer additional pathways for cations with higher mobility in the aqueous electrolytes. The DFT calculation revealed that the surface electron density tends to increase on the Mn vacancies, which helps reduce the reaction energy barrier and accelerate the electron transfer dynamics (Figure 6.6g). In this respect, the reaction pathway of the $MnO_2$ dissolution with the assistance of Mn vacancies, which involves four major steps, proceeds in promoted reaction dynamics with lower overpotentials (Figure 6.6h). Recently, the Ni-catalyzed $Mn^{2+}/MnO_2$ electrolysis kinetics process was carefully explored by Qiao and coworkers with the help of synchrotron spectra techniques and DFT calculations.[116] The significant role of Ni introduction in powering electrolytic reactivity is proposed via enhanced active electron states, charge delocalization and facilitated charge transfer around strongly electronegative Ni.

## 6.5 Towards Large-scale Energy Storage Applications

Owing to the remarkable advantages of low cost, good safety, high operation voltage, long lasting and reliable shelf life, high capacity and energy density, Mn-based batteries with the cathode $Mn^{2+}/MnO_2$ chemistry are promising for next-generation large-scale energy storage.[47,49] Figure 6.7a shows a conceptual schematic of the representative future grid-scale energy storage based on the rechargeable Mn-based batteries with the $Mn^{2+}/MnO_2$ chemistry. By virtue of integrating renewable yet intermittent energy resources such as solar, wind and hydro-electricity into electrical grids, the centralized electrolytic Zn-MnO$_2$ battery system with capacity of MWh and even GWh can be used to power individual households, factories, buildings and transportations like electric vehicles.[139] The electrolytic Zn-MnO$_2$ battery system plays a critical role as an indispensable supplement to the present electrical grid and could even replace it completely to achieve 100% green electricity in our society. In addition, the dispatchable electrolytic Zn-MnO$_2$ batteries can be functionalized as distributed utility energy resources and found applications wherever they are able to.

The realization of the proposed fascinating spectacle of the future sustainable energy storage applications becomes feasible when taking into account mass production of the electrolytic Zn-MnO$_2$ batteries. As depicted in Figure 6.7a, the electrolytic Zn-MnO$_2$ battery pack can be assembled on the basis of multiple stacks of batteries that are constructed by connecting a number of individual Mn cells in series and in parallel to deliver high current and voltage for practical applications. It can be achieved through mass production of the Mn-based batteries with progressive technology readiness levels from research lab to industry. In order to demonstrate the scalability of Mn-based batteries with the $Mn^{2+}/MnO_2$ chemistry, a prototype lab-scale membrane-free electrolytic Zn-MnO$_2$ flow battery with a capacity of 1200 mAh has been fabricated by increasing the area of the cathode carbon substances, which demonstrated excellent electrochemical behaviors with stable cycle performance (Figure 6.7b).[49] Considering that it is a primary scale-up cell design that has not been optimized for maximum electrochemical performance, the Mn-based battery is believed to meet the requirements of large-scale energy storage after further improvement. In addition, Mn-based batteries with the $Mn^{2+}/MnO_2$ chemistry are feasible to be greatly scaled up by various cell designs such as pouch and prismatic cell packs, providing high flexibility to the future Mn cell design and optimization of packaging efficiency at the battery level. For example, Hu and coworkers[115] fabricated a decoupled Zn-MnO$_2$ battery with a high discharge voltage of ~2.75 V by scaling up its capacity to 3330 mAh. As shown in Figure 6.7c, when connecting such two single cells in series, the output voltage of the battery was

(a)

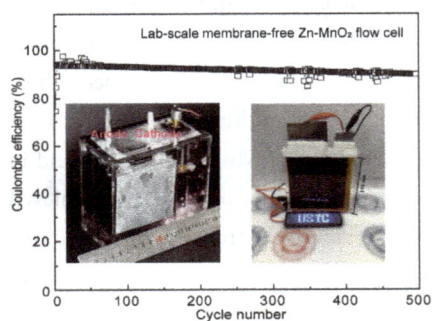

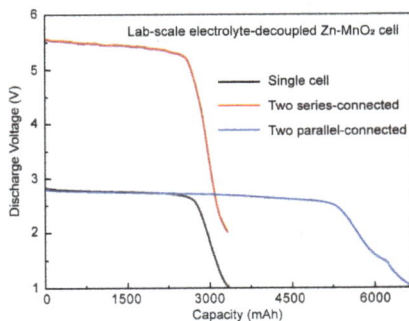

(b)                                                      (c)

**Figure 6.7.** The grid-scale energy storage applications of electrolytic Zn-MnO$_2$ batteries with Mn$^{2+}$/MnO$_2$ chemistry. (a) A conceptual schematic illustration of the future grid-scale energy storage applications of the rechargeable aqueous electrolytic Zn-MnO$_2$ batteries. It is achieved by the integration of renewable solar-, wind- and hydro-powered electricity with the electrical grid, as well as the transformation of the electrolytic Zn-MnO$_2$ battery technology with large scaling up and mass production from the research lab to industry. The grid-scale energy storage applications of the electrolytic Zn-MnO$_2$ batteries will benefit our society from individual households, transportations, factories and buildings to the entire community. (b) The long-term stability of a prototype membrane-free Zn-MnO$_2$ flow battery over 500 cycles. The battery was charged at 2 V to 1200 mAh and then discharged at 1000 mA to 1 V for 500 consecutive times. (c) The discharge behaviors of the lab-scale decoupled Zn-MnO$_2$ batteries with two different cell connections, e.g., two series-connected and two parallel-connected cells. The single unit cell is a decoupled Zn-MnO$_2$ battery with a theoretical capacity of 3370 mAh. Reproduced with permission from Ref. [115], Springer Nature.

doubled, reaching ~5.44 V without sacrificing the high capacity. While connecting two individual cells in parallel, the resulting battery showed capacity of ~6660 mAh with similar discharge potential of ~2.7 V. Furthermore, the battery pack, which was formed by connecting several decoupled Zn-MnO$_2$ cells in parallel and series, was integrated with a wind-driven generator and photovoltaic hybrid power generation systems to demontrate its practical energy storage capability. Previous successful integration of the renewable energy resources and practical applications of the stored energy to power LED panel, cellphone, and vehicle model have demonstrated the significant potential of Mn-based batteries with the Mn$^{2+}$/MnO$_2$ chemistry for large-scale energy storage.[47,49,50,116]

## 6.6 Opportunities and Future Perspectives

The continued pursuit of renewable battery technologies for large-scale energy storage will accelerate the exploration of state-of-the-art rechargeable aqueous electrolytic Zn-MnO$_2$ batteries with the Mn$^{2+}$/MnO$_2$ chemistry. However, further research is required to effectively enable their practical applications for the benefit of society. Critical challenges include, but are not limited to, understanding battery charge storage mechanism, achieving high electrochemical performance that complies with the best practice for measurements, incorporating theoretical calculation and simulation, optimizing battery system integration, and applying the electrolytic Zn-MnO$_2$ batteries in large-scale applications via scaling up and commercialization (Figure 6.8). In this endeavor, it is highly desirable to develop future electrolytic Zn-MnO$_2$ batteries by considering the above-mentioned aspects in a holistic strategy.

**Charge storage mechanism.** Although the charge storage mechanism of the cathode Mn$^{2+}$/MnO$_2$ reactions has been primarily revealed by different characterization techniques, simulation and theoretical calculations, it is still necessary to understand the intermediate steps involving the complex Mn$^{2+}$/MnO$_2$ reactions toward two-electron charge transfer in the charge and discharge processes from a battery-level perspective. It was generally accepted that the Mn$^{2+}$/MnO$_2$ reactions are involved in multiple electrochemical and chemical redox reactions that are highly dependent on the proton reactivity of the electrolytes and the charge/discharge parameters. These reactions include the electrochemical oxidation of Mn$^{2+}$ to Mn$^{3+}$, the chemical hydrolysis of Mn$^{3+}$ to MnOOH or the disproportionation of Mn$^{3+}$ to form Mn$^{2+}$ and Mn$^{4+}$, and

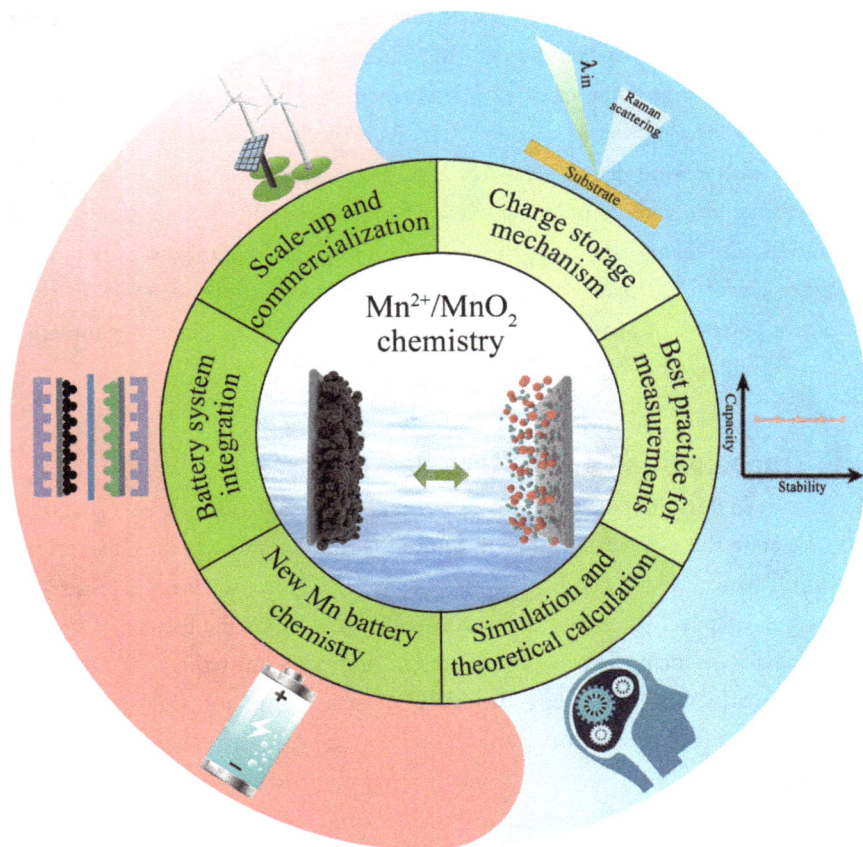

**Figure 6.8.** New opportunities for developing electrolytic Zn-MnO$_2$ batteries with the Mn$^{2+}$/MnO$_2$ chemistry, including charge storage mechanism, best practice for measurements, simulation and theoretical calculation, new Mn-based battery chemistry, battery system integration, and scale-up and commercialization for large-scale energy storage applications.

the further electrochemical oxidation of MnOOH to MnO$_2$ or the hydrolysis of Mn$^{4+}$ to MnO$_2$.[94,140] Due to the complex cathode Mn$^{2+}$/MnO$_2$ reactions paired with various anode reactions in the respective battery systems, there has been controversy regarding the Mn$^{2+}$/MnO$_2$ deposition/stripping mechanism on the reaction pathways and electron transfer. Therefore, some advanced *in situ*/operando characterization tools with high spatial and temporal resolutions should be applied to unveil the intrinsic storage mechanism of the cathode Mn$^{2+}$/MnO$_2$ reactions.[115,141,142] A number of *in situ*/operando techniques may provide powerful capabilities for understanding the energy storage mechanism of Mn-based batteries by addressing key issues of the $Mn^{2+}/MnO_2$ reactions regarding the identity and configuration of possible intermediates, preferred reaction pathways and impact of reaction conditions. The various techniques include *in situ*/operando infrared and Raman spectroscopies, electrochemical impedance spectroscopy, X-ray absorption spectroscopy, scanning probe microscopy, scanning electrochemical microscopy, liquid-phase transmission electron microscopy, cryogenic electron microscopy and their rational combinations.[143,144] Besides proton reactivity, the introduction of both other cations ($Ni^{2+}$ and $Zn^{2+}$) and anions (sulfate and acetate) may result in different electrolysis kinetics.[30,116] These catalysis/passivation mechanisms with different solutes and solvents are very interesting and should be carefully investigated to improve the electrochemical performance of Mn-based batteries further.

**Best practice for measurements.** A challenge to evaluate the electrochemical performance of Mn-based batteries with the cathode $Mn^{2+}/MnO_2$ reactions is the lack of standardized criteria regarding calculations and measurements. As a consequence, it is difficult and unreliable to compare the reported results from different studies, which will lead to unhealthy growth in this area. For the cathode $Mn^{2+}/MnO_2$ reactions which are typically conducted in $MnSO_4$ electrolyte, high theoretical specific capacity of ~616 mAh $g^{-1}$ is often reported based on the two-electron charge transfer reactions. Actually, defects cannot be eliminated in the derived Mn oxides via the electrodeposition process,[50,53] which makes the intrinsic valence state of Mn oxides lower than 4+ and results in a lower capacity than the theoretical value of 616 mAh $g^{-1}$. Moreover, when it comes to the concentration of the $MnSO_4$ electrolyte by considering the electrolyte mass and volume, the theoretical specific capacity of the $Mn^{2+}/MnO_2$ reactions drops to ~141 mAh $g^{-1}$ (volumetric capacity ~214 Ah $L^{-1}$) with 100% DoD at a concentration of ~4 M of $MnSO_4$ at room temperature.[47] As the cathode $Mn^{2+}/MnO_2$ reactions involve liquid/solid conversion, one is highly suggested to include the solute and solvent as a whole into consideration for the evaluation of battery performance such as gravimetric and volumetric capacity, energy and power densities. One general practice we urge is to clearly state the parameters of batteries (including the cathode and anode, electrolyte volume, composition and concentration), the measurements (including charge/discharge techniques, rates, DoC, DoD) and the specific calculation equations (including voltage, capacity, energy density, power density, self-discharge, *etc.*) on how these parameters are determined. It is a matter of utmost urgency for the Mn-based battery community to establish widely accepted guidelines for standardizing calculations and measurements. In addition, the Mn-based batteries with extraordinary electrochemical performance and multi-functions such as the demonstrated ultra-low temperature operation of the $MnO_2$-hydronium battery,[113] flexible and wearable devices, and self-charging properties are exciting directions to be explored for unique applications.

**Simulation and theoretical calculation.** In combination with experimental observations by advanced characterization tools, *ab initio* kinetic simulation, theoretical calculations, as well as machine learning are powerful tools to gain fundamental insights into the principles of electrolytic $Zn-MnO_2$ batteries with the $Mn^{2+}/MnO_2$ chemistry. For example, *ab initio* kinetic simulation of the Mn-based batteries can provide detailed information about behaviors of the redox reactions on a molecular level. Numerical simulations such as the finite element method can assist in failure analysis and lifespan prediction of the batteries.[145,146] In addition, first-principle calculations can be used to predict the battery redox reactions at an atomic level.[147] Through DFT calculation of the adsorption energies of the intermediates, we can analyze the electrode surface species and the preferred reaction pathways in specific electrolytes, thus gaining comprehensive insight into the energy storage mechanism.[148] Furthermore, artificial intelligence and machine learning will play invaluable roles in the prediction and refinement of the most rational material combinations, cell manufacturing and operation.[149]

**New Mn-based battery chemistry.** Despite the recent progress in the development of Mn-based batteries with the $Mn^{2+}/MnO_2$ mechanism, there remain some critical challenges that require further investigations, and therefore call for the development of new Mn-based battery chemistry with enhanced performance. The cathode $Mn^{2+}/MnO_2$ reactions with two-electron charge transfer give rise to theoretical capacity of ~616 mAh g$^{-1}$ based on the deposited $MnO_2$ solely. However, the pursuit of cathode reactions with multiple (>2) electron transfers and superior output voltage is attractive yet very challenging. A recent study implies that the $Mn^{2+}/MnO_4^-$ reactions may be reversible in a Mn-based battery with acidic $MnO_4^-$ in the cathode electrolyte.[114] Hence, more effort should be devoted to the development of new Mn cathode with superior performance. On the other hand, the matching anodes of Mn-based batteries, which are generally metallic materials, have potential risks of HER and dendrite formation. As HER accelerates on the metal anode in acidic electrolytes, while the cathode $Mn^{2+}/MnO_2$ reactions perform well with a certain acidity, it is important to optimize the acidity of the electrolyte for both the cathode and anode reactions to maximize the overall electrochemical performance. Meanwhile, dendrite formation due to uneven and irregular metal plating is detrimental to the cycling stability and safety of the battery. In this regard, the employment of non-metals such as polymers,[113] carbon materials[53] and other materials as alternative anodes provides additional opportunities to enrich the Mn-based battery chemistry. Exploring new cathodes and anodes with proper functionalities will contribute to the development of new Mn-based battery chemistry.

**Battery system integration.** The battery system integration refers to the overall considerations of electrolytic $Zn$-$MnO_2$ battery as a delicate system from various aspects with respect to battery key components (current collector, cathode, anode, separator, electrolyte, as well as cell configuration, packing and manufacturing), cost, performance, and real-life applications in order to maximize the techno-economic benefits. As the capacity of both cathode and anode are expected to increase, their current collectors are required to have improved mass loadings for active materials, which is a challenge to the current state-of-the-art electrolytic $Zn$-$MnO_2$ battery. Future strategy should be focused on exploring current collectors that are lightweight, with high surface area, good electrical conductivity, and hydrophilicity. The adoption of cathode, anode and electrolyte needs to be considered simultaneously to comply with standard guidelines for the electrochemical performance, while maintaining low cost for practical applications. Special attention needs to be paid to the optimization of $Mn^{2+}$, $H^+$ and other cations/anions in the electrolyte, so as to understand their incorporated impacts on the electrochemical performance in terms of discharge voltage, capacity, charge/discharge rates, long-term stability and self-discharge. In addition, the cost of Mn-based battery should be evaluated when relatively expensive separators/ion-exchange membranes are applied to the battery. Overall, the system-level concerns and solutions in the development of Mn-based batteries will help translate the research lab-based battery designs to the industry with scalable production in a cost-effective manner.[150]

**Scale-up and commercialization.** Aqueous electrolytic Zn-MnO$_2$ batteries with the cathode Mn$^{2+}$/MnO$_2$ chemistry have shown the advantages of low cost, high-level safety, environmental friendliness, good electrochemical performance and long-term stability, representing promising candidates for future large-scale energy storage applications. By integration with the electric grid, the electrolytic Zn-MnO$_2$ batteries can make up for the randomness and intermittence of the renewable energy sources, hence effectively improving the performance of the power grid.[151] Low cost, scalable manufacturing and long-term shelf life are among the most prominent challenges that need to be overcome for the Mn-based batteries toward large-scale energy storage applications. Despite the recent developments where lab-scale prototype batteries have shown very attractive features, the electrolytic Zn-MnO$_2$ batteries remain challenging for large-scale commercialization.[115] We believe that the gradual shift from the lab research to industry, and the further development of the electrolytic Zn-MnO$_2$ batteries in the industry with standard guidelines and targeted applications, will enable commercially viable projects to fulfill the practical energy storage solutions.

## Acknowledgements

W.C. acknowledges the startup funds from University of Science and Technology of China (KY2060000150) and the Fundamental Research Funds for the Central Universities (WK2060000040).

## References

[1] C. McGlade, P. Ekins, *Nature* **2015**, 517, 187–190.

[2] R. F. Service, *Science* **2018**, 362, 508–509.

[3] S. Chu, A. Majumdar, *Nature* **2012**, 488, 294–303.

[4] Z. Zhang, C. Shi, G. Xiao, L. Kai, J. Yue, *Ceram. Int.* **2017**, 43, 10052–10056.

[5] H. Li, L. Ma, C. Han, Z. Wang, Z. Liu, Z. Tang, C. Zhi, *Nano Energy* **2019**, 62, 550–587.

[6] S. Chu, Y. Cui, N. Liu, *Nat. Mater.* **2016**, 16, 16–22.

[7] Y. Meng, M. Wang, Z. Zhu, T. Jiang, Z. Liu, N. Chen, C. Shen, Q. Peng, W. Chen, *ACS Appl. Energy Mater.* **2021**, 4, 12927–12934.

[8] M. Wang, Y. Meng, K. Li, T. Ahmad, N. Chen, Y. Xu, J. Sun, M. Chuai, X. Zheng, Y. Yuan, C. Shen, Z. Zhang, W. Chen, *eScience* **2022**, 2, 509–517.

[9] H. J. Herfurth, *Nature* **2008**, 451, 652–657.

[10] P. Gu, M. Zheng, Q. Zhao, X. Xiao, H. Xue, H. Pang, *J. Mater. Chem. A* **2017**, 5, 7651–7666.

[11] J. Y. Hwang, S.-T. Myung, Y.-K. Sun, *Chem. Soc. Rev.* **2017**, 46, 3529–3614.

[12] K. Cao, H. Liu, X. Chang, Y. Li, Y. Wang, L. Jiao, *Adv. Mater. Technol.* **2017**, 2, 1600221.

[13] J. Hu, Y. Xiao, H. Tang, H. Wang, Z. Wang, C. Liu, H. Zeng, Q. Huang, Y. Ren, C. Wang, W. Zhang, F. Pan, *Nano Lett.* **2017**, 17, 4934–4940.

[14] M. Wang, J. Ma, Y. Meng, J. Sun, Y. Yuan, M. Chuai, N. Chen, Y. Xu, X. Zheng, Z. Li, W. Chen, *Angew. Chem. Int. Ed.* **2023**, 62, e202214966.

[15] F. Wang, O. Borodin, M. S. Ding, M. Gobet, J. Vatamanu, X. Fan, T. Gao, N. Edison, Y. Liang, W. Sun, *Joule* **2018**, 2, 927–937.

[16] J. Zhang, Y. Shi, Y. Ding, W. Zhang, G. Yu, *Nano Lett.* **2016**, 16, 7276–7281.

[17] W. Mao, G. Ai, Y. Dai, Y. Fu, Y. Ma, S. Shi, R. Soe, X. Zhang, D. Qu, Z. Tang, V. S. Battaglia, *J. Power Sources* **2016**, 310, 54–60.

[18] M. Song, H. Tan, D. Chao, H. J. Fan, *Adv. Funct. Mater.* **2018**, 28, 1802564.

[19] N. Alias, A. A. Mohamad, *J. Power Sources* **2015**, 274, 237–251.

[20] T. Liu, X. Cheng, H. Yu, H. Zhu, N. Peng, R. Zheng, J. Zhang, M. Shui, Y. Cui, J. Shu, *Energy Storage Mater.* **2019**, 18, 68–91.

[21] S. Lian, C. Sun, W. Xu, W. Huo, Y. Luo, K. Zhao, G. Yao, W. Xu, Y. Zhang, Z. Li, K. Yu, H. Zhao, H. Cheng, J. Zhang, L. Mai, *Nano Energy* **2019**, 62, 79–84.

[22] C. Guo, Q. Zhou, H. Liu, S. Tian, B. Chen, J. Zhao, J. Li, *Electrochim. Acta* **2019**, 324, 134867.

[23] X. Fan, K. Ni, J. Han, S. Wang, L. Gou, D.-L. Li, *Funct. Mater. Lett.* **2019**, 12, 1950073.

[24] E. Iruin, A. R. Mainar, M. Enterria, N. Ortiz-Vitoriano, J. A. Blazquez, L. C. Colmenares, T. Rojo, S. Clark, B. Horstmann, *Electrochim. Acta* **2019**, 320, 134557.

[25] S. Khamsanga, R. Pornprasertsuk, T. Yonezawa, A. A. Mohamad, S. Kheawhom, *Sci. Rep.* **2019**, 9, 8441.

[26] Y. Tang, S. Zheng, Y. Xu, X. Xiao, H. Xue, H. Pang, *Energy Storage Mater.* **2018**, 12, 284–309.

[27] Y. Zhang, C. Yuan, K. Ye, X. Jiang, J. Yin, G. Wang, D. Cao, *Electrochim. Acta* **2014**, 148, 237–243.

[28] B. Li, J. Chai, X. Ge, T. An, P. C. Lim, Z. Liu, Y. Zong, *ChemNanoMat* **2017**, 3, 401–405.

[29] Y. Zhang, S. Deng, M. Luo, G. Pan, Y. Zeng, X. Lu, C. Ai, Q. Liu, Q. Xiong, X. Wang, X. Xia, J. Tu, *Small* **2019**, 15, 1905452.

[30] C. Xie, T. Li, C. Deng, Y. Song, H. Zhang, X. Li, *Energy Environ. Sci.* **2020**, 13, 135–143.

[31] C. Ling, R. Zhang, T. S. Arthur, F. Mizuno, *Chem. Mater.* **2015**, 27, 5799–5807.

[32] K. Zhang, X. Han, Z. Hu, X. Zhang, Z. Tao, J. Chen, *Chem. Soc. Rev.* **2015**, 44, 699–728.

[33] W. Wei, X. Cui, W. Chen, D. G. Ivey, *Chem. Soc. Rev.* **2011**, 40, 1697–1721.

[34] D.-S. Li, Q.-L. Gao, H. Zhang, Y.-F. Wang, W.-L. Liu, M.-M. Ren, F.-G. Kong, S.-J. Wang, J. Chang, *Appl. Surf. Sci.* **2020**, 510, 145458.

[35] D. Aurbach, Z. Lu, A. Schechter, Y. Gofer, E. Levi, *Nature* **2000**, 407, 724–727.

[36] Y. Liu, X. Chi, Q. Han, Y. Du, J. Huang, Y. Liu, J. Yang, *J. Power Sources* **2019**, 443, 227244.

[37] G. Wang, L. Zhang, J. Zhang, *Chem. Soc. Rev.* **2012**, 41, 791–828.

[38] B. Lee, S. Y. Chong, H. R. Lee, K. Y. Chung, H. O. Si, *Sci. Rep.* **2014**, 4, 6066.

[39] J. Mcbreen, *Electrochem. Acta* **1975**, 20, 221–225.

[40] C. Wang, Y. Zeng, X. Xiao, S. Wu, G. Zhong, K. Xu, Z. Wei, W. Su, X. Lu, *J. Energ. Chem.* **2020**, 43, 182–187.

[41] D. Wang, L. Wang, G. Liang, H. Li, Z. Liu, Z. Tang, J. Liang, C. Zhi, *ACS Nano* **2019**, 13, 10643–10652.

[42] A. R. Armstrong, P. G. Bruce, *Nature* **1996**, 381, 499–500.

[43] Y. Jiang, D. Ba, Y. Li, J. Liu, *Adv. Sci.* **2020**, 7, 190295.

[44] M. Salehi, Z. Shariatinia, *Electrochim. Acta* **2016**, 188, 428–440.

[45] X. Wu, A. Markir, Y. Xu, C. Zhang, D. P. Leonard, W. Shin, X. Ji, *Adv. Funct. Mater.* **2019**, 29, 1900911.

[46] L. Chen, Z. Yang, F. Cui, J. Meng, Y. Jiang, J. Long, X. Zeng, *Mater. Chem. Front.* **2020**, 4, 213–221.

[47] W. Chen, G. Li, A. Pei, Y. Li, L. Liao, H. Wang, J. Wan, Z. Liang, G. Chen, H. Zhang, J. Wang, Y. Cui, *Nat. Energy* **2018**, 3, 428–435.

[48] C. Liu, X. Chi, Q. Han, Y. Liu, *Adv. Energy Mater.* **2020**, 10, 1903589.

[49] G. Li, W. Chen, H. Zhang, Y. Gong, F. Shi, J. Wang, R. Zhang, G. Chen, Y. Jin, T. Wu, Z. Tang, Y. Cui, *Adv. Energy Mater.* **2020**, 10, 1902085.

[50] D. Chao, W. Zhou, C. Ye, Q. Zhang, Y. Chen, Lin Gu, K. Davey, S.-Z. Qiao, *Angew. Chem. Int. Ed.* **2019**, 58, 7823–7828.

[51] J. Huang, Z. Guo, X. Dong, D. Bin, Y. Wang, Y. Xia, *Sci. Bull.* **2019**, 64, 1780–1787.

[52] L. Wei, L. Zeng, M. C. Wu, H. R. Jiang, T. S. Zhao, *J. Power Sources* **2019**, 423, 203–210.

[53] Y. Feng, Q. Zhang, S. Liu, J. Liu, Z. Tao, J. Chen, *J. Mater. Chem. A* **2019**, 7, 2922–2922.

[54] J. Huang, L. Yan, D. Bin, X. Dong, Y. Wang, Y. Xia, *J. Mater. Chem. A* **2020**, 8, 5959–5967.

[55] Z. Ma, T. Zhao, *Electrochim. Acta* **2016**, 201, 165–171.

[56] G. Liang, F. Mo, Q. Yang, Z. Huang, X. Li, D. Wang, Z. Liu, H. Li, Q. Zhang, C. Zhi, *Adv. Mater.* **2019**, 31, 1905873.

[57] Y. H. Lu, J. B. Goodenough, Y. Kim, *J. Am. Chem. Soc.* **2011**, 133, 5756–5759.

[58] A. M. Kannan, S. Bhavaraju, F. Prado, M. M. Raja, A. Manthirama, *J. Electrochem. Soc.* **2000**, 149, A483–A492.

[59] C. Xu, B. Li, H. Du, F. Kang, *Angew. Chem. Int. Ed.* **2012**, 51, 933–935.

[60] A. Era, Z. Takehara, S. Yoshizawa, *Electrochim. Acta* **1967**, 12, 1199–1212.

[61] F. R. McLarnon, E. J. Cairns, *J. Electrochem. Soc.* **1991**, 138, 645–656.

[62] B. J. Hertzberg, A. Huang, A. Hsieh, M. Chamoun, G. Davies, J. K. Seo, Z. Zhong, M. Croft, C. Erdonmez, Y. S. Meng, *Chem. Mater.* **2016**, 28, 4536–4545.

[63] V. K. Nartey, L. Binder, A. Huber, *J. Power Sources* **2000**, 87, 205–211.

[64] J. F. Parker, C. N. Chervin, E. S. Nelson, D. R. Rolison, J. W. Long, *Energy Environ. Sci.* **2014**, 7, 1117–1124.

[65] Y. Shen, K. Kordesch, *J. Power Sources* **2000**, 87, 162–166.

[66] Manickam Minakshi, Pritam Singh, M. Carter, K. Prince, *Electrochem. Solid Lett.* **2008**, 11, 515–582.

[67] X. Zhu, X. Wu, T. N. L. Doan, Y. Tian, H. Zhao, P. Chen, *J. Power Sources* **2016**, 326, 498–504.

[68] J. Yang, J. Wang, L. Zhu, Q. Gao, W. Zeng, J. Wang, Y. Li, *Ceram. Int.* **2018**, 44, 23073–23079.

[69] G. G. Yadav, J. W. Gallaway, D. E. Turney, M. Nyce, J. Huang, X. Wei, S. Banerjee, *Nat. Commun.* **2017**, 8, 14424.

[70] C. Mondoloni, M. Laborde, J. Rioux, E. Andoni, C. Levyclement, *J. Electrochem. Soc.* **1992**, 139, 954–959.

[71] B. Hertzberg, L. Sviridov, E. A. Stach, T. Gupta, D. Steingart, *J. Electrochem. Soc.* **2014**, 161, A835–A840.

[72] N. D. Ingale, J. W. Gallaway, M. Nyce, A. Couzis, S. Banerjee, *J. Power Sources* **2015**, 276, 7–18.

[73] L. Bat, D. Y. Qu, B. E. Conway *J. Electrochem. Soc.* **1993**, 140, 884.

[74] C. G. Castledine, B. E. Conway *J. Appl. Electrochem.* **1995**, 25, 701–705.

[75] J. Lee, J. B. Ju, W. I. Cho, B. W. Cho, S. H. Oh, *Electrochim. Acta* **2013**, 112, 138–143.

[76] M. H. Alfaruqi, J. Gim, S. Kim, J. Song, D. T. Pham, J. Jo, Z. Xiu, V. Mathew, J. Kim, *Electrochem. Commun.* **2015**, 60, 121–125.

[77] J. Huang, Z. Guo, Y. Ma, D. Bin, Y. Wang, Y. Xia, *Small Methods* **2019**, 3, 1800272.

[78] T. Yamamoto, T. Shoji, *Cheminform* **1986**, 17, 017.

[79] T. Shoji, M. Hishinuma, T. Yamamoto, *J. Appl. Electrochem.* **1988**, 18, 521–526.

[80] D. Xu, B. Li, C. Wei, Y.-B. He, H. Du, X. Chu, X. Qin, Q.-H. Yang, F. Kang, *Electrochim. Acta* **2014**, 133, 254–261.

[81] I. Stosevski, A. Bonakdarpour, F. Cuadra, D. P. Wilkinson, *Chem. Commun.* **2019**, 55, 2082–2085.

[82] N. Qiu, H. Chen, Z. Yang, S. Sun, Y. Wang, *Electrochem. Acta* **2018**, 272, 154–160.

[83] Y. Li, S. Wang, J. R. Salvador, J. Wu, B. Liu, W. Yang, J. Yang, W. Zhang, J. Liu, J. Yang, *Chem. Mater.* **2019**, 31, 2036–2047.

[84] H. Pan, Y. Shao, P. Yan, Y. Cheng, K. S. Han, Z. Nie, C. Wang, J. Yang, X. Li, P. Bhattacharya, K. T. Mueller, J. Liu, *Nat. Energy* **2016**, 1, 16039.

[85] H. Wei, X. Hu, X. Zhang, Z. Yu, T. Zhou, Y. Liu, Y. Liu, Y. Wang, J. Xie, L. Sun, M. Liang, P. Jiang, *Energy Technol.* **2019**, 7, 1800912.

[86] W. Sun, F. Wang, S. Hou, C. Yang, C. Wang, *J. Am. Chem. Soc.* **2017**, 139, 9775–9775.

[87] C. Mylad, B. W. R., T. Cheuk-Wai, K. Gunder, N. Dag, *Energy Storage Mater.* **2018**, 15, 251–360.

[88] Q. Zhao, X. Chen, Z. Wang, L. Yang, R. Qin, J. Yang, Y. Song, S. Ding, M. Weng, W. Huang, J. Liu, W. Zhao, G. Qian, K. Yang, Y. Cui, H. Chen, F. Pan, *Small* **2019**, 15, 1904545.

[89] N. Zhang, F. Cheng, J. Liu, L. Wang, X. Long, X. Liu, F. Li, J. Chen, *Nat. Commun.* **2017**, 8, 405.

[90] N. Zhang, F. Cheng, Y. Liu, Q. Zhao, K. Lei, C. Chen, X. Liu, J. Chen, *J. Am. Chem. Soc.* **2016**, 138, 12894–12901.

[91] M. Li, Q. He, Z. Li, Q. Li, Y. Zhang, J. Meng, X. Liu, S. Li, B. Wu, L. Chen, Z. Liu, W. Luo, C. Han, L. Mai, *Adv. Energy Mater.* **2019**, 9, 1901469.

[92] S. Wang, Q. Wang, W. Zeng, M. Wang, L. Ruan, Y. Ma, *Nano-Micro Lett.* **2019**, 11, 70.

[93] Y. Huang, J. Mou, W. Liu, X. Wang, L. Dong, F. Kang, C. Xu, *Nano-Micro Lett.* **2019**, 11, 49.

[94] M. Chotkowski, Z. Rogulski, A. Czerwiński, *J. Electroanal. Chem.* **2011**, 651, 237–242.

[95] P. G. Perret, P. R. L. Malenfant, C. Bock, B. MacDougall, *J. Electrochem. Soc.* **2012**, 159, A1554–A1561.

[96] A. J. Gibson, B. Johannessen, Y. Beyad, J. Allen, S. W. Donne, *J. Electrochem. Soc.* **2016**, 163, H305–H312.

[97]  G. Liang, F. Mo, H. Li, Z. Tang, Z. Liu, D. Wang, Q. Yang, L. Ma, C. Zhi, *Adv. Energy Mater.* **2019**, 91, 1901838.

[98]  M. Wang, N. Chen, Z. Zhu, Y. Meng, C. Shen, X. Zheng, D. Liang, W. Chen, *Small* **2021**, 17, 2103921.

[99]  M. Wang, Y. Meng, N. Chen, M. Chuai, C. Shen, X. Zheng, Y. Yuan, J. Sun, Y. Xu, W. Chen, *ACS Mater. Lett.* **2021**, 3, 1558–1565.

[100] A. Kozawa., R. A. Powers., *J. Electrochem. Soc.* **1966**, 113, 830.

[101] C. Guo, J. Li, Y. Chu, H. Li, H. Zhang, L. Hou, Y. Wei, J. Liu, S. Xiong, *Dalton Trans.* **2019**, 48, 7403–7412.

[102] M. Manickam, P. Singh, T. B. Issa, S. Thurgate, R. D. Marco, *J. Power Sources* **2004**, 138, 319–322.

[103] J. K. Seo, J. Shin, H. Chung, P. Y. Meng, X. Wang, Y. S. Meng, *J. Phys. Chem. C* **2018**, 122, 11177–11185.

[104] F. Y. Cheng, J. Chen, X. L. Gou, P. W. Shen, *Adv. Mater.* **2005**, 17, 2753–2756.

[105] Q. Wang, J. Pan, Y. Sun, Z. Wang, *J. Power Sources* **2012**, 199, 355–359.

[106] G. G. Yadav, X. Wei, J. Huang, J. W. Gallaway, D. E. Turney, M. Nyce, J. Secor, S. Banerjee, *J. Mater. Chem. A* **2017**, 5, 15845–15854.

[107] G. G. Yadav, J. Cho, D. Turney, B. Hawkins, X. Wei, J. Huang, S. Banerjee, M. Nyce, *Adv. Energy Mater.* **2019**, 9, 1902270.

[108] B. Lee, H. R. Lee, H. Kim, K. Y. Chung, B. W. Cho, H. O. Si, *Chem. Commun.* **2015**, 51, 9265–9268.

[109] M. H. Alfaruqi, S. Islam, J. Gim, J. Song, S. Kim, D. T. Pham, J. Jo, Z. Xiu, V. Mathew, J. Kim, *Chem. Phy. Lett.* **2015**, 550, 64–68.

[110] M. H. Alfaruqi, J. Gim, S. Kim, J. Song, J. Jo, S. Kim, V. Mathew, J. Kim, *J. Power Sources* **2015**, 288, 320–327.

[111] B. Wu, G. Zhang, M. Yan, T. Xiong, L. Mai, *Small* **2018**, 14, 1703850.

[112] J. Huang, Z. Wang, M. Hou, X. Dong, Y. Liu, Y. Wang, Y. Xia, *Nat Commun.* **2018**, 9, 2906.

[113] Z. Guo, J. Huang, X. Dong, Y. Xia, L. Yan, Z. Wang, Y. Wang, *Nat. Commun.* **2020**, 11, 959.

[114] G. G. Yadav, D. Turney, J. Huang, X. Wei, S. Banerjee, *ACS Energy Lett.* **2019**, 4, 2144–2146.

[115] C. Zhong, B. Liu, J. Ding, X. Liu, Y. Zhong, Y. Li, C. Sun, X. Han, Y. Deng, N. Zhao, W. Hu, *Nat. Energy* **2020**, 5, 440–449.

[116] D. Chao, C. Ye, F. Xie, W. Zhou, Q. Zhang, Q. Gu, K. Davey, L. Gu, S.-Z. Qiao, *Adv. Mater.* **2020**, 32, 2001894.

[117] W. Chen, Y. Jin, J. Zhao, N. Liu, Y. Cui, *Proc. Natl. Acad. Sci.* **2018**, 115, 11694–11699.

[118] Q. Li, X. Cui, Q. Pan, *ACS Appl. Mater. Interfaces* **2019**, 11, 38762–38770.

[119] N. Ma, P. Wu, Y. Wu, D. Jiang, G. Lei, *Funct. Mater. Lett.* **2019**, 12, 1930003.

[120] X. Guo, J. Zhou, C. Bai, X. Li, G. Fang, S. Liang, *Mater. Today Energy* **2020**, 16, 100396.

[121] L. Wang, Q. Wu, A. Abraham, P. J. West, L. M. Housel, G. Singh, N. Sadique, C. D. Quilty, D. Wu, E. S. Takeuchi, A. C. Marschilok, K. J. Takeuchi, *J. Electrochem. Soc.* **2019**, 166, A3575–A3584.

[122] X. Zheng, R. Luo, T. Ahmad, J. Sun, S. Liu, N. Chen, M. Wang, Y. Yuan, M. Chuai, Y. Xu, T. Jiang, W. Chen, *Energy Environ. Mater.* **2022**, e12433.

[123] J. Lei, Y. Yao, Z. Wang, Y.-C. Lu, *Energy Environ. Sci.* **2021**, 14, 4418–4426.

[124] X. Zheng, Y. Wang, Y. Xu, T. Ahmad, Y. Yuan, J. Sun, R. Luo, M. Wang, M. Chuai, N. Chen, T. Jiang, S. Liu, W. Chen, *Nano Lett.* **2021**, 21, 8863–8871.

[125] M. Chuai, J. Yang, M. Wang, Y. Yuan, Z. Liu, Y. Xu, Y. Yin, J. Sun, X. Zheng, N. Chen, W. Chen, *eScience* **2021**, 1, 178–185.

[126] M. Chuai, J. Yang, R. Tan, Z. Liu, Y. Yuan, Y. Xu, J. Sun, M. Wang, X. Zheng, N. Chen, W. Chen, *Adv. Mater.* **2022**, 34, 2203249.

[127] M. Wang, W. Wang, Y. Meng, Y. Xu, J. Sun, Y. Yuan, M. Chuai, N. Chen, X. Zheng, R. Luo, K. Xu, W. Chen, *Energy Storage Mater.* **2023**, 56, 424–431.

[128] Y. Xu, Z. Liu, X. Zheng, K. Li, M. Wang, W. Yu, H. Hu, W. Chen, *Adv. Energy Mater.* **2022**, 12, 2103352.

[129] J. Sun, Z. Liu, K. Li, Y. Yuan, X. Zheng, Y. Xu, M. Wang, M. Chuai, H. Hu, W. Chen, *ACS Appl. Mater. Interfaces* **2022**, 14, 51900–51909.

[130] Y. Yuan, J. Yang, Z. Liu, R. Tan, M. Chuai, J. Sun, Y. Xu, X. Zheng, M. Wang, T. Ahmad, N. Chen, Z. Zhu, K. Li, W. Chen, *Adv. Energy Mater.* **2022**, 12, 2103705.

[131] J. Sun, X. Zheng, K. Li, G. Ma, T. Dai, B. Ban, Y. Yuan, M. Wang, M. Chuai, Y. Xu, Z. Liu, T. Jiang, Z. Zhu, J. Chen, H. Hu, W. Chen, *Energy Storage Mater.* **2023**, 54, 570–578.

[132] M. Liu, Q. Zhao, H. Liu, J. Yang, X. Chen, L. Yang, Y. Cui, W. Huang, W. Zhao, A. Song, Y. Wang, S. Ding, Y. Song, G. Qian, H. Chen, F. Pan, *Nano Energy* **2019**, 64, 103942.

[133] S. Zhang, W. Guo, F. Yang, P. Zheng, R. Qiao, Z. Li, *Batteries & Supercaps* **2019**, 2, 627–637.

[134] G. L. Soloveichik, *Chem. Rev.* **2015**, 115, 11533–11558.

[135] M. F. Dupont, S. W. Donne, *Electrochim. Acta* **2014**, 120, 219–225.

[136] S. W. Donne, F. H. Feddrix, R. Glo¨ckner, S. Marion, T. Norby, *Solid State Ionics* **2002**, 152–153, 695–701.

[137] G. J. Browning, S. W. Donne, *J. Appl. Electrochem.* **2005**, 35, 437–443.

[138] C. J. Clarke, G. J. Browning, S. W. Donne, *Electrochim. Acta* **2006**, 51, 5773–5784.

[139] S. Hameer, J. L. Van Niekerk, *Int. J. Energy Res.* **2015**, 39, 1179–1195.

[140] M. P. Owen, G. A. Lawrance, S. W. Donne, *Electrochim. Acta* **2007**, 52, 4630–4639.

[141] C. Zhan, Z. Yao, J. Lu, L. Ma, V. A. Maroni, L. Li, E. Lee, E. E. Alp, T. Wu, J. Wen, Y. Ren, C. Johnson, M. M. Thackeray, M. K. Y. Chan, C. Wolverton, K. Amine, *Nat. Energy* **2017**, 2, 963–971.

[142] A. M. Simes, A. C. Bastos, M. G. Ferreira, Y. González-García, S. González, R. M. Souto, *Corro. Sci.* **2007**, 49, 726–739.

[143] A. D. Handoko, F. Wei, Jenndy. B. S. Yeo, Z. W. Seh, *Nat. Catal.* **2018**, 1, 922–934.

[144] D. Liu, Z. Shadike, R. Lin, K. Qian, H. Li, K. Li, S. Wang, Q. Yu, M. Liu, S. Ganapathy, X. Qin, Q.-H. Yang, M. Wagemaker, F. Kang, X.-Q. Yang, B. Li, *Adv. Mater.* **2019**, 31, 1806620.

[145] K. Jang, K. Joo, K. Kim, *Nanosci. Nanotech. Lett.* **2017**, 9, 1165–1174.

[146] J.-S. Park, J. H. Jo, Y. Aniskevich, A. Bakavets, G. Ragoisha, E. Streltsov, J. Kim, S.-T. Myung, *Chem. Mater.* **2018**, 30, 6777–6787.

[147] M. H. Alfaruqi, S. Islam, J. Lee, J. Jo, V. Mathew, J. Kim, *J. Mater. Chem. A* **2019**, 7, 26966–26974.

[148] Y. Y. Birdja, E. Perez-Gallent, M. C. Figueiredo, A. J. Gottle, F. Calle-Vallejo, M. T. M. Koper, *Nat. Energy* **2019**, 4, 732–745.

[149] P. M. Attia, A. Grover, N. Jin, K. A. Severson, T. M. Markov, Y.-H. Liao, M. H. Chen, B. Cheong, N. Perkins, Z. Yang, P. K. Herring, M. Aykol, S. J. Harris, R. D. Braatz, S. Ermon, W. C. Chueh, *Nature* **2020**, 578, 397–402.

[150] D. L. Chao, W. H. Zhou, X. F. X , C. Ye, Li H., M. Jaroniec, S. Z. Qiao, *Sci. Adv.* **2020**, 6, aba4098.

[151] B. Li, M. Gu, Z. Nie, Y. Shao, Q. Luo, X. Wei, X. Li, J. Xiao, C. Wang, V. Sprenkle, W. Wang, *Nano Lett.* **2013**, 13, 1330–1335.

© 2024 World Scientific Publishing Company
https://doi.org/10.1142/9789811278327_0007

Chapter 7

# Toward High-Voltage Aqueous Batteries: Super- or Low-Concentrated Electrolyte?[1]

Dongliang Chao,* Shi-Zhang Qiao*

*School of Chemical Engineering & Advanced Materials,
The University of Adelaide, Adelaide, SA 5005, Australia*

## 7.1 Fundamentals

Recent incidents in organic media-based Li-ion batteries (LIBs), such as the Boeing 787 battery fires in 2013, Samsung Note 7 explosions in 2016, and the Tesla Model S combustions in 2019, have caused severe threats to human health and life. This is an alert that safety is a vital prerequisite for next-generation rechargeable batteries for both consumable electronics and electric vehicles. In this context, aqueous batteries (ABs) offer tremendous competitiveness in terms of safe operation, low cost, facile manufacturing and recycling, together with high power density and tolerance against misuse. We have witnessed the prosperity of various types of rechargeable ABs in the market, such as primary Zn-Mn alkaline, valve-regulated lead-acid, nickel-cadmium, and nickel-metal hydride batteries. These batteries possess competitiveness in either economical profit, safety, or environmental influence. However, the market share has been continuously shrinking due to the inadequate energy density of ABs compared with the mainstream LIBs.

Fundamentally, water has the thermodynamic potentials for oxygen evolution reaction (OER) and hydrogen evolution reaction (HER) (Figure 7.1 and Eqs. (7.1)–(7.4)). The electrochemical stability window (ESW, potential difference between OER and HER) of water is ~1.23 V, which restrains the operating output voltage of ABs and leads to insufficient energy density. Moreover, the generation of gases ($H_2$, $O_2$, etc.) consumes

---

* Corresponding authors: dongliang.chao@adelaide.edu.au; s.qiao@adelaide.edu.au
[1] Adapted with permission from D. Chao, S.-Z. Qiao, *Joule* **2020**, 4, 1846.

(177)

Toward High-Voltage Aqueous Batteries: Super- or Low-Concentrated Electrolyte?

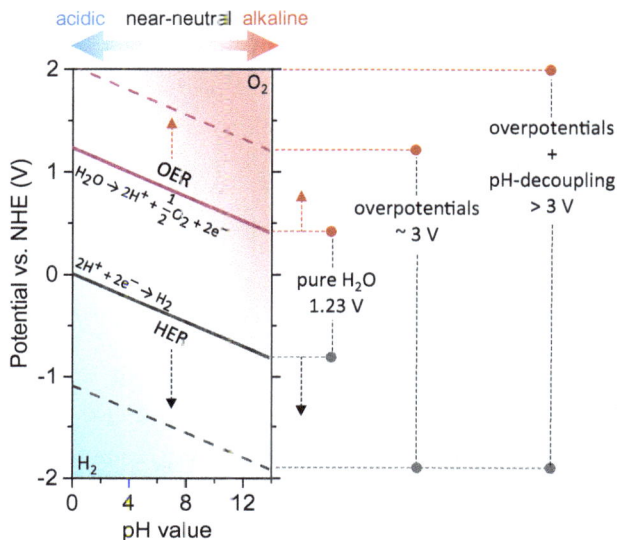

**Figure 7.1.** Pourbaix diagram of water and electrochemical principles in tuning the water activity. The overpotentials are associated with OER and HER polarization associated with altered electrodes and/or electrolytes. The pH decoupling refers to the extended voltage window by coupling separated anolyte and catholyte with different pH.

electrons and electrolyte, destroys the structure of the electrodes, and results in battery swell and inferior cycling stability. To overcome these issues and achieve better performance of ABs, it is crucial to suppress the water splitting and extend the ESW. Practically, the output voltage (E) is generally higher than 1.23 V (see Figure 7.1) because of the sluggish kinetics of HER (in acidic media: $2H^+ + 2e^- \rightarrow H_2$; in neutral or alkaline media: $2H_2O + 2e^- \rightarrow H_2 + 2OH^-$) and OER (in acidic media: $H_2O \rightarrow 2H^+ + \frac{1}{2}O_2 + 2e^-$; in neutral or alkaline media: $2OH^- \rightarrow H_2O + \frac{1}{2}O_2 + 2e^-$) and the corresponding overpotentials ($\eta$), including the anode ($\eta_a$) and cathode ($\eta_c$) overpotentials from the intrinsic activation barriers and other overpotentials ($\eta_o$) from the solution and interface resistances, etc. Taking water activity into consideration (especially in the concentrated electrolyte, the equilibrium constant would vary with the water activity), the corresponding environmental potentials according to Nernst equations and related output voltage can be expressed as:

$$\phi_{HER} : \phi_{H^+/H_2} = \phi^\Theta_{H^+/H_2} + \frac{RT}{nF} \ln\left(\frac{\alpha^2_{H^+}}{P_{H_2}/P_0}\right) \tag{7.1}$$

$$\phi_{H_2O/H_2} = \phi_{H_2O/H_2}^{\ominus} + \frac{RT}{nF} \ln\left(\frac{\alpha_{H_2O}^2}{\alpha_{OH^-}^2 P_{H_2}/P_0}\right) \tag{7.2}$$

$$\phi_{OER} : \phi_{H_2O/O_2} = \phi_{H_2O/O_2}^{\ominus} + \frac{RT}{nF} \ln\left(\frac{\alpha_{H_2O}^2}{\alpha_{H^+}^4 P_{O_2}/P_0}\right) \tag{7.3}$$

$$\phi_{OH^-/O_2} = \phi_{OH^-/O_2}^{\ominus} + \frac{RT}{nF} \ln\left(\frac{\alpha_{OH^-}^4}{\alpha_{H_2O}^2 P_{O_2}/P_0}\right) \tag{7.4}$$

Output voltage: $E = (\phi_{OER} - \phi_{HER}) + \eta_a + \eta_c + \eta_o$ (7.5)

where the $\phi_{H^+/H_2}^{\ominus}$ and $\phi_{H_2O/H_2}^{\ominus}$ are the standard potentials of HER; $\phi_{H_2O/O_2}^{\ominus}$ and $\phi_{OH^-/O_2}^{\ominus}$ are the standard potentials of OER; $\phi_{H^+/H_2}$, $\phi_{H_2O/H_2}$, $\phi_{H_2O/O_2}$ and $\phi_{OH^-/O_2}$ represent the corresponding environmental potentials; $\alpha_{H_2O}$, $\alpha_{H^+}$ and $\alpha_{OH^-}$ correspond to the activities of water, $H^+$ and $OH^-$ in electrolyte, respectively; $P_{H_2}/P_0$ and $P_{O_2}/P_0$ are partial pressures of hydrogen and oxygen, respectively; R, T, F and n are thermodynamic constant, temperature, Faraday constant, and number of electrons, respectively.

Widening the ESW of water has not only significantly boosted the viability of ABs in the energy storage industry, but also brought untouched chemistries and broadened the scope of materials selection. Super-concentrated electrolytes (SCEs) have proved effective in extending the ESW, but at the expense of high cost and confined reaction kinetics. Recent achievements in introducing competitive redox couples, optimizing current collectors and applying pH decoupling with ion-selective membranes provide new opportunities in low-concentrated electrolytes (LCEs). Herein, we summarize the strategies towards high-voltage ABs from a distinct perspective of different battery components. Critical appraisal is first raised regarding being SCE or LCE in solving the voltage issue. We stress that SCE or LCE for high-voltage ABs should be studied together with the affordability, stability and gravimetric/volumetric energy and power densities, when large-scale practical applications are considered. Finally, we render insights for future development of high-voltage ABs.

## 7.2 To be Super-concentrated Electrolyte

In the past few years, advanced H/Li/Na/K/Zn-based ABs have garnered enormous interests worldwide.[1] According to the Nernst equations and the electrochemical

(179)

Toward High-Voltage Aqueous Batteries: Super- or Low-Concentrated Electrolyte?

principles of water decomposition, the overpotentials are tunable via the temperature, the pH value and concentration of electrolytes, and the interface effects (Figure 7.1 and Eqs. (7.1)–(7.5)). Several strategies have been implemented effectively in inhibiting water activity and extending the ESW, such as redox-active additives, "water-in-salt" (WIS) electrolyte, hydrated eutectic electrolyte, and molecular crowding strategy.[1a,2] Herein, we summarize the key strategies towards high-voltage ABs from a special perspective of different battery components (Figure 7.2). As shown in the schematic, an aqueous cell consists of four main components, including electrode (cathode and anode redox couples), electrolyte (solvent and solute), separator, and current collector. It is believed that all these components can be designed to contribute to high voltage output in ABs. For example, from the perspective of the solute of electrolyte, Wang and co-workers proposed an effective SCE of WIS approach to extend the ESW of aqueous LIBs.[3] By super-concentrated solute of lithium bis(trifluoromethane sulfonimide) (LiTFSI, 21 mol $kg^{-1}$ (abbreviated as m hereafter)), a broadened ESW of ~3.0 V has been realized in a $Mo_6S_8//LiMn_2O_4$ full cell (Figure 7.3).[4] Compared to traditional aqueous electrolytes, a highly concentrated solute can decrease the number of free water molecules due to strengthened interaction between anions/ cations and water. However, the use of super-concentrated fluorinated solutes, including LiTFSI, lithiumtrifluoromethane sulfolate (LiOTF), and lithium bis(pentafluoroethanesulfonyl)imide (LiBETI), will significantly increase the toxicity and cost on account of the excess reaction-irrelevant ions, which compromises the

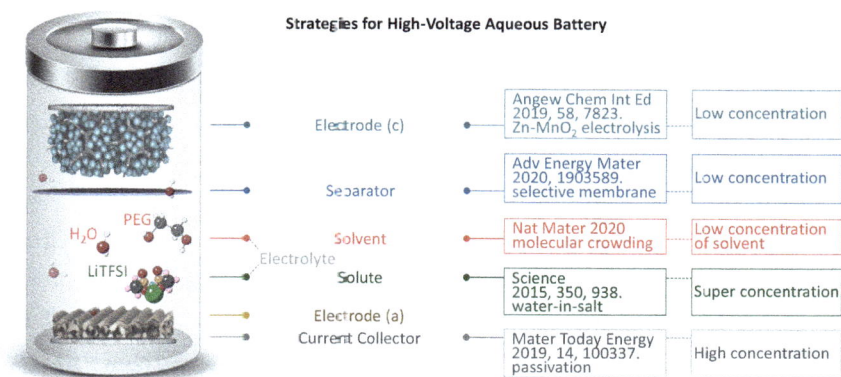

**Figure 7.2.** AB components and key strategies towards high voltage.

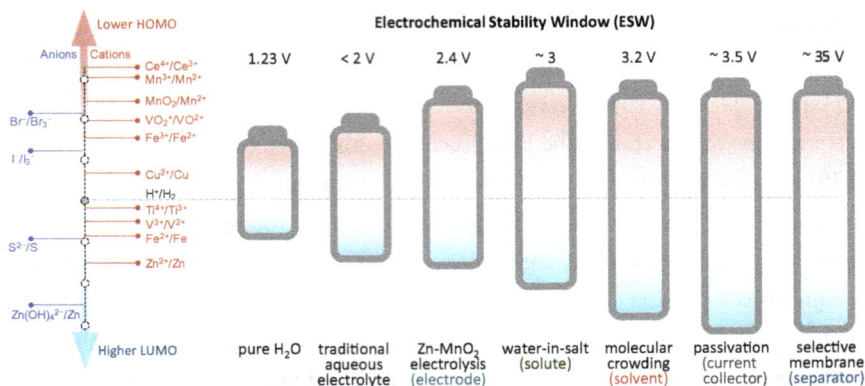

**Figure 7.3.** Extension of the ESW of ABs. (Left) Redox potentials of available redox couples that may be suitable for new ABs with extended ESW. (Right) The competitive redox couples refer to Zn/Zn²⁺ and MnO₂/Mn²⁺ redox couples at anode and cathode, respectively, in 1 mol/L ZnSO₄ + 1 mol/L MnSO₄ electrolyte.[7] The "water-in-salt" corresponds to a super-concentrated solute of 21 m LiTFSI electrolyte in Mo₆S₈//LiMn₂O₄ cell,[3a] or super-concentrated solute of sugar in sodium Prussian blue//activated carbon cell.[4] The molecular crowding is Li₄Ti₅O₁₂//LiMn₂O₄ cell in an electrolyte of 2 m LiTFSI–94%PEG–6%H₂O.[6] The passivation can be observed in Al₂O₃-coated Al and TiO₂-coated Ti current collectors in 15 m NaClO₄ electrolyte.[5] The pH decoupling was achieved with ion-selective membrane in alkaline Zn/Zn(OH)₄²⁻ and acidic MnO₂/Mn²⁺ dual-electrolyte system.[8]

economic benefits, reduces the specific energy and plagues the large-scale viability of ABs. The search for other low-cost and eco-friendly approaches to extend the ESW is thus indispensable.

Some relatively cheap non-fluorinated solutes such as sodium perchlorate (NaClO₄), zinc chloride (ZnCl₂), lithium acetate (CH₃COOLi), and potassium acetate (CH₃COOK) have been explored recently.[1] Super-concentrated sugar as a low-cost hydrogen bond-regulated solute proved effective in extending the ESW up to ~2.8 V.[4] Although sugar is naturally inert in the electrolyte, the SCE with sugar solute is effective in reducing the free water molecules, destroying the tetrahedral structure, and breaking the hydrogen bonds of water molecules. In addition to solutes, it is also possible to stabilize the aqueous electrolyte by passivating the current collectors. Hou *et al.* found that films of Al₂O₃ at the cathode and TiO₂ at the anode current collectors exhibit an evident passivation effect which increases the energy barrier of water splitting.[5] By regulating the thickness of oxide films and the concentration of the WIS electrolyte,

(181)

Toward High-Voltage Aqueous Batteries: Super- or Low-Concentrated Electrolyte?

an ESW of ~3.5 V is achieved in a full cell coupling TiS$_2$//Prussian blue in a relatively high concentration of 15 m NaClO$_4$ aqueous electrolyte (Figure 7.3). Despite the great success in widening ESW from high-concentration solutes, the high viscosity and desolvation energy, decreased wettability and therefore longer wetting (electrode stack infiltration) time of the SCEs further complicate the structure and assembly of ABs in their practical applications. Hence, the large-scale application of SCEs has been beset. Other strategies to achieve wide ESW with low-cost and moderate solute concentrations are urgently needed.

## 7.3 To be Low-concentrated Electrolyte

In addition to solutes, engineering the solvent also provides new opportunities. Recently, Lu and coworkers reported a molecular crowding electrolyte with a low solute concentration of 2 m LiTFSI in a mixed solvent of 94% poly(ethylene glycol) (PEG) and 6% H$_2$O.[6] Uisng the conventional electrode materials, the constructed Li$_4$Ti$_5$O$_{12}$// LiMn$_2$O$_4$ cell shows an ESW of 3.2 V via confining water molecules in a crowding agent of PEG network through hydrogen bonding. The PEG solvent is water-miscible, non-toxic, inert and inexpensive. Different from the previous WIS solute SCE, electrolyte with a molecular crowding solvent endows a lower solution density of ca. 1.10 g cm$^{-3}$ than that of the 21 m LiTFSI WIS electrolyte (ca. 1.65 g cm$^{-3}$) and the Li(TFSI)$_{0.7}$(BETI)$_{0.3}$ hydrate melt electrolyte (ca. 1.78 g cm$^{-3}$). Hence, this adds benefit to the gravimetric energy density when the total weight of all cell components is counted. This work offers a new avenue to enlarge the ESW especially by suppressing HER and broadens the choice of negative electrode materials (Figure 7.3). Nevertheless, there are also shortfalls of this new low-solute and high-solvent concentration electrolyte. Firstly, the water-retaining property and electric/ionic conductivity need to be further improved to demonstrate its viability in high-rate and long-cycling energy storage applications. Another drawback of this work is its limited gravimetric cell capacity of less than 50 mAh g$^{-1}$, which is confined by the theoretical capacity of electrodes for both Li$_4$Ti$_5$O$_{12}$ (175 mAh g$^{-1}$) and LiMn$_2$O$_4$ (148 mAh g$^{-1}$). Hopefully this is not an intrinsic limitation and can be mitigated by choosing concomitant high-capacity and high-voltage redox couples.

From the perspective of electrodes, introduction of competitive redox couples that show faster kinetics than those of HER or OER is considered a feasible approach to enhance both the output voltage and the capacity of ABs. As sketched in Figure 7.3, suitable candidates of the redox couples should exhibit high potential for cathode, such as $Ce^{3+}/Ce^{4+}$ (1.74 V $vs.$ standard hydrogen potential (SHE)), $Mn^{2+}/Mn^{3+}$ (1.54 V $vs.$ SHE) and $Mn^{2+}/MnO_2$ (1.23 V $vs.$ SHE), or low potential for the anode, such as $Zn(OH)_4^{2-}/Zn$ (−1.21 V $vs.$ SHE) and $Zn^{2+}/Zn$ (−0.76 V $vs.$ SHE). Recently, we introduced a high-voltage redox couple of $Mn^{2+}/MnO_2$ (Figure 7.3) in conventional Zn-ion battery via a facile tuning of the proton activity.[7] Without using the SCEs with concentrated solutes or solvents, the electrolytic $Zn-MnO_2$ system presents an ultra-flat output voltage as high as 1.95 V in a low-cost LCE (1 mol/L $ZnSO_4$ + 1 mol/L $MnSO_4$). The voltage is significantly higher than that of conventional Zn//Mn battery (1.2–1.4 V).

The ESW of the electrolytic $Zn-MnO_2$ system in LCE can be further widened from 2.4 V to 3.5 V when an ion-selective membrane and more negative redox couple of alkaline $Zn(OH)_4^{2-}/Zn$ were incorporated (Figure 7.3).[8] The rationale is that the redox potentials of HER and OER are pH dependent. Hence, according to the Pourbaix diagrams, individual control in the pH of catholyte and anolyte by ion-selective membranes is possible. In addition to high output voltage, given the high theoretical capacity from both Zn anode (820 mAh $g^{-1}$) and Mn cathode (616 mAh $g^{-1}$), a high gravimetric cell capacity of ~270 mAh $g^{-1}$ can be achieved on the basis of the total weight of positive and negative active materials.[8a] On one hand, the usage of an ion-selective membrane, such as anion/cation exchange membranes and bipolar membrane, will activate most of the cathode and anode redox couples and broaden the ESW of ABs. On the other hand, addition of membranes complicates the battery structure and poses extra challenges on battery assembly, welding, and housing. It should also be noted that, although the issues of cost and kinetics in SCEs can be eliminated in LCEs, the anodic dissolution and side reaction problems of the anode materials, cathode materials or current collectors may be critical in achieving ultrastable aqueous energy storage in diluted systems.

## 7.4 Outlook

To conclude, the fast progress in rechargeable ABs is encouraging and exciting. A rational choice and integration strategies from electrolytes, redox couples, membranes and current collector materials will endow not only affordable high-voltage ABs but also high energy and power densities. Although significant advances have been made

(183)

Toward High-Voltage Aqueous Batteries: Super- or Low-Concentrated Electrolyte?

in this area, unremitting efforts are still required to meet the practical metrics for the next stage of development of ABs. In the future design of high-voltage ABs, the following directions are believed to be of potential:

1. Exploring new competitive redox couples in LCEs that are stable, highly reversible and of faster reaction kinetics compared with that of OER/HER, such as solving the reversibility problem of the ultra-high voltage redox couple of $MnO_4^-/Mn^{2+}$ (1.51 V $vs.$ SHE) and ultra-low voltage redox couple of $Al^{3+}/Al$ (−1.66 V $vs.$ SHE).

2. Developing untouched high-capacity electrode materials in novel SCEs that are affordable and of high specific energy, such as low-cost solutes of $NaClO_4$, $LiCH_3COO$, $KCH_3COO$, $ZnCl_2$ and sugar, solvents of PEG and succinonitrile, and high-capacity redox couples of $S_6/S^{2-}$ or $S/S^{2-}$.

3. Inventing novel low-cost electrolytes with suppressed water activity and density, such as using lightweight molecular crowding solvent and hydrous salts directly with/without solvents.

4. Manipulating the interfacial reactions on cathode, anode, current collector, and separator to fundamentally understand the voltage-widening mechanisms.

5. Optimizing the device design in pH-decoupling ABs and constructing relevant membranes with low cost and high ion selectivity.

6. Evaluating the performances of ABs in a more practical metric, such as at high gravimetric and volumetric loadings, with starved electrolyte-based device design, and the transient power release; self-discharge and wide-temperature performances.

## Acknowledgments

Financial supports from the Australian Research Council through Discovery Projects (FL170100154 and DE200101244) are appreciated.

## References

[1] a) D. Chao, W. Zhou, F. Xie, C. Ye, H. Li, M. Jaroniec, S. Qiao, *Sci. Adv.* **2020**, 6, eaba4098; b) Z. Liu, Y. Huang, Y. Huang, Q. Yang, X. Li, Z. Huang, C. Zhi, *Chem. Soc. Rev.* **2020**, 49, 180.

[2] W. Yang, X. Du, J. Zhao, Z. Chen, J. Li, J. Xie, Y. Zhang, Z. Cui, Q. Kong, Z. Zhao, C. Wang, Q. Zhang, G. Cui, *Joule* **2020**, 4, 1557.

[3] a) L. Suo, O. Borodin, T. Gao, M. Olguin, J. Ho, X. Fan, C. Luo, C. Wang, K. Xu, *Science* **2015**, 350, 938; b) O. Borodin, J. Self, K. A. Persson, C. Wang, K. Xu, *Joule* **2020**, 4, 69; c) C. Yang, J. Chen, X. Ji, T. P. Pollard, X. Lu, C. J. Sun, S. Hou, Q. Liu, C. Liu, T. Qing, Y. Wang, O. Borodin, Y. Ren, K. Xu, C. Wang, *Nature* **2019**, 569, 245.

[4] H. Bi, X. Wang, H. Liu, Y. He, W. Wang, W. Deng, X. Ma, Y. Wang, W. Rao, Y. Chai, H. Ma, R. Li, J. Chen, Y. Wang, M. Xue, *Adv. Mater.* **2020**, 32, 2000074.

[5] Z. Hou, X. Zhang, H. Ao, M. Liu, Y. Zhu, Y. Qian, *Mater. Today Energy* **2019**, 14, 100337.

[6] J. Xie, Z. Liang, Y. C. Lu, *Nat. Mater.* **2020**, 19, 1006.

[7] D. Chao, W. Zhou, C. Ye, Q. Zhang, Y. Chen, L. Gu, K. Davey, S. Z. Qiao, *Angew. Chem. Int. Ed.* **2019**, 58, 7823.

[8] a) D. Chao, C. Ye, F. Xie, W. Zhou, Q. Zhang, Q. Gu, K. Davey, L. Gu, S.-Z. Qiao, *Adv. Mater.* **2020**, 32, 2001894; b) C. Liu, X. Chi, Q. Han, Y. Liu, *Adv. Energy Mater.* **2020**, 10, 1903589.

https://doi.org/10.1142/9789811278327_0008

Chapter 8

# Opportunity for Eutectic Mixtures in Metal-Ion Batteries[1]

Mingming Han,[a] Jiang Zhou,[b] Hong Jin Fan[a,*]

*aSchool of Physical and Mathematical Sciences,
Nanyang Technological University, Singapore 637371, Singapore*
*bSchool of Materials Science and Engineering,
Central South University, Changsha 410083, China*

Breakthroughs in super-concentrated electrolytes have pushed the aqueous solution to the forefront of high-safety battery devices. An ideal electrolyte system should be cost-effective and stable in a wide electrochemical window. In recent years, eutectic mixtures have emerged as a green, safe, low-cost, and electrochemically stable electrolyte system for rechargeable metal-ion batteries (MIBs). In this chapter, the fundamental understanding of the formation mechanisms, physio-chemical properties, and composition-structure-property relationships of eutectic mixtures are summarized. Our focus is their advanced function and applications in MIBs. Considering that eutectic mixtures in MIBs are still at an early stage, we provide the challenges and perspectives which hopefully may guide the rational design of advanced eutectic mixtures for different electrochemical energy storage and conversion systems.

## 8.1 The Emergence of Eutectic Mixtures

Effort towards global decarbonization has led to research on renewable electrochemical energy storage and conversion systems (EESCs) to secure a sustainable future. State-of-the-art lithium-ion batteries (LIBs) are approaching their energy density limit yet are challenged by ever-increasing costs and safety concerns.[1] Better energy management systems beyond the horizon of LIBs are urgently needed to meet the energy demands

---

* Corresponding author: fanhj@ntu.edu.sg
[1] Adapted with permission from M. Han, J. Zhou, H. J. Fan, *Trends Chem.* **2023**, 5, 214.

of future generations.[2] Water-based technologies such as metal-ion batteries (MIBs), metal-O$_2$/S batteries, redox-flow batteries, and electrochemical supercapacitors are competitive alternatives due to their relatively low cost and ease of operation.[3] However, the relatively low energy density and insufficient cycling stability stemming from the narrow operating voltage range and severe side reactions of the aqueous system exclude them from "high-performance rechargeable batteries".

An expanded voltage window to 3 V and beyond has been achieved in super-concentrated and aqueous/organic hybrid (>21 m LiTFSI, mol kg$^{-1}$; 14 M LiTFSI/dimethyl carbonate, mol L$^{-1}$) electrolytes. But these electrolyte systems have high cost and limited reaction kinetics.[4] Quasi- and all-solid-state electrolytes offer an avenue to enlarge the operation window and avoid side reactions, but struggle with low conductivity and high interfacial impedance.[5] Solvent engineering, such as molecular crowding and surfactant/organic molecule addition, gives common low-concentrated electrolytes a chance to suppress solvent decomposition.[6] Nonetheless, they still suffer from cathode dissolution and side reactions. Electrolyte-decoupling strategy separates the electrolyte solution on the cathode from that on the anode by using special ion-selective membranes for charge balance.[7] This enables the optimal redox chemistry of cathode and anode and broadens the operating voltage, but introduces additional challenges in manufacturing, welding, and housing. Molten salt electrolytes of ionic liquids (ILs) overcome safety concerns and widen the operating voltage. They have been widely applied in EESCs because of their thermal stability, fire-retardant properties, and widened electrochemical window.[8] In the past 20 years, IL-derived eutectic mixtures have attracted significant attention because they share many characteristics with ILs. Eutectic mixture-based rechargeable MIBs may provide opportunities for next-generation EESCs (Figure 8.1). However, it is suggested that these two types of liquids are vastly different especially in their formation mechanisms and physio-chemical properties.[9]

This chapter presents a systematic discussion of the underlying fundamentals and advanced functions of eutectic mixtures as well as their applications in MIBs. Progress in feasible design, tunable composition, thermal/electrochemical stability, and safety issue of eutectic mixtures are analyzed. Compared to prior reviews on eutectic electrolytes in battery applications,[9b,10] here we focus on advanced functions of the

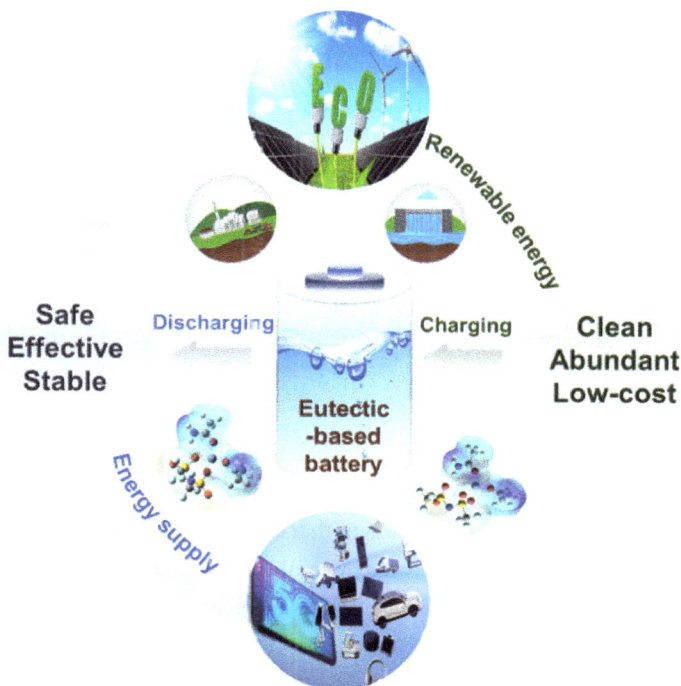

**Figure 8.1.** Eutectic mixture-based MIBs as potential EESCs.

eutectic mixtures in MIBs, including expanding the electrochemical stability window, stabilizing electrode/electrolyte interfaces, and improving the charge conductance. Additionally, the present perspective complements existing reviews by providing a comprehensive summary on the current challenges and explicit suggestions for future research. Specifically, we propose that a judicious design of eutectic mixtures should consider the solvation structure, ion transfer mechanism, and electrochemical reaction pathway, for which advanced experimental characterizations combined with multiscale simulations will be helpful.

## 8.2 Fundamentals of Eutectic Mixtures

ILs are known as room-temperature molten salts that contain organic cations and anions, such as $Cl^-$, $I^-$, $BF_4^-$, and $CF_3SO_3^-$. They share similar features with simple and high-melting-point salts such as NaCl. In the early stages, room-temperature ILs were prepared by mixing metal chlorides ($ZnCl_2$, $AlCl_3$, $SnCl_2$, $FeCl_3$) with quaternary ammonium salts.[8a] In the following years, ILs combining imidazolium-based organic cations with different types of anions ($Cl^-$, $BF_4^-$, $PF_6^-$, $NTf_2^-$) emerged because of the increased interests in metal-free green chemistry. These are classified as "green solvents" because of the possibility of recycling.[10a] Abbott and colleagues first discovered a series of eutectic mixtures by combining urea with choline chloride at room temperature.[11] ILs still face challenges of toxicity, poor biodegradability, and high cost. In addition, their synthesis is not environmentally friendly as large amounts of salts and solvents are required to completely exchange the anions. These drawbacks limit their large-scale applications.

Deep eutectic solvents (DESs) are a new subclass of eutectic mixtures, which can be synthesized by simply mixing two or more low-cost and bio-friendly components. For example, ChCl is a widely used component that can be extracted from biomass. Many types of eutectic mixtures can be prepared by combining with urea, renewable carboxylic acids (such as oxalic, citric, succinic) or polyols (such as glycerol and carbohydrates). DESs share some physio-chemical features with ILs (e.g., thermally/electrochemically stable, low vapor pressure, composition tunability, and wide operation voltage), but their low ecological footprint and cost-effectiveness render more opportunities as the ion/charge transfer media in EESC devices.

## 8.2.1 Definition and Formation Mechanism of Eutectic Mixtures

A eutectic mixture is composed of two or more components, which associate with each other when the intermolecular interactions are sufficiently strong compared to the internal reactions. The type, structure, dielectric constants, dipole moments and molar ratios of the components determine the intermolecular interactions.[12] Fundamentally, a eutectic system has a lower freezing point ($T_f$) than that of an ideal mixture, as shown in the binary phase diagram (see Figure 8.2a). $\Delta T_f$ is the difference between the lowest freezing point and the theoretical value of the ideal mixture.[9b] A larger $\Delta T_f$ means stronger intermolecular interactions. Generally, eutectic occurs at the lowest freezing point and the intermolecular interaction is stronger than that in individual components.

Up to now, the well-accepted intermolecular reactions in eutectic mixtures include van der Waals interaction, hydrogen bond interaction, and Lewis acid-base interaction

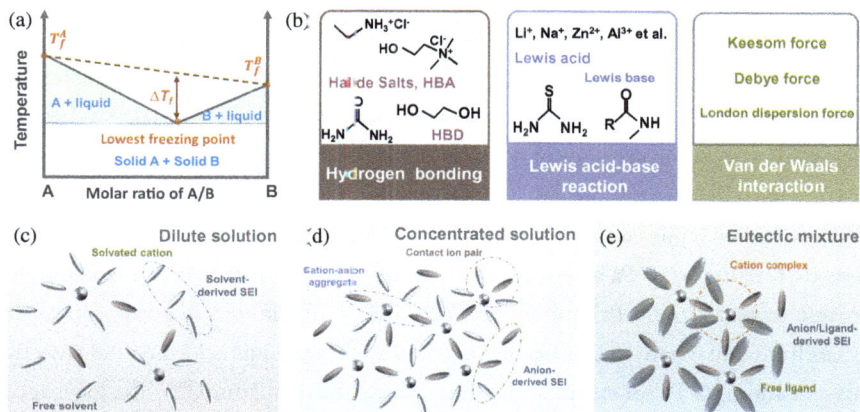

**Figure 8.2.** Fundamentals of eutectic mixtures. (a) Typical binary phase diagram to define a two-component eutectic system. (b) Formation mechanisms. (c–e) Solvation structures of dilute, concentrated and eutectic mixtures.

(Figure 8.2b).[9a] Particularly, van der Waals interaction is a kind of weak chemical force that exists between any two molecules. While the formation of eutectic mixture involves multiple types of interactions, van der Waals interaction dominates the effect of tuning the physio-chemical property of eutectic mixtures. Hydrogen bonding effect takes place by delocalizing charge densities between hydrogen bond acceptor (HBA) and hydrogen bond donor (HBD).[13] The HBAs are usually metal halides, choline chloride (ChCl), or quaternary ammonium salts; HBDs include carboxylic acids, urea, sugar-derived polyols, amide and so on.[14] In the Lewis acid-base theory, the component that accepts electrons is the Lewis acid, and that which donates electrons is the Lewis base. Hence, all the cations and electron-deficient chemicals (e.g., $BF_3$, $PF_3$) can be defined as Lewis acids; anions and electron-rich chemicals (e.g., $NH_3$, $PH_3$, and $C_6H_5CH_2NH_2$) are Lewis bases. It is believed that only the adjacent chemical environment of the eutectic components can be altered, which will not be accompanied by formation of new chemical bonds. Different from the solvated structures in dilute solutions and ionic aggregates in concentrated solutions,[15] large cation-anion coordinated complexes dominate eutectic mixtures (Figure 8.2c–e). Various metal ion-based eutectic mixtures have been applied to batteries and supercapacitors. Note that the Lewis base ligands can enter the solvation shells of metal ions due to the large mutual binding energy.

## 8.2.2 *Vapor Pressure and Non-flammability*

In general, the higher the saturated vapor pressure of a eutectic mixture, the higher the volatilization rate at the liquid/vapor interface.[16] While there are few studies on measuring the vapor pressure of eutectic mixtures, it is accepted that the eutectic mixture has a lower vapor pressure ($P_{vap}$) than common water or organic solvents. For example, it was reported that the $P_{vap}$ value of LiTFSI/N-methylacetamide (NMAc) eutectic mixture is 50 Pa at 40°C, which is much lower than that of the common organic solvent (10200 Pa) and water (7380 Pa).[17] In addition, the vapor pressure differs from one system to another, being 540.9 Pa for decanoic acid-menthol eutectic mixture and 55.5 Pa for decanoic acid-lidocaine eutectic mixture.[16b] The low vapor pressure, combined with the high thermal stability, assures the non-flammability of the eutectic mixture.

## 8.2.3 *Viscosity and Conductivity*

The viscosity of eutectic mixtures is usually higher than that of ILs and molecular-based solvents due to the intrinsic large complex radius and almost no free volume.[18] Generally, the viscosity is largely dependent on intermolecular interactions and temperature. For example, the viscosity in LiTFSI/NMAc (78.38 mPa·s) is lower than that in LiNO$_3$/NMAc (107.19 mPa·s) because of the larger ionic size and more delocalized bond of Li$^+$–TFSI$^-$ than Li$^+$–NO$_3^-$. Replacing NMAc by acetamide (Ace), the viscosity increases because of the stronger hydrogen bonding interaction between Li$^+$ and Ace.[19] While low viscosity is achievable at high temperatures, it is suggested that the viscosity of eutectic mixtures could be as high as 1000 Pa·s when the hole size is smaller than the ion size at low temperatures.[9a] Hence, the rational design of eutectic mixtures with low viscosity for battery applications under room-temperature condition is urgently needed.

Ionic conductivity is another fundamental character of eutectic mixtures. In the hole theory for eutectic mixtures, it is hypothesized that ionic substances contain holes with irregular sizes, and ions with smaller sizes than the holes will move into the empty spaces.[20] Generally, the hole space increases with temperature. It is reported that the ionic motion in Zn(TFSI)$_2$/Ace eutectic mixture is temperature dependent in the range of 25–80°C.[21] Despite the fact that eutectic mixtures generally deliver slow ion diffusion kinetics due to the high viscosity, large cation species and special transfer mechanism, the rate performance of MIBs is reasonably good. Moreover, molecular interactions cannot be ignored for the ionic mobility property. It is proved that the molar ratios, structure, and polarity of eutectic components determine the ionic conduction properties. For example, LiTFSI-Ace eutectic mixture delivers an ionic conductivity of 1.07 mS cm$^{-1}$,[19] but a decreased value of 0.37 mS cm$^{-1}$ in Zn(TFSI)$_2$-Ace mixture was observed due to the stronger Coulombic interaction among Zn$^{2+}$, Ace and TFSI$^-$ than that among Li$^+$, Ace and TFSI$^-$.[21] However, it increases to 2.85 mS cm$^{-1}$ in LiTFSI-SN mixture due to the higher polarity of SN which facilitates the dissociation of LiTFSI.[10b] In general, low surface tension, large hole voids, and low viscosity are favorable factors for ionic conductivity.

## 8.3 Functionality of Eutectic Mixtures for Metal-ion Batteries

Owing to the above merits, eutectic mixtures offer opportunities for the development of MIBs with various metal cations (such as Li$^+$, Zn$^{2+}$, Mg$^{2+}$, Al$^{3+}$). A range of metal-plating processes (e.g. Zn, Sn, Cu, Ni, Al) in eutectic mixtures have been reported, and nanomaterials with different morphologies synthesized in eutectic mixtures have been used as efficient electrocatalysts in air batteries or fuel cells.[22] While eutectic mixtures have the overall advantages of high energy densities, safety, and cycling stability,[9b] the performance of organic/aqueous-based MIBs are still not sufficiently good for commercialization due to one or more reasons including use of organic solvent, dendrite formation, and water decomposition. The following section discusses the functions of eutectic mixtures in MIBs, which includes expanding the electrochemical stability window (ESW), stabilizing the interface, and improving the ion conductivity (Figure 8.3).

### 8.3.1 *High Level of Safety*

Traditional organic electrolytes have fire hazard due to the flammable and volatile solvents.[23] A eutectic mixture uses non-flammable substances, and the strong cation-anion-ligand molecular interaction reduces the volatility. For example, LiTFSI/NMAc eutectic mixture can keep the liquid phase in the temperature range from –72 to 150°C.[17] The solution shows obviously improved ionic conductivity with increasing temperature (1.61 at room temperature and 28.4 at 150°C), and the assembled LiFePO$_4$//Li battery maintains the capacity at 80°C. More evidence is provided by the Cui group,[24] who studied the SN-based dual-anion (LiDFOB and LiTFSI) eutectic mixture (D-DES) for Li metal battery. It is demonstrated that there is a negligible weight loss of the eutectic mixture at 150°C, and both the liquid and vapor phase D-DES are non-flammable. Furthermore, the anion-derived solid electrolyte interphase (SEI) could circumvent parasitic side reactions and avoid potential short circuit. A significantly improved interface stability during Li plating in D-DES can be achieved through forming SEI layer, which restricts fatal side reactions associated with dendrites, gas release and Li metal corrosion.

### 8.3.2 Electrochemical Stability Window and Energy Density of Metal-ion Batteries

Aqueous electrolytes intrinsically have narrow ESW especially when free water is involved. That is the reason aqueous-based batteries usually correspond to low energy densities. Based on the chosen aqueous electrolytes, it may also be challenging to search for suitable anodic electrodes with low electrochemical potentials and high capacities.[25] As illustrated in Figure 8.3a, eutectic mixtures with wide ESWs could be an ideal selection to address this challenge, as proposed also by Mai, Zhou and coworkers.[9b] For example, LiTFSI-MAc and LiNO$_3$-MAc eutectic mixtures show ESWs of 5.3 and 4.7 V (vs. Li), respectively (passivated Al collector).[17] LiTFSI-methyl carbamate and LiPF$_6$-methyl carbamate eutectic mixtures at 1:5 molar ratios give an ESW of 3.2 and 4 V, respectively.[26] The explanation for the 3.5 V-window Li$_4$Ti$_5$O$_{12}$//LiMn$_2$O$_4$ full cell rendered by the methylsulfonylmethane (MSM)-lithium perchlorate-water (MSM/LiClO$_4$/H$_2$O) eutectic mixture[27] was that all water molecules participate in constructing the primary solvation sheath, leading to weakened activity of water and increased overpotentials of hydrogen/oxygen evolution. Yamada and coworkers[28] put forward a suggestion that the water molecules in Li(TFSI)$_{0.7}$(BETI)$_{0.3}$·2H$_2$O hydrate-melt electrolyte are quite close to the state of crystalline hydrates. The Lewis acid-base reaction between water and Li$^+$ limits the possible hydrogen bonding reaction, giving rise to expansion of the voltage window of 2.4 V in Li$_4$Ti$_5$O$_{12}$//LiCoO$_2$ and 3.1 V in Li$_4$Ti$_5$O$_{12}$//LiNi$_{0.5}$Mn$_{1.5}$O$_4$ full cells. The corresponding Li$_4$Ti$_5$O$_{12}$//LiCoO$_2$ full cell

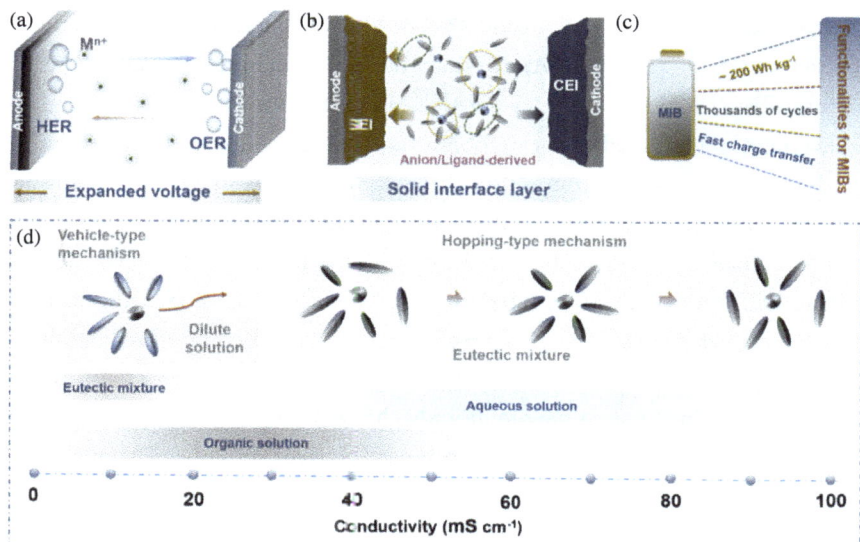

**Figure 8.3.** Functionality of eutectic mixtures in MIBs: (a) Expansion of the ESW. (b) Formation of the solid interface layer on the cathode and anode surfaces. (c) Battery performance enhancement by eutectic mixtures. (d) A summary of ionic transfer mechanism and ionic conductivity in different electrolyte systems.

delivers an actual energy of 130 Wh kg$^{-1}$ (55.3 Ah kg$^{-1}$ and 2.35 V on average), which is comparable to the commercial non-aqueous Li$_4$Ti$_5$O$_{12}$//LiMn$_2$O$_4$ battery with theoretical energy density of 160 Wh kg$^{-1}$. Very recently, Wang *et al.* used dilute CO(NH$_2$)$_2$ to prepare LiTFSI-KCH-CO(NH$_2$)$_2$-H$_2$O ternary eutectic mixture,[4a] and found that both the LiTFSI and CO(NH$_2$)$_2$ can reduce water activity and stabilize H$_2$O molecules due to the strong hydrogen bonding interactions. Thus, the ESW of LiMn$_2$O$_4$//Li$_4$Ti$_5$O$_{12}$ 18650-type cell is expanded to 3.3 V with an energy density of 103 Wh kg$^{-1}$ for 2.5 mA h cm$^{-2}$.

## 8.3.3 *Electrochemical Stability*

Generally, eutectic mixtures, including the case of hydrated eutectic or water-in-eutectic mixtures, exhibit better long-term cycling stability than aqueous counterparts. It is reported that the Li//Li cell could cycle more than 2,200 hours and the Zn//LMO full cell could keep running for 600 cycles in LZ-DES-2H$_2$O eutectic mixture.[29] In another report, Zn//NVO full cell cycled for 800 hours in the Zn(ClO$_4$)$_2$·6H$_2$O/MSM/H$_2$O eutectic mixture.[30] Apart from Li- and Zn-based batteries, recent research reports the long-term cycling stability of Al metal anode in chloroaluminate-based eutectic mixtures. The Al//graphite cell using the AlCl$_3$-urea eutectic mixture shows a charging/discharging Coulombic efficiency of 99.7% and cycling stability up to 200 cycles.[31] In addition to the common insertion/extraction reactions, irreversible electrode/electrolyte interface reactions also occur in eutectic-based MIBs, leading to an SEI layer that enhances the electrochemical stability.[4b] The anion-ligand complex-derived SEI and CEI (cathode electrode interface) layers are helpful to mitigate the issues of metal anode corrosion, volume changes of electrodes, and electrolyte decomposition (Figure 8.3b).[32] For example, the TFSI$^-$ anion could be reduced to ZnF$_2$ below 0.37 V (vs. Zn/Zn$^{2+}$) due to the significantly changed coordination environment in the Zn(TFSI)$_2$-Ace eutectic mixture. An inner LiF layer could be generated from TFSI$^-$ anion at high potential, while the CO(NH$_2$)$_2$ polymerizes to polyurea at the outer surface at low potential in LiTFSI-KOH-CO(NH$_2$)$_2$-H$_2$O eutectic mixture.[4] Anode interfacial layers composed of Li$_x$BF$_y$O$_z$, LiF, Li$_3$N, Li$_2$N$_2$O$_2$ and sulfides could be formed at lithium metal side due to the interactions between TFSI$^-$/DFOB$^-$ anions and lithium metal. Similarly, cathode interfacial layers comprising B–F, B–O, N- and S-rich components resulting from the reactions among TFSI$^-$ anion, DFOB$^-$ anion and SN ligand have also been verified in D-DES.[24] The *in situ* formed SEI and CEI layers, which are mechanically robust and ionically permeable, do not necessarily deteriorate the ionic diffusion in eutectic mixtures and charge transfer at the electrode/electrolyte interface. Instead, the fast ionic diffusion kinetics is an advantage offered by eutectic mixtures (Figure 8.3d).

## 8.4 Suggestions for Future Development

Although the concept of eutectic mixtures is seemly straightforward, the development of green, low-cost, and high-performance eutectic mixtures is still challenging. The large coordination species, high viscosity and low ionic conductivity of eutectic mixtures may make dissociation and diffusion kinetically sluggish. As a rule of thumb, the design of optimal eutectic mixtures should consider the solvation structure, ion transfer mechanism and electrochemical reaction pathway, as elaborated in the following sections.

## 8.4.1 *Explore a Diversity of Salts and Ligands*

In contrast to common dilute aqueous/organic systems where a fixed composition is required, the diverse composition of eutectic mixtures is a major uniqueness.[33] Future design should carefully consider the charge density and molecular weight of metal cations, the functional group and spatial structure of ligands, and the ratio between metal salt and ligand combinations.[34] The ternary liquidus phase diagrams of Na/K(PTFSI)$_x$(TFSI)$_y$(OTf)$_z$/H$_2$O systems (Figure 8.4a and 8.4b) suggest that the ratios between the mixed components have a major effect on the physio-chemical properties.[35] Hence, the ternary mixture should be optimized to achieve better solubility and lower water content than those in single or binary mixtures. In addition, the ESWs of eutectic mixtures vary from amide-ligand based system to nitrile-, urea- and MSM-based systems, and from TFSI$^-$ anion-based system to (CF$_3$SO$_2$)C$^-$, NO$_3^-$, ClO$_4^-$, and OAc$^-$ anion-based systems. One can introduce common additives such as H$_2$O, dimethyl sulfoxide, ethylene carbonate, dimethyl carbonate, fluoethylene carbonate, and ethylene glycol

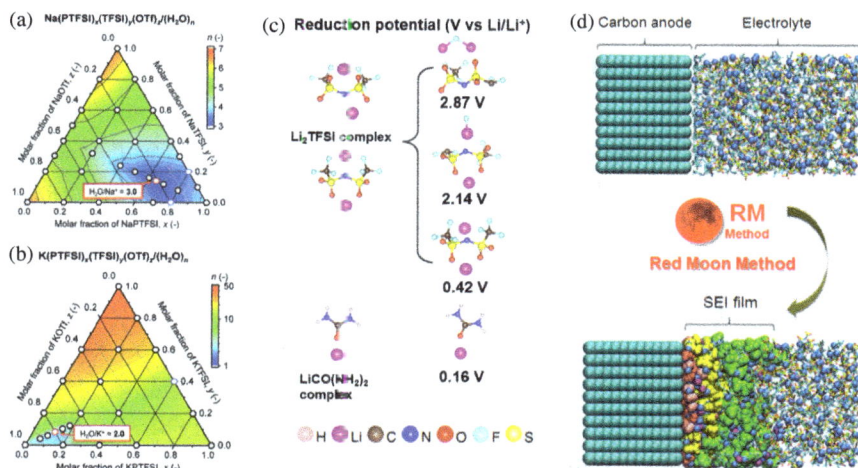

**Figure 8.4.** Examples of current studies on eutectic mixtures. (a, b) Preparation of the Na/K(PTFSI)$_x$(TFSI)$_y$(OTf)$_z$/H$_2$O eutectic mixtures, where $x + y + z = 1$, $n$ represents the molar ratio between water and salt, and the color gradation represents the water content. Adapted with permission from Ref. [35]. (c) Predicted reduction potential of the Li$_2$TFSI complex by quantum chemistry calculations. Adapted with permission from Ref. [4a]. (d) Typical snapshots of the SEI formation process using a hybrid Monte Carlo/molecular dynamics reaction method. Adapted with permission from Ref. [40].

to modulate the viscosity, conductivity and ESW by altering the molecular interaction in a eutectic mixture.[9a,19,36] In principle, metal salts and organic molecules that are inactive in conventional electrolytes could be suitable components in eutectic mixtures.

A large variety of metal salt-ligand combinations and additives should be employed to achieve eutectic mixtures with various functions.[12,37] By designing cosolvent, hydrated, and solid-state eutectic solvent mixtures, one may achieve non-flammability, high conductivity, stable interface, and mechanical strength.[38] For example, the water molecules can enter the $Zn^{2+}$ sheath, leading to a reduced nucleation overpotential and homogeneous Zn deposition. Cui *et al.* prepared a solid-state eutectic mixture through solidifying $Zn(TFSI)_2$-Ace eutectic mixture using nano-$TiO_2$.[39] Due to the preferential Lewis acid-base reaction between $TFSI^-$ and $TiO_2$, the ionic associations in liquid eutectic mixtures are weakened, resulting in $Zn^{2+}$-conducting channels on $TiO_2$ surfaces. The obtained solid-state eutectic mixture shows relatively high $Zn^{2+}$ conductivity (0.0378 mS cm$^{-1}$ at 30°C) and compatibility with Zn metal anode.

## 8.4.2 *Deepen the Fundamental Understanding*

Research on eutectic mixtures is still in its early stage since the first publication in 2001. The fundamentals of solution structure, electrical structure, charge transfer mechanism and interface chemistry are still not well understood. In common dilute solutions, the metal salts can be fully dissolved and thus metal cations are solvated with solvents. However, in most cases, a eutectic mixture is composed of large cation-anion complexes. For example, the $ZnCl_2$-urea eutectic mixture can have six types of ions, including $[ZnCl_3]^-$, $[Zn_2Cl_5]^-$ and $[Zn_3Cl_7]^-$ anions, and $[ZnCl(urea)]^+$, $[ZnCl(urea)_2]^+$, and $[ZnCl(urea)_3]^+$ cations.[11] And the $Zn^{2+}$ species in the $Zn(TFSI)_2$-Ace mixture include $[ZnTFSI(Ace)_2]^+$, $[ZnTFSI(Ace)_2]^+$, $[ZnTFSI(Ace)_3]^+$ and $[ZnTFSI(Ace)_4]^+$.[21] A comprehensive understanding of the solution structure may require multiscale computational methodologies in combination with advanced experimental characterizations techniques. Nuclear magnetic resonance spectroscopy, neutron scattering, fluorescence spectroscopy, X-ray scattering and dynamic light scattering are important techniques for characterizing the micro-environmental polarity, structure, interactions, solvation and size distributions in eutectic mixtures. In addition, simulation and modeling can be used, which may include *ab initio* molecular dynamics

(MD), density-functional tight-binding MD, classical MD, machine learning, and the combination of Monte Carlo and molecular dynamics (MC/MD).[15]

In addition to solvation structure, fundamental studies on the ionic conduction mechanism in eutectic mixtures are also needed. The frontier orbital interaction and the electronic structure, which correlate closely with electrochemical stability and interface reactions, have been extensively employed for dilute or concentrated aqueous/organic solutions.[41] However, they may not exactly apply to eutectic mixtures. In theory, solvated metal cations can move independently in traditional dilute solution.[35] The "super-ionicity" feature of both $Na(PTFSI)_{0.65}(TFSI)_{0.14}(OTf)_{0.21}\cdot 3H_2O$ and $K(PTFSI)_{0.12}(TFSI)_{0.08}(OTf)_{0.8}\cdot 2H_2O$ suggests a high ionic conductivity of the eutectic mixture. While the conventional vehicle-type mechanism and the hopping-type mechanism offer guidelines to the ionic transport property in dilute and concentrated electrolytes, respectively (Figure 8.3d), a clear picture of the ionic transport mechanism of eutectic mixtures is not yet available and should be investigated.

The third key point is the electrode-electrolyte interface chemistry. According to the frontier molecular orbital concept, SEI forms when the ESW is not sufficient to cover the redox potential of the cathode and anode.[42] As for the eutectic mixtures, the interface chemistry is different and unclear. First, it is not definite if the LUMO (the lowest unoccupied molecular orbital)-HOMO (highest occupied molecular orbital) law is applicable in eutectic mixtures. The current *in situ* experimental detection and simulation techniques for theoretical prediction are insufficient to address this problem. Recently, the SEI formation mechanism has been discussed by density functional theory calculation (Figure 8.4c). Mixed simulation and modeling methods, such as MC/MD (Figure 8.4d),[40] could be used to unravel the SEI formation process. In summary, some fundamental questions about the generality of eutectic structure, theories of charge transport and interface reactions remain unsolved and should be the focus of future research.

### 8.4.3 *Effect of Water Upon Eutectic Structure*

The fundamental formation mechanism of hydrogen bond interaction of eutectic mixture makes it strongly water-miscible and hygroscopic. Generally, latent absorbed

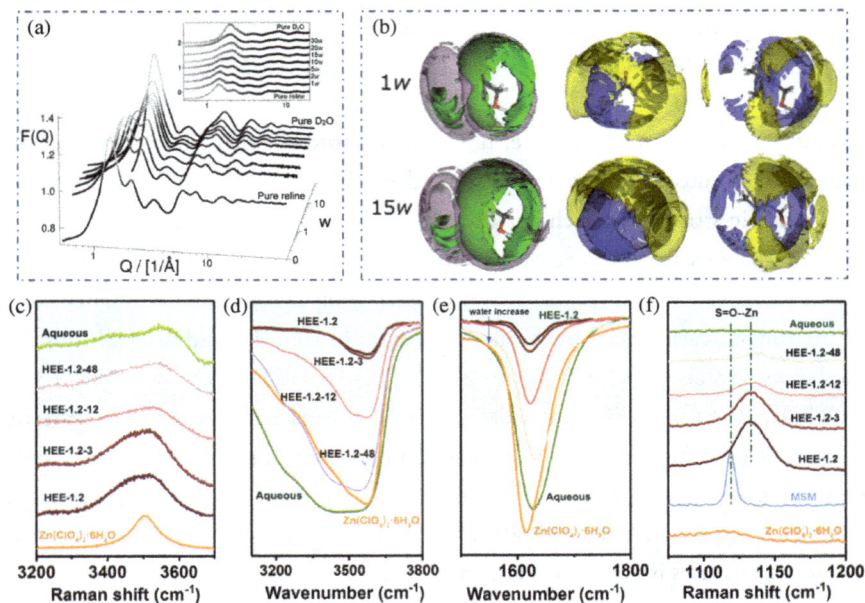

**Figure 8.5.** The effect of water on the eutectic structure. (a) Experimental neutron diffraction results in 3D and 2D plots of the perdeuterated eutectic mixtures. (b) SDF plots describing 3D nanostructure of reline-1w and 15w. Iso-surfaces denote chloride (green), urea (lilac), choline (yellow) and water (blue) molecules at the 7.5% probability level. Adapted with permission from Ref. [43] (c, f) Raman and (d, e) FTIR spectra of the HEE system. Adapted with permission from Ref. [44].

water under natural environment is unavoidable and the water content impacts their physico-chemical properties such as viscosity, ion conductivity, freezing point and solvation structure. It is proved that there is an intermolecular-scale nanostructural transition from eutectic state to aqueous state in choline chloride-urea mixture when the extra water level increases.[43] Neutron diffraction data (Figure 8.5a) highlights a contraction in the major intermolecular interaction length from 4.3 Å in the pure DES to 3.1 Å, the value found for water. This process gradually goes up to 10w, and then suddenly to 15w (xw is defined as the molar ratio of eutectic mixture to water). 3D spatial density function (SDF) plots (Figure 8.5b) demonstrate that the orientations of urea and choline are retained at 1w but choline-choline structure is affected by the strong water interaction and the hydration of eutectic mixture increases with water volume. Furthermore, the urea molecule has a saturated first hydration shell with increase in crowding choline at 15w. Thus, the breakdown of the eutectic network is related to the point where the eutectic mixture-water interaction is stronger than the original eutectic interaction. In 2022, Han *et al.* systematically studied the physico-chemical behaviors of $Zn(ClO_4)_2 \cdot 6H_2O/MSM/H_2O$ hydrated eutectic electrolyte (HEE) with different water content. It is shown that the ion conductivity improves with water content and reaches 12.13 mS cm$^{-1}$ when the molar ratio of $Zn(ClO_4)_2 \cdot 6H_2O$, MSM and $H_2O$ is 1.2:3.6:48. Raman and Fourier transform infrared (FTIR) data (Figure 8.5c–e) show that the hydrogen bond proportion in HEE is obviously lower than that in aqueous solution. Meanwhile, there is almost no bulk water molecule in the HEE-1.2-3 (molar ratio: $Zn(ClO_4)_2 \cdot 6H_2O:MSM:H_2O$ = 1.2:3.6:3); instead they are isolated from each other forming the $Zn-H_2O$ coordinated ionic phases. It is noteworthy that the peak intensity associated with $Zn^{2+}$–S=O bond decreases gradually with water content (Figure 8.5f), suggesting the extra water molecules could disturb the original Zn-MSM coordination structure. Overall, the influence of water content on eutectic mixture remains poorly understood and hard to control.

## 8.5 Conclusions

Eutectic mixtures may provide solutions to many challenging issues associated with the conventional aqueous or organic dilute/concentrated electrolyte systems. The merits of eutectic mixtures include high thermal stability, realizing high-safety operation at high temperature, low vapor pressure without flammable potential, and wide voltage window up to 5 V (vs. Li). These advantageous features enable the use of high-voltage cathodes, side reaction suppression, and cycle life (Figure 8.6a). The extension of this concept to aqueous battery technologies may lead to the development of high-energy-density batteries, particularly the aqueous lithium battery towards an energy density up to 200 Wh kg$^{-1}$. High power performance is also possible as fast kinetics in electrode reaction process in eutectic mixtures is achievable despite the fact that the ionic transport mechanism is still unclear. Apart from the reported vehicle- and hopping-type mechanisms, there might be alternative mechanisms to describe the charge transfer process more accurately. Recently, researchers have demonstrated that the anion-derived SEI layer can not only expand the operational voltage window, but also accelerate the charge transfer and uniformize the metal deposition (e.g., Li, Zn and Al).[12,21,27] Hence, the charge transfer and SEI formation mechanism associated with eutectic mixtures should be deeply investigated using advanced characterization techniques and theoretical calculations.

(a)

(b)

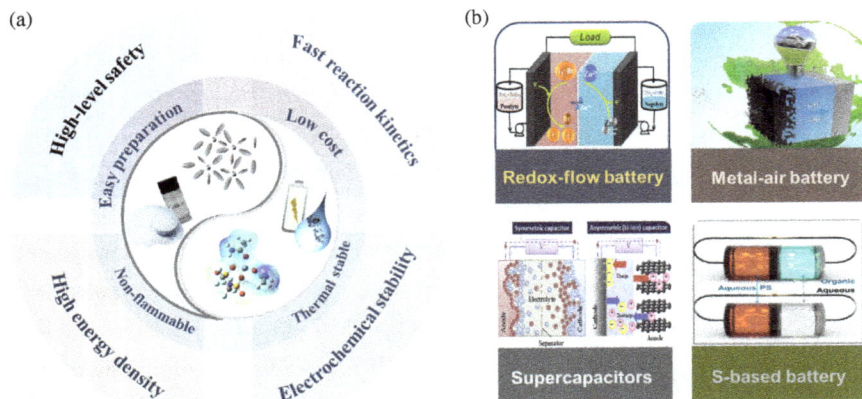

**Figure 8.6.** Conclusion and perspectives on eutectic mixtures in EESCs. (a) Advantages and functionalities of eutectic mixtures for MIBs. (b) Potential application of eutectic mixtures in other battery and supercapacitor technologies. Adapted with permission from Ref. [46].

Eutectic mixtures have not yet been widely applied in various MIBs (e.g., Na⁺, K⁺, Mg²⁺, Ca²⁺). Although there are some attempts in using Na salts containing eutectic mixtures in electrochemical capacitors,[45] their application in Na-ion based batteries has not been reported. NaTFSI and KTFSI are potential salts to constitute eutectic mixtures and the TFSI⁻ anion could boost the formation of F-containing SEI layer. Eutectic mixtures based on Na- and K-salts and ligand molecules are suggested for Na/K ion batteries with high stability. Recently, multi-valent MIBs have attracted broad attentions due to their potential in providing high energy densities. In these batteries, metals can be directly used as the anode. However, an issue for Mg, Ca and Al metal anodes is that the surfaces are possibly passivated due to their high activity with atmospheric components, protic solvents, and salt anions. The intrinsic chemical coordination environment and functional interface chemistry of eutectic mixtures could benefit the multi-valent metal ion battery in terms of operational stability. Additionally, it is possible that the eutectic mixture provides a more effective and greener medium for MIB recycling under mild conditions, as compared to the conventional recycling methods from organic solvents.

In conclusion, these low-cost, facile-synthesis and high-performance eutectic mixtures with special intrinsic advantages enrich the electrolyte family, which offer opportunities for not only MIBs but also other energy storage systems including air batteries, redox-flow batteries, sulfur-based batteries, and supercapacitors (Figure 8.6b).

# References

[1] a) Z. Zhu, A. Kushima, Z. Yin, L. Qi, K. Amine, J. Lu, J. Li, *Nat. Energy* **2016**, 1, 16111; b) Z. Hou, X. Zhang, H. Ao, M. Liu, Y. Zhu, Y. Qian, *Mater. Today Energy* **2019**, 14, 100337; c) P. Ruan, S. Liang, B. Lu, H. J. Fan, J. Zhou, *Angew. Chem. Int. Ed.* **2022**, 61, e202200598.

[2] X. Lin, G. Zhou, M. J. Robson, J. Yu, S. C. T. Kwok, F. Ciucci, *Adv. Funct. Mater.* **2021**, 32, 2109322.

[3] a) H. Gao, B. M. Gallant, *Nature Rev. Chem.* **2020**, 4, 566; b) T. H. Gu, D. A. Agyeman, S. J. Shin, X. Jin, J. M. Lee, H. Kim, Y. M. Kang, S. J. Hwang, *Angew. Chem. Int. Ed.* **2018**, 57, 15984; c) T. Ma, Z. Pan, L. Miao, C. Chen, M. Han, Z. Shang, J. Chen, *Angew. Chem. Int. Ed.* **2018**, 57, 3158.

[4] a) J. Xu, X. Ji, J. Zhang, C. Yang, P. Wang, S. Liu, K. Ludwig, F. Chen, P. Kofinas, C. Wang, *Nat. Energy* **2022**, 7, 186; b) L. Suo, O. Borodin, T. Gao, M. Olguin, J. Ho, X. Fan, C. Luo, C. Wang, K. Xu, *Science* **2015**, 350, 938.

[5] a) Z. Wang, H. Li, Z. Tang, Z. Liu, Z. Ruan, L. Ma, Q. Yang, D. Wang, C. Zhi, *Adv. Funct. Mater.* **2018**, 28, 1804560; b) M. Wu, M. Zhang, L. Xu, C. Yang, M. Hong, M. Cui, B. C. Clifford, S. He, S. Jing, Y. Yao, L. Hu, *Matter* **2022**, 5, 3402.

[6] a) Y. Lv, Y. Xiao, L. Ma, C. Zhi, S. Chen, *Adv. Mater.* **2021**, 34, 2106409; b) C. Yang, J. Chen, T. Qing, X. Fan, W. Sun, A. von Cresce, M. S. Ding, O. Borodin, J. Vatamanu, M. A. Schroeder, N. Eidson, C. Wang, K. Xu, *Joule* **2017**, 1, 122.

[7] a) D. Chao, C. Ye, F. Xie, W. Zhou, Q. Zhang, Q. Gu, K. Davey, L. Gu, S. Z. Qiao, *Adv. Mater.* **2020**, 32, e2001894; b) C. Zhong, B. Liu, J. Ding, X. Liu, Y. Zhong, Y. Li, C. Sun, X. Han, Y. Deng, N. Zhao, W. Hu, *Nat. Energy* **2020**, 5, 440.

[8] a) M. Watanabe, M. L. Thomas, S. Zhang, K. Ueno, T. Yasuda, K. Dokko, *Chem. Rev.* **2017**, 117, 7190; b) S. Tang, J. Feng, R. Su, M. Zhang, M. Guo, *ACS Sustain. Chem. Eng.* **2022**, 10, 8423; c) T. Xuan, L. Wang, *Energy Storage Mater.* **2022**, 48, 263.

[9] a) E. L. Smith, A. P. Abbott, K. S. Ryder, *Chem. Rev.* **2014**, 114, 11060; b) L. Geng, X. Wang, K. Han, P. Hu, L. Zhou, Y. Zhao, W. Luo, L. Mai, *ACS Energy Lett.* **2022**, 7, 247.

[10] a) X. Lu, E. J. Hansen, G. He, J. Liu, *Small* **2022**, 18, e2200550; b) C. Zhang, L. Zhang, Y. Ding, X. Guo, G. Yu, *ACS Energy Lett.* **2018**, 3, 2875.

[11] A. P. Abbott, G. Capper, D. L. Davies, H. L. Munro, R. K. Rasheed, V. Tambyrajah, *Chem. Commun.* **2001**, 19, 2010.

[12] J. Kim, B. Koo, J. Lim, J. Jeon, C. Lim, H. Lee, K. Kwak, M. Cho, *ACS Energy Lett.* **2021**, 7, 189.

[13] M. K. Tran, M.-T. F. Rodrigues, K. Kato, G. Babu, P. M. Ajayan, *Nat. Energy* **2019**, 4, 339.

[14] A. P. Abbott, J. C. Barron, K. S. Ryder, D. Wilson, *Chemistry* **2007**, 13, 6495.

[15] Y. Yamada, J. Wang, S. Ko, E. Watanabe, A. Yamada, *Nat. Energy* **2019**, 4, 269.

[16] a) D. R. MacFarlane, M. Forsyth, P. C. Howlett, M. Kar, S. Passerini, J. M. Pringle, H. Ohno, M. Watanabe, F. Yan, W. Zheng, S. Zhang, J. Zhang, *Nat. Rev. Mater.* **2016**, 1, 15005; b) C. H. J. T. Dietz, J. T. Creemers, M. A. Meuleman, C. Held, G. Sadowski, M. van Sint Annaland, F. Gallucci, M. C. Kroon, *ACS Sustain. Chem. Eng.* **2019**, 7, 4047.

[17] A. Boisset, S. Menne, J. Jacquemin, A. Balducci, M. Anouti, *Phys. Chem. Chem. Phys.* **2013**, 15, 20054.

[18] J. Song, Y. Si, W. Guo, D. Wang, Y. Fu, *Angew. Chem. Int. Ed.* **2021**, 60, 9881.

[19] J. Wu, Q. Liang, X. Yu, Q. F. Lü, L. Ma, X. Qin, G. Chen, B. Li, *Adv. Funct. Mater.* **2021**, 31, 2011102.

[20] a) A. P. Abbott, *Chemphyschem* **2004**, 5, 1242; b) A. W. Taylor, P. Licence, A. P. Abbott, *Phys. Chem. Chem. Phys.* **2011**, 13, 10147.

[21] H. Qiu, X. Du, J. Zhao, Y. Wang, J. Ju, Z. Chen, Z. Hu, D. Yan, X. Zhou, G. Cui, *Nat. Commun.* **2019**, 10, 5374.

[22] a) Q. Zhang, K. De Oliveira Vigier, S. Royer, F. Jerome, *Chem. Soc. Rev.* **2012**, 41, 7108; b) S. Y. Zhang, Q. Zhuang, M. Zhang, H. Wang, Z. Gao, J. K. Sun, J. Yuan, *Chem. Soc. Rev.* **2020**, 49, 1726; c) D. Lee, H. I. Kim, W. Y. Kim, S. K. Cho, K. Baek, K. Jeong, D. B. Ahn, S. Park, S. J. Kang, S. Y. Lee, *Adv. Funct. Mater.* **2021**, 31, 2103850.

[23] Y. Sun, G. Zheng, Zhi W. Seh, N. Liu, S. Wang, J. Sun, Hye R. Lee, Y. Cui, *Chem* **2016**, 1, 287.

[24] Z. Hu, F. Xian, Z. Guo, C. Lu, X. Du, X. Cheng, S. Zhang, S. Dong, G. Cui, L. Chen, *Chem. Mater.* **2020**, 32, 3405.

[25] a) X. Li, M. Li, Z. Huang, G. Liang, Z. Chen, Q. Yang, Q. Huang, C. Zhi, *Energy Environ. Sci.* **2021**, 14, 407; b) S. D. Han, N. N. Rajput, X. Qu, B. Pan, M. He, M. S. Ferrandon, C. Liao, K. A. Persson, A. K. Burrell, *ACS Appl. Mater. Interfaces* **2016**, 8, 3021.

[26] N. Z. Hardin, Z. Duca, A. Imel, *ChemElectroChem* **2022**, 9, e202200628.

[27] P. Jiang, L. Chen, H. Shao, S. Huang, Q. Wang, Y. Su, X. Yan, X. Liang, J. Zhang, J. Feng, Z. Liu, *ACS Energy Lett.* **2019**, 4, 1419.

[28] Y. Yamada, K. Usui, K. Sodeyama, S. Ko, Y. Tateyama, A. Yamada, *Nat. Energy* **2016**, 1, 1.

[29] J. Zhao, J. Zhang, W. Yang, B. Chen, Z. Zhao, H. Qiu, S. Dong, X. Zhou, G. Cui, L. Chen, *Nano Energy* **2019**, 57, 625.

[30] M. Han, J. Huang, S. Liang, L. Shan, X. Xie, Z. Yi, Y. Wang, S. Guo, J. Zhou, *iScience* **2020**, 23, 100797.

[31] M. Angell, C. J. Pan, Y. Rong, C. Yuan, M. C. Lin, B. J. Hwang, H. Dai, *Proc. Natl. Acad. Sci.* **2017**, 114, 834.

[32] C.-C. Su, M. He, J. Shi, R. Amine, J. Zhang, K. Amine, *Angew. Chem. Int. Ed.* **2020**, 59, 18229.

[33] T. Liang, R. Hou, Q. Dou, H. Zhang, X. Yan, *Adv. Funct. Mater.* **2021**, 31, 2006749.

[34] K. Kim, Y. Ando, A. Sugahara, S. Ko, Y. Yamada, M. Otani, M. Okubo, A. Yamada, *Chem. Mater.* **2019**, 31, 5190.

[35] Q. Zheng, S. Miura, K. Miyazaki, S. Ko, E. Watanabe, M. Okoshi, C. P. Chou, Y. Nishimura, H. Nakai, T. Kamiya, T. Honda, J. Akikusa, Y. Yamada, A. Yamada, *Angew. Chem. Int. Ed.* **2019**, 58, 14202.

[36] J. Shi, T. Sun, J. Bao, S. Zheng, H. Du, L. Li, X. Yuan, T. Ma, Z. Tao, *Adv. Funct. Mater.* **2021**, 31, 2102035.

[37] A. P. Abbott, J. C. Barron, G. Frisch, K. S. Ryder, A. F. Silva, *Electrochim. Acta* **2011**, 56, 5272.

[38] a) H. Sano, H. Sakaebe, H. Matsumoto, *J. Electrochem. Soc.* **2011**, 158, A316; b) X. Zeng, J. Mao, J. Hao, J. Liu, S. Liu, Z. Wang, Y. Wang, S. Zhang, T. Zheng, J. Liu, P. Rao, Z. Guo, *Adv. Mater.* **2021**, 33, e2007416.

[39] G. Cui, H. Qiu, R. Hu, X. Du, Z. Chen, J. Zhao, G. Lu, M. Jiang, Q. Kong, Y. Yan, J. Du, X. Zhou, *Angew. Chem. Int. Ed.* **2021**, 61, e202113086.

[40] N. Takenaka, T. Fujie, A. Bouibes, Y. Yamada, A. Yamada, M. Nagaoka, *J. Phys. Chem. C* **2018**, 122, 2564.

[41] L. Miao, L. Liu, Z. Shang, Y. Li, Y. Lu, F. Cheng, J. Chen, *Phys. Chem. Chem. Phys.* **2018**, 20, 13478.

[42] T. Zhang, Y. Tang, S. Guo, X. Cao, A. Pan, G. Fang, J. Zhou, S. Liang, *Energy Environ. Sci.* **2020**, 13, 4625.

[43] O. S. Hammond, D. T. Bowron, K. J. Edler, *Angew. Chem. Int. Ed.* **2017**, 56, 9782.

[44] M. Han, J. Huang, X. Xie, T. C. Li, J. Huang, S. Liang, J. Zhou, H. J. Fan, *Adv. Funct. Mater.* **2022**, 32, 2110957.

[45] J. Feng, Y. Wang, Y. Xu, Y. Sun, Y. Tang, X. Yan, *Energy Environ. Sci.* **2021**, 14, 2859.

[46] a) G.-M. Weng, Z. Li, G. Cong, Y. Zhou, Y.-C. Lu, *Energy Environ. Sci.* **2017**, 10, 735; b) Y. Li, H. Dai, *Chem. Soc. Rev.* **2014**, 43, 5257; c) H. D. Yoo, E. Markevich, G. Salitra, D. Sharon, D. Aurbach, *Mater. Today* **2014**, 17, 110; d) C. Liu, W. Xu, C. Mei, M.-C. Li, X. Xu, Q. Wu, *Chem. Eng. J.* **2021**, 405, 126737.

Chapter 9

# Hydrogels Enable Smart Aqueous Batteries[1]

Jin-Lin Yang,[a] Peihua Yang,[b] Jia Li,[c]
Seok Woo Lee,[c,d] Hong Jin Fan[a,*]

[a]*School of Physical and Mathematical Science,
Nanyang Technological University,
Singapore 637371, Singapore*
[b]*The Institute of Technological Sciences, Wuhan University,
Wuhan 430072, China*
[c]*Rolls-Royce@NTU Corporate Lab, Nanyang Technological
University, Singapore 639798, Singapore*
[d]*School of Electrical and Electronic Engineering,
Nanyang Technological University,
Singapore 639798, Singapore*

The growing trend of intelligent devices ranging from wearables and soft robots to artificial intelligence has set a high demand for smart batteries. Hydrogels provide opportunities for smart aqueous batteries to self-adjust their functions according to the operating conditions. Despite the progress in hydrogel-based aqueous batteries, a gap remains between the designable functions of diverse hydrogels and the expected performance of batteries. In this chapter, we first briefly introduce the fundamentals of hydrogels, including formation, structure and characteristics of the internal water and ions. Batteries that operate under unusual mechanical and temperature conditions enabled by hydrogels are highlighted. Challenges and opportunities for further development of hydrogels are outlined to propose future research in smart batteries towards all-climate power sources and intelligent wearables.

* Corresponding author: fanhj@ntu.edu.sg
[1] Adapted with permission from P. Yang, J.-L. Yang, K. Liu, H. J. Fan, *ACS Nano* **2022**, 16, 15528.

## 9.1 Introduction

The rapid development of portable and wearable electronics, from smartphones to virtual reality, requires smaller and lighter energy storage devices with higher energy and power density.[1,2] Aqueous rechargeable batteries are promising candidates due to their low cost and environmental friendliness.[3,4] However, the aqueous solution electrolytes create technical difficulties for large-scale integration and encapsulation especially for wearable devices. In addition, it is also necessary that these batteries are intelligent and can work under some extreme conditions such as high and low temperatures and repeated mechanical deformations.[5] For this purpose, it is of significant importance to explore advanced materials and battery design to render high chemical stability, excellent temperature adaptability and strong mechanical properties.

Hydrogels are made from water-saturated and cross-linked polymer networks. Hence, hydrogels are intrinsically compatible to aqueous batteries. The polymer networks make the hydrogel an elastomer, which is beneficial for device flexibility; and the water molecules make the hydrogel an ionic conductor, which guarantees ion transport in the battery during charging and discharging.[6] In addition, more functions could be generated by exploiting gelation physics and chemistry with versatile molecular building blocks and functional additives,[7,8] laying a solid foundation for smart batteries. Here, smart batteries refer to the batteries that can adapt spontaneously to external environmental variations, such as temperature and mechanical stimuli. Therefore, hydrogels will play an important role in developing aqueous smart batteries for future energy management components and systems.

This chapter aims to discuss the ways in which hydrogels can be used as a key electrolyte component in smart batteries. As comprehensive reviews on hydrogels and their broad applications have been available,[9,10] this chapter will only briefly introduce the basic properties of gel formation, network structure, and water/ion behaviors, and will place the focus on analyzing the advantages brought by hydrogel electrolytes as alternatives to inorganic and non-aqueous ionic conductors. We highlight the advantage brought by tuning the functions and extending the application of hydrogels as smart electrolytes in aqueous batteries. Our opinions on present challenges and future opportunities are also provided, which hopefully will convince the reader that hydrogels, when judicially designed, will have great potentials in smart batteries, energies and beyond.

## 9.2 Fundamentals of Hydrogels

Hydrogels have broad application prospects in flexible electronics due to their superior softness, stretchability, and intrinsic ionotronic properties. Ionic hydrogels contain mobile ions or fixed ionic groups. These charges exhibit specific ionic effects in electric, force and chemical potential fields, endowing hydrogels with special functions such as ionic conductivity, electrostatic adhesion, and controlled swelling degree.

### 9.2.1 Hydrogel Formation

Hydrogels form by cross-linking of polymer chains in an aqueous environment. Gelation can be achieved through a variety of mechanisms, including physical entanglement, electrostatic interactions, and covalent chemical cross-linking.[11] Take the chemical cross-linking as an example (Figure 9.1a); a more stable hydrogel matrix can be realized with substantially improved flexibility and spatiotemporal accuracy compared with physical methods, since stronger covalent bonds are formed between the chains during polymerization. By introducing double network structure with two interpenetrating polymer networks, in which one is stiff and brittle and the other is soft and flexible, one can realize robust and tough hydrogels.[12] To further increase the mechanical

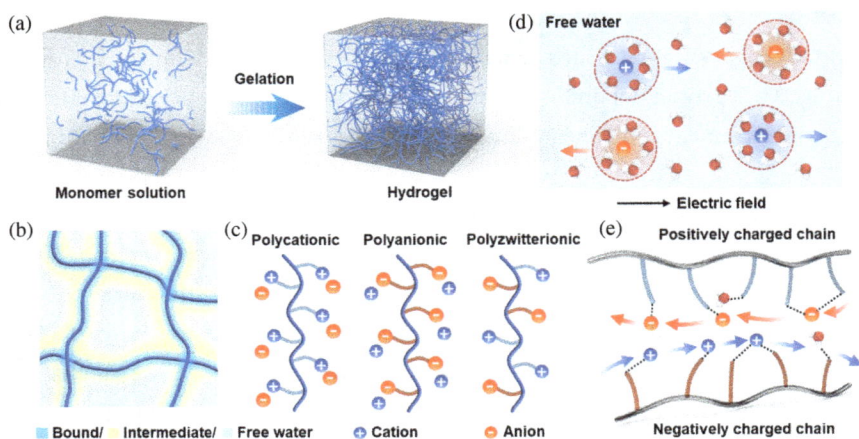

**Figure 9.1.** Hydrogel formation and basic properties. (a) Polymer networks crosslinking from monomers. (b) Water states in hydrogel. (c) Polyionic hydrogel chains. (d) Diffusion of solvated ions in free water. (e) Interaction of ions with polyionic polymer chains.

strength of hydrogels for soft robotics and wearable applications, hybridization strategies by mixing multiple components such as inorganic nanoparticles and cellulose have been attempted.[13,14]

In addition to conventional hydrogel synthesis in designed molds, 3D printing is amenable to smart polymers with arbitrary configuration,[15] including direct ink writing, stereolithography, and fused deposition modeling. Although 3D printing can in principle provide on-demand precision fabrication, more efforts are needed to develop advanced inks with desirable mechanical, physical and chemical properties.

### 9.2.2 Structure and Water Properties

Macroscopically, hydrogels behave like both solids and liquids, and can be considered as a type of quasi-solid materials between liquids and solids. High water content makes hydrogels similar to liquid water, including permeability to a wide range of chemicals and transparency to light. On the other hand, due to the constraint of the polymer network, water molecules within the hydrogel cannot flow freely like liquid water. Hence, hydrogels can maintain a certain shape like a solid, which is the basis for its application in flexible and stretchable devices.

The water in hydrogels can be divided into bound water, intermediate water and free water (Figure 9.1b).[16] Free water has a similar structure and mobility to liquid water, while bound water is physically bound to polymer chains through strong hydrogen bonds. Between free and bound water, intermediate water interacts weakly with the polymer chain and surrounding free water molecules. The theory of water-water interaction and water-polymer interaction in hydrogels has been largely enriched in recent years.[17–20] The fundamental properties of these water molecules and polymer chains will determine the environmental and electrochemical stability of hydrogels, which is crucial in energy storage devices. For example, when expanding the electrochemical potential window of the batteries containing hydrogel electrolyte, it is unclear whether the water molecules or the polymer chains decompose first. Water decomposition is closely associated with hydrogen evolution in aqueous batteries, so it is necessary to unveil the stability of water in hydrogels under electric fields.

### 9.2.3 *Conductivity Principles*

Introducing ions, including free ions and immobile ionic groups, into the hydrogel can affect the interactions between the components inside the hydrogel, thereby regulating the physicochemical properties of hydrogels to realize their ionotronic applications. According to the source of ions, ionic hydrogels can be divided into two categories: free ionic hydrogels and polyionic hydrogels (Figure 9.1c). In free ionic hydrogels, the ions are derived from the dissociated salts in the aqueous solution, and they are not bound by the polymer chains and can move freely. In contrast, the ions in polyionic hydrogels originate from ionic groups on the polymer chain and counterions with opposite charges are close to the ionic groups. The ionic groups bound to the polymer chains cannot move freely, while the counterions are mobile, but their movement is affected by the electrostatic interaction of ionic groups. Depending on the charged properties of ionic groups, we may subdivide polyionic hydrogels into polycationic hydrogels, polyanionic hydrogels, and polyzwitterionic hydrogels (Figure 9.1c). In polyzwitterionic hydrogels, an equal number of anionic groups and cationic groups distribute on the polymer chain, which can retain water and provide ion migration channels. Therefore, polyzwitterionic electrolyte is regarded as promising in energy storage applications.[21–23]

The charge conduction, contributed by ions and electrons, have different principles at interface and in bulk. There are two charge conduction mechanisms at the electrode-hydrogel interface: capacitive charge conduction and Faradaic charge conduction.[24] The capacitive mechanism is prevalent at the interface of all ionic and electronic conductors. The ions accumulate at the interface and do not carry a continuous current. When an electrochemical reaction happens, for example, during charging and discharging of a battery, a redox reaction with Faradaic charge conduction occurs at the electrode-hydrogel interface. Such an ion conduction is accompanied by continuous electron transfer across the electrode.

In contrast to the interface conduction, charge conduction in the bulk of an ionic hydrogel is based on the migration of charge carriers under an electric field. Since free ionic hydrogels contain a large amount of aqueous electrolyte solution, the electrochemical properties of free ionic hydrogels are close to that of aqueous electrolyte solutions. The charge conduction in such hydrogels is dominated by vehicular diffusion (Figure 9.1d). However, the presence of the polymer networks usually reduces the diffusion rate of free ions, and thus lowers the ionic conductivity of the hydrogel. Additionally, charged functional groups on the network in polyionic hydrogels can become hopping sites for ion migration (Figure 9.1e), which is known as Grotthuss diffusion. In the actual situation of hydrogel electrolytes, these two mechanisms can coexist. By choosing appropriate mobile ions and functional groups, highly conductive hydrogels with ionic conductivity magnitude of 100 mS cm$^{-1}$ can be obtained,[25] which is roughly the same as aqueous electrolytes with similar ion concentration. It is important to note that hydrogels can be electronically conductive when the polymer chains contain conjugated units or when conductive nanoparticles are percolated in the hydrogel matrix.[26] These conductive hydrogels are not suitable as electrolytes but may be used for electrode materials in energy storage devices.

## 9.3 Application of Hydrogels in Smart Aqueous Batteries

Batteries are becoming indispensable in daily life. Demand for high-energy-density batteries keeps increasing in portable electronics, electric vehicles and household power sources. Aqueous batteries are prospected as a promising next-generation technology owing to their non-toxicity and low-cost metrics. Battery usage scenarios vary with the electronics it powers and occasionally must be adaptive to extreme mechanical and temperature conditions. In the following, basic requirements for future smart batteries are elaborated with a focus on aqueous batteries.

### 9.3.1 Zn Anode Protection

In addition to high capacity, the chemical stability of electrode materials is a formidable issue in its operation. The metal anode in Zn aqueous batteries inevitably form dendrites due to limited nucleation sites and inhomogeneous plating (Figure 9.2a), which may finally lead to inner short and battery failure. In addition, the Zn dendrite formation is also accompanied by hydrogen evolution and electrode surface passivation, resulting in battery bulge and capacity decay, respectively. Numerous strategies have been developed to mitigate this issue, including anode structural design (3D host), anode interfacial protection, and electrolyte engineering.[27,28] Their main purpose is to stabilize metal stripping/plating and limit side reactions.

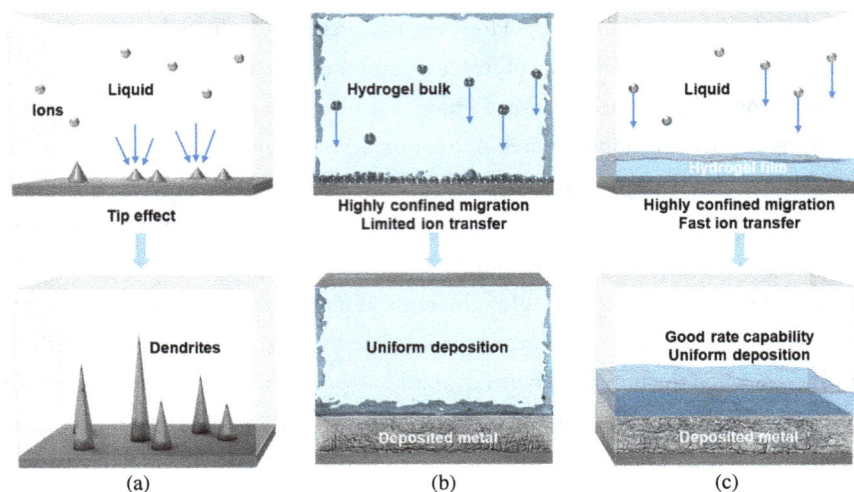

**Figure 9.2.** Hydrogel stabilizes metal anode in aqueous electrolytes. (a) Anode metal plating in aqueous electrolyte and generation of detrimental dendrites. (b) Hydrogel electrolyte regulates ion flux and promotes uniform metal deposition. (c) Hydrogel film protection for metal anode.

Hydrogel electrolyte with 3D polymer network is proven effective in suppressing dendrite formation during battery operation, resulting in a homogeneous and smooth anode surface (Figure 9.2b). This is consistent with our results in zinc batteries with polyacrylamide hydrogel electrolytes.[29] Such phenomenon can be attributed to electro-chemo-mechanical multi-field combination rather than just one factor.[30] Firstly, the polymers with high shear modulus and appropriate stiffness on the anode have been successfully applied to suppress the metal dendrite growth.[31,32] Secondly, the hydrogel polymer networks can uniformize ion flux in the bulk and induce lateral ion flux near the interface,[33] which can be helpful in regulating the metal ion deposition in a homogenous way and avoiding dendrite formation. Moreover, by introducing polyanions into the polymer network, such as carboxylate ($-COO^-$)[34,35] and sulfonate ($-SO_3^-$) groups,[36–38] the ion-confinement capability in hydrogel electrolytes may restrict the movement of cations and further stabilize the metal anode. These hydrogel strategies in aqueous batteries are equally applicable to ionogels in non-aqueous batteries, such as lithium and sodium batteries.[39]

We need to consider two aspects of hydrogel electrolytes in aqueous batteries. First, the ion conductivity of hydrogel is still several times lower than that of the liquid electrolyte, so the internal resistance of batteries with bulk hydrogel can be considerable. Therefore, it is urgent to develop highly conductive hydrogels or engineer the configuration of hydrogels for batteries that require high-rate capabilities. The other aspect is to create robust adhesion of hydrogel with metal anode. This will stabilize the hydrogel/electrode interface and subsequently regulate metal deposition and suppress hydrogen evolution.

Combining the interfacial protection concept with hydrogels provides a distinctive idea for stabilizing aqueous batteries. We used a polyanionic hydrogel thin film on zinc metal anode as the protective layer (Figure 9.2c).[40] A polyanionic hydrogel film with negative charge ($-SO_3^-$) was fabricated with a strong hydrogel-solid adhesion (denoted as Zn-SHn). The zincophilic $-SO_3^-$ group can efficiently facilitate $Zn^{2+}$ desolvation and repel anions during the plating process, and a uniform ion flux can be maintained by the hydrogel skeleton to avoid high local electric field density (schematically illustrated in Figure 9.3a). With the assistance of saline coupling agent (3-methacryloxypropyltrimethoxysilane), Zn-O bonds form at the hydrogel-Zn interface. Experiments and calculation reveal that such chemical bonding effectively prevents hydrogen evolution, guides 3D Zn nucleation and inhibits hydrogel detaching from the Zn surface. Consequently, dendrite-free Zn metal anode with high Zn plating/stripping reversibility has been achieved (Figure 9.3c and 9.3d). Specifically, the Zn-SHn electrodes show cycling stability over 3500 h at 1 mA cm$^{-2}$/1 mAh cm$^{-2}$ and 1000 h at 10 mA cm$^{-2}$/5 mAh cm$^{-2}$ in symmetric cells.

Different from the bulk hydrogel electrolyte, the micron scale hydrogel-modified electrode does not change the aqueous environment and can retain the capacity advantages and rate capability of aqueous batteries. Similar result may be obtained by employing zwitterionic polymers,[41] in which it is still polyanionic groups that exhibit high bonding energy with cations and ensure uniform metal deposition. Although the hydrogel film protection achieves excellent battery performance, it is more suitable for coin/pouch cells and static energy storage rather than flexible devices because of the liquid electrolytes.

Despite the considerable progress made on metal anode stabilization, high-performance anodes do not always translate to high-performance devices; the stability of cathodes should not be neglected. Luckily, hydrogel does help improve cathode stability. For example, the dissociated sodium ions in alginate hydrogels can alleviate

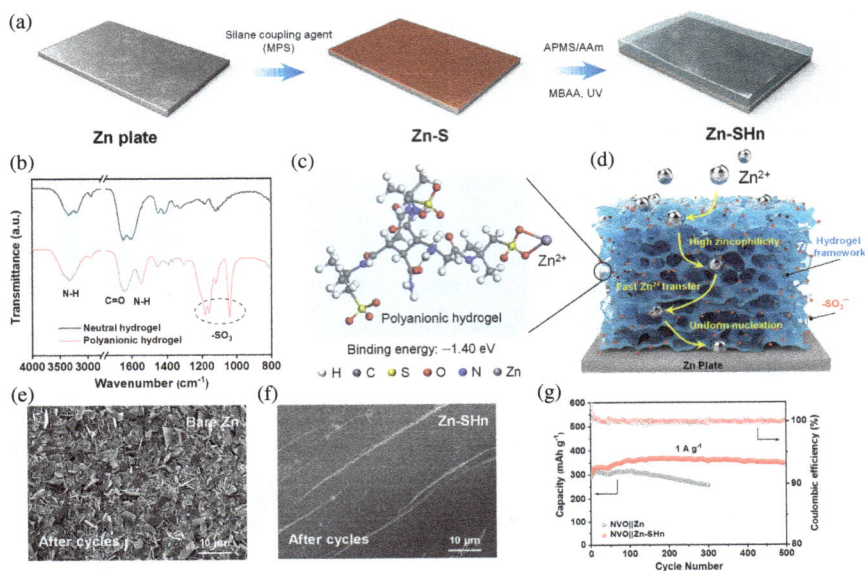

**Figure 9.3.** Hydrogel for Zn anode protection. (a) Schematic of the preparation of Zn-SHn (S denotes saline coupling agent, H for hydrogel, n for negatively charged). (b) Fourier Transform infrared spectra of Zn-SH and Zn-SHn. (c) Optimized configuration of the $Zn^{2+}$ adsorption on $-SO_3^-$ group. (d) Schematic of $Zn^{2+}$ transport in the polyanionic hydrogel solid-electrolyte interphase layer. Drawing is not to scale. Top-view scanning electron microscope (SEM) image of (e) bare Zn and (f) Zn-SHn at the end of 2,400 cycles at a current density of 5 A g$^{-1}$. (g) Cycling performance comparison of full cells. Reprinted with permission from Wiley.[40]

the collapse of the crystal structure of $Na_{0.65}Mn_2O_4$ cathode during charging process,[42] thus offering a superior stable battery. More work is needed to reinforce this conclusion. Different from the dendrites and side reactions on the anode, the main issue for cathodes is structure instability of the materials. Little has been known about how the hydrogel structure and polyionic groups affect the cathode material stability.

## 9.3.2 Hetero-Polyionic Hydrogels for Zn-I$_2$ Battery

Aqueous zinc-iodine (Zn-I$_2$) batteries work in different mechanisms from the currently popular Zn-MnO$_2$ batteries. The conversion-type cathode provides relatively longer cycle life and higher capacity. I$_2$ exhibits low solubility in water (approximately

0.29 g L$^{-1}$), making it suitable in aqueous batteries. Additionally, iodine offers several advantages: abundant natural reserves (approximately 50–60 $\mu$g L$^{-1}$ in the ocean and 0.0075% in the Earth's crust), a high specific capacity of approximately 211 mAh g$^{-1}$ and a notable high energy density of 750 Wh kg$^{-1}$. The downfall of Zn-I$_2$ battery, however, is that it suffers from a serious shuttle effect of polyiodides, sluggish iodine species conversion kinetics, and detrimental Zn dendrite growth. These are the hindrances to widescale adoption of the Zn-I$_2$ battery. In previous studies, researchers either used catalyst or host design for iodine cathode (such as single atom catalyst, porous carbon), or applied various protection strategies for the Zn metal anode. However, there has been little effort in designing electrolytes to mitigate issues from both sides.

Recently, a hetero-polyionic hydrogel electrolyte is designed which can synchronously stabilize the Zn anode and accelerate the redox kinetics of the iodine cathode, leading to evident enhancement of Zn-I$_2$ battery cycle life (Figure 9.4).[43] On the Zn anode side, the presence of sulfonate groups in the hydrogel hinders the transport of H$_2$O, polyiodides, and SO$_4^{2-}$, which is helpful in preventing the formation of deprotonated hydroxyl species and surface corrosion on the Zn anode. On the cathode side, iodophilic polycationic hydrogel (PCH) effectively alleviates the shuttle effect and facilitates redox kinetics of the iodine species. Meanwhile, polyanionic hydrogel (PAH) towards Zn metal anode uniformizes Zn$^{2+}$ flux and prevents surface corrosion by electrostatic repulsion of polyiodides. Consequently, the Zn symmetric cells with PAH electrolyte demonstrate remarkable cycling stability over 3,000 h at 1 mA cm$^{-2}$ (1 mAh cm$^{-2}$) and 800 h at 10 mA cm$^{-2}$ (5 mAh cm$^{-2}$). The Zn-I$_2$ full cell with our PAH-PCH hetero-hydrogel exhibits outstanding capacity retention with an ultralow decay rate of 0.008% per cycle for 18,000 cycles at 8 C (drops from initial capacity of 159.4 mAh g$^{-1}$ to 135.1 mAh g$^{-1}$). The paper battery (single piece) can reach an areal capacity around 2.2 mAh cm$^{-2}$. It is possible to increase the output voltage of 3.5 V and areal capacity over 6 mAh cm$^{-2}$ by parallel connection of multiple pieces. This strategy is believed to shed light on the design of multifunctional hydrogel electrolytes for long-life aqueous batteries.

There are also lingering issues to overcome for this hydrogel-based Zn-I$_2$ battery. *First*, to enhance the energy density of the battery, it is imperative to further reduce the thickness of the hydrogel (or the weight of cellulose separator). This should be

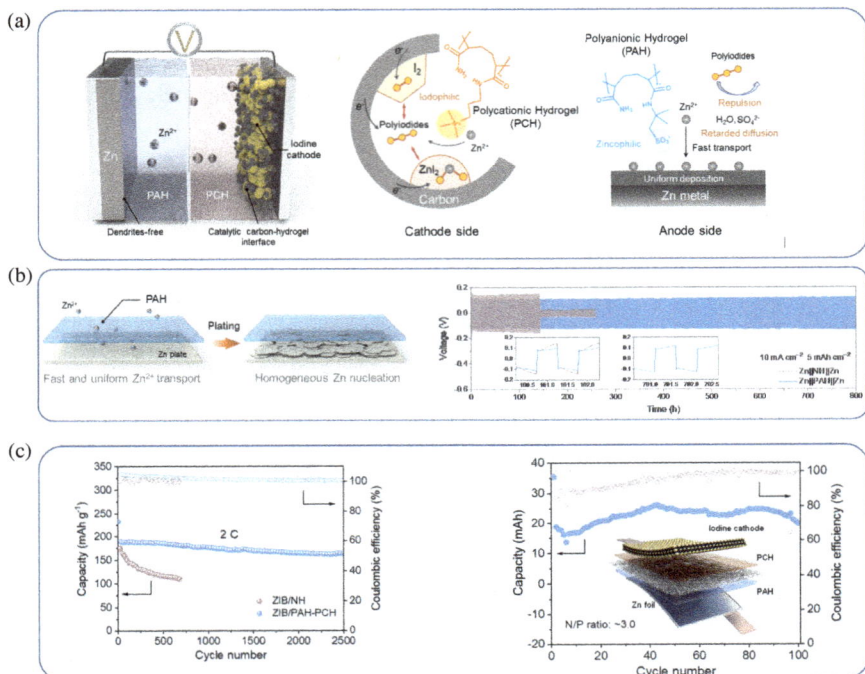

**Figure 9.4.** Hydrogel for Zn anode protection. (a) Schematic of the Zn-I$_2$ battery structure (left), the catalytic effect of PCH on iodine cathode side (middle), and PCH protection on Zn metal anode side (right). (b) Schematics of the Zn plating on the PAH-covered Zn surface (left), and comparison of cyclic performance of two symmetric cells (right). (c) Full battery performance comparison. Coin cell (left) and pouch cell (right). 1 C = 211 mA g$^{-1}$. Reprinted with permission from Wiley.[43]

considered in future work. *Second*, to maximize the interfacial catalytic effect, it is essential to increase the contact area between the hydrogel and the iodine cathode. However, achieving gelation in porous cathodes poses a challenge due to the high redox activity of iodine. And immersion of the hydrogel in zinc sulfate solution is also a time-consuming process in our current study. *Third*, the monomer of the PCH contains Cl$^-$, which can be detrimental to the Zn plating and stripping process due to its corrosion property. Therefore, exploring alternative Cl$^-$-free monomers is imperative. *Finally*, for certain applications under extreme conditions, further improvement is needed to warrant the mechanical flexibility, anti-freezing properties, and thermal stability of the hetero-hydrogel.

### 9.3.3 *Temperature Adaptability*

The principles of batteries are essentially chemical reactions governed by electrochemical thermodynamics and kinetics, both of which depend on temperature. Therefore, the battery performance is inevitably affected by temperature. Generally, batteries will not operate normally in extremely cold or hot climates. At sub-zero temperatures, the electrolyte (both aqueous solution and hydrogel) will have freezing problems leading to reduction in ionic conductivity and subsequently capacity (Figure 9.5a). It is possible to apply external thermal insulation or heating elements to keep the temperature above 0°C. Alternatively, an all-climate battery with internal short circuit design has also been reported,[43] which can heat itself from temperatures below 0°C. Despite being an innovative concept, this approach complicates the battery design and increases the fabrication cost and safety risk.

To achieve low-temperature operation of batteries, it is more feasible to engineer electrolytes with low freezing point and decent ionic conductivity.[44] By adding salt/organic additives or adjusting the structure of cross-linked polymer chains/functional groups (Figure 9.5b), hydrogel electrolytes can realize a wide range of operating temperature (−50 to 100°C).[44,45] The introduced free ions and additives can change the hydrogen bonds in water and reduce the interaction of water molecules under sub-zero temperatures. Although the addition of anti-freeze is proven to be effective, the solute, especially organic small molecule additives, may escape from the hydrogel network. In contrast, additives grafted on the polymer chains present a more stable freezing tolerance. For example, a high-strength and anti-freezing hydrogel can be achieved with ethylene glycol anchored on the polymer network,[46] where the hydrogen bonds between ethylene glycol, water molecules and the polymer matrix hinder the formation of ice crystals. The combination of intrinsic cold endurance and mechanical robustness means such hydrogel electrolyte has great potential for flexible batteries and wearable electronics in cold environments.

In contrast to low temperatures, high temperatures will usually increase ion conductivity which favors the rate capability of batteries. However, concomitant side reactions in batteries at elevated temperatures may occur and cause permanent capacity fading or risk of fire (when organic electrolytes are employed). Hydrogels ward off fires due to the intrinsic endothermic effect of water evaporation and the oxygen barrier effect of

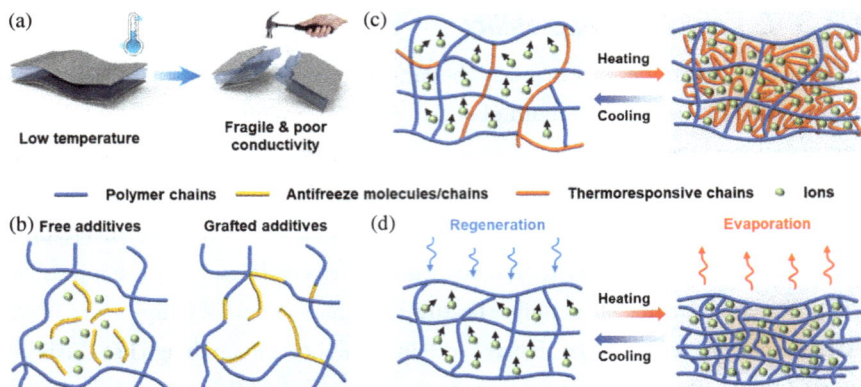

**Figure 9.5.** Temperature adaptability of aqueous batteries enabled by hydrogels. (a) Performance deterioration due to freezing of aqueous electrolyte at low temperatures. (b) Anti-freezing by introducing additives to hydrogel. (c) Phase transition of hydrogel electrolyte that regulates the migration of ions. (d) Reversible water evaporation and regeneration in hygroscopic hydrogel electrolyte.

inner salts,[47,48] but the hydrogels are still limited by the instability of the electrolyte/ electrode interface. External thermal management system cools down the batteries by fans or flowing liquids, which increases the complexity and footprint of the entire system. Therefore, it is of great significance to develop smart batteries that stop working at high temperatures, i.e., a function of thermal self-protection.

Thermoresponsive polymers with ionic and electronic modulation offer viable opportunities.[49,50] For example, poly($N$-isopropylacrylamide) hydrogel is a typical thermoresponsive polymer, in which the internal chains present swollen (ion-conductive) and shrunk (ion-resistive) states below and above phase transition temperature, respectively. Figure 9.5c illustrates the principle of smart batteries based on poly($N$-isopropylacrylamide) electrolyte. When the temperature increases, the hydrogel transforms from hydrophilic to hydrophobic state, and the transition is reversed after cooling down. Such hydrogel can also be modified on porous separators to enable thermal self-protection of batteries.[51] However, these thermoresponsive polymers still allow notable ionic mobility in the shrunk state, so the battery is not completely deactivated at high temperatures. This imperfection needs to be addressed by engineering the polymer or the porous electrode structure.

Apart from the use of phase-change hydrogels, we have demonstrated an original thermal self-protection approach for flexible batteries recently (Figure 9.6).[29] Hygroscopic hydrogels are employed as smart electrolytes. When the temperature increases, a large amount of water in the hydrogel evaporates quickly like sweating. Consequently, the ion transport in the hydrogel electrolyte is gradually restricted until it is completely cut off, and the overheated battery thus shuts down automatically. When the battery cools down to room temperature, the dehydrated hydrogel spontaneously absorbs moisture from the surrounding air and regenerates itself, and the battery returns to its original state. The thermal switch is intensive and repeatable. Such reversible moisture recycling lays a solid foundation for sustainable thermal self-protection of the batteries. However, as shown in Figure 9.6b and 9.6c, a major drawback

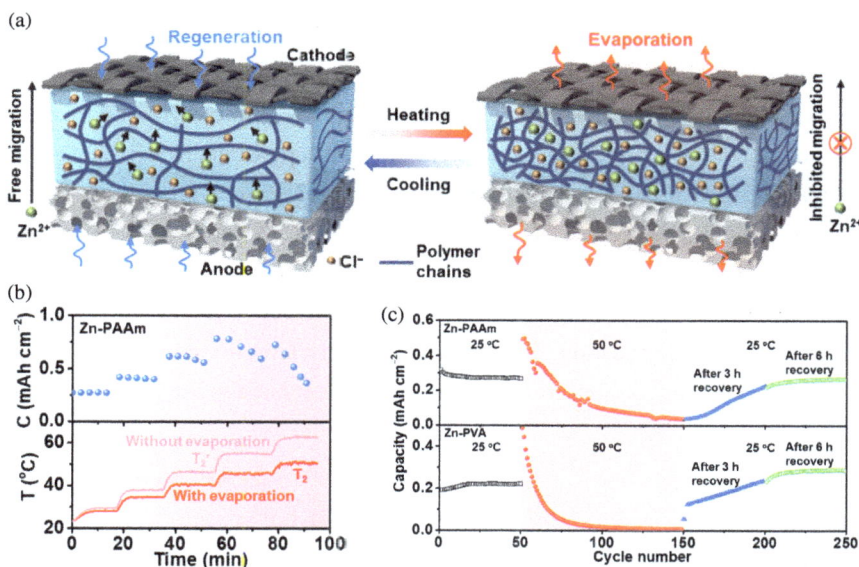

**Figure 9.6.** Thermal self-protective zinc-ion batteries based on hygroscopic hydrogel electrolyte. (a) Illustration of the working principle. (b) The capacity variation of battery based on Zn-polyacrylamide (PAAm) hydrogel electrolyte and temperature evolution of $MnO_2$ cathodes under transient heating with and without evaporation. The temperature increased from 30 to 70°C with an interval of 10°C. (c) The thermal-responsive reversibility of the batteries based on two types of hydrogel electrolytes: Zn-PAAm hydrogel (top panel) and Zn-polyvinyl alcohol (PVA) hydrogel (bottom panel). The capacity is derived from galvanostatic discharge measurement with a current density of 12 mA cm$^{-2}$. Reprinted with permission from Wiley.[29]

of this approach is the slow recovery (a few hours). For practical application of such thermal protection method, it is necessary to combine battery design and polymer engineering to further improve thermal regulation capability and response speed.

## 9.3.4 *Paper Zn Batteries and Soft Batteries by Mechanical Regulation*

Deformable batteries are needed to power wearable electronics that interface tightly with irregular or non-smooth surfaces including human skin.[52] In a hydrogel-electrolyte battery (Figure 9.7a), the hydrogels play dual roles of electrolyte and separator. It may also provide a solution to avoid aqueous electrolyte leakage under tensile strain. Based on the intrinsic mechanical advantages of hydrogel electrolytes, stretchable batteries can be assembled through hydrogel pre-stretching.[53,54] Specifically, flat electrodes are paved on each side of the pre-stretched hydrogel, thus the battery demonstrates a wavy structure at relaxed state. In this way, hydrogel electrolytes with suitable monomers and networks endow excellent stretchable batteries. For example, zinc-air batteries based on dual-network hydrogel electrolyte presented super-stretchable capability up to 800% of their original size.[55] In contrast to the electrolyte pre-strain strategy, electrode film shrinkage can also create curved-surface electronics for localized stretchable applications.[56] The negative side of this strategy is that delamination of electrode materials and hydrogel electrolyte under reiterant mechanical deformation will cause capacity degradation. This issue has received relatively less attention. Efforts can be devoted to not only electrode architecture design, but also interfacial adhesion

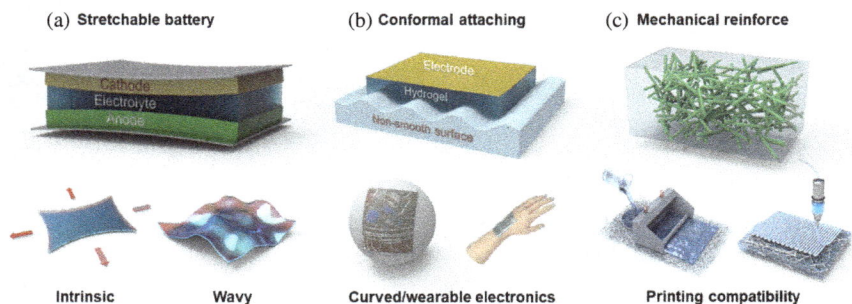

**Figure 9.7.** Mechanical regulation of batteries with hydrogel electrolytes. (a) Stretchable battery enabled by wavy structures. (b) Conformal design for structural batteries and wearable electronics. (c) Schematic of framework-reinforced hydrogel electrolyte.

strength, for example, by directly cross-linking hydrogel electrolyte on electrodes. In addition to the inherent stretchability of hydrogels, batteries may also be made stretchable by macrostructure design. As the hydrogel-based batteries are shape tailorable into desirable structures (e.g., zigzag, honeycomb), it is possible to design zinc batteries into a honeycomb structure with a maximum tensile strain up to 3,500%.[57]

Most wearables do not sustain frequent large tensile strains. One possible solution is that the functional device conforms to the irregular surfaces of the substrate,[58,59] which may guarantee the consistency of deformation (Figure 9.7b). To achieve this, the entire device must be very thin. Kotov and co-workers designed functionalized PVA film with thickness around 10 $\mu$m, and fabricated biomorphic structural zinc batteries to stick on robotics.[60] In general, the volumetric capacity of such types of thin and conformal batteries can be much greater compared with a standalone Li-ion battery. However, due to the limited active materials, the batteries may present much lower areal capacity. Hence, they are suitable for low power consumption or emergency power supply. Achieving high capacity is urgent for conformal batteries but has remained challenging. For this purpose, researchers need to exploit structural configurations and high-capacity electrode materials to accelerate the success of reliable conformal batteries for future skin-attachable devices and wearable electronics.

Another challenge in flexible and wearable batteries is to maintain the operation upon extreme mechanical stimuli, such as pressing, bending, and squeezing. Reinforcing the mechanical properties of hydrogel electrolytes by constructing frameworks helps to improve the overall strength of the device (Figure 9.7c). These frameworks are generally hydrophilic skeletons, which are compatible with hydrogel formation, such as polyacrylonitrile networks,[61] chitin nanofibers,[62] and cellulose paper.[63] In our study of the paper battery,[63] we fabricated hydrogel-cellulose paper composite which presents favourable mechanical and conductivity properties that are suitable to its function as the separator and electrolyte for quasi-solid zinc batteries. The anode and cathode materials are printed on the front and back of the hydrogel-cellulose paper, respectively (see Figure 9.8a and 9.8b). This is different from previous nano-cellulose/hydrogel composites where the composites were employed only as the separator for energy devices, but the electrode materials cannot be printed on the composite. The fabricated paper batteries are flexible during operation and long-term cycling (Figure 9.8c and 9.8d). Given that cellulose fibers can be obtained from various biomass waste materials, there are enormous opportunities for large-scale fabrication of hydrogel-cellulose composites.

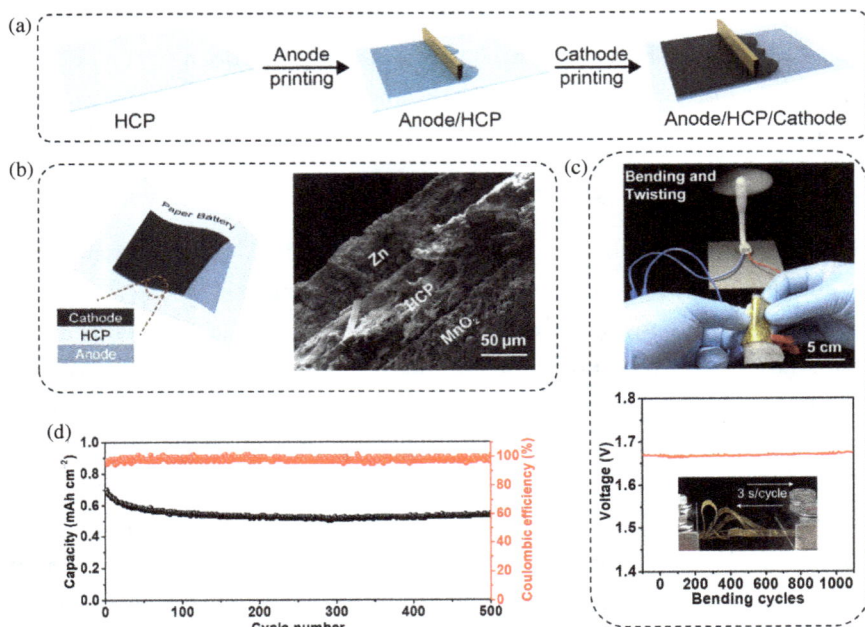

**Figure 9.8.** Hydrogel-reinforced cellulous papers for zinc paper batteries. (a) Schematic of the printing process. (b) Illustration of and cross-sectional SEM image of a Mn-Zn paper battery. (c) Mechanical test. Top: 4 cm × 4 cm paper battery powering a mini-electric fan. Bottom: Voltage profile of battery while undergoing 1,000 bending cycles. Inset illustrates the battery under repeated 180° bending cycles at a speed of 3 seconds per cycle. (d) Charge and discharge profiles of the printed Mn-Zn battery (1 cm × 1 cm) connected in series from one to three cells at a fixed current of 5 mA. Reprinted with permission from Wiley.[63]

The softness of hydrogel-composite electrode allows us to fabricate thin and soft batteries that may find application in versatile electronic skins and next-generation wearable electronics. For example, by employing electrochromic cathode materials, we obtained dual-function batteries for skin-interfaced wearable electronics using a scalable transfer printing method, featuring a thickness of less than 50 $\mu$m (Figure 9.9).[64] The soft battery can be attached conformally on arbitrary surfaces (Figure 9.9b) and show optical transparency modulation at different discharge states (Figure 9.9c and 9.9d). The high flexibility of the ultrathin batteries endows them with good conformity on arbitrarily shaped surfaces, including polyhedrons and elastic human skin.

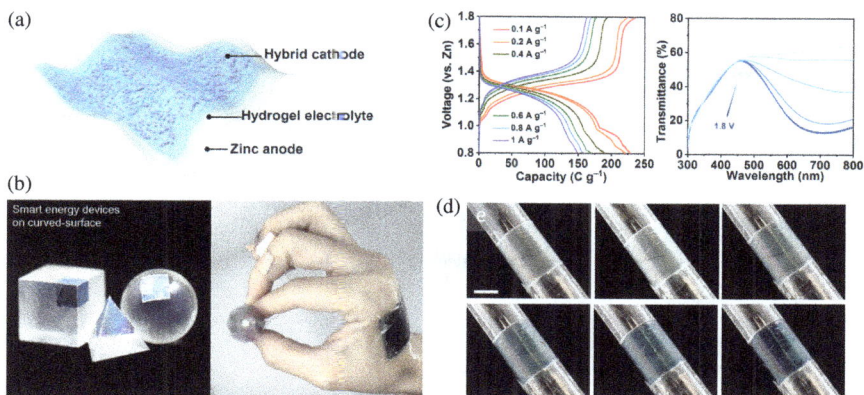

**Figure 9.9.** Ultrathin smart battery for skin-interfaced wearable electronics. (a) Configuration of the battery. (b) Soft hydrogel Zn batteries attached on various surfaces and human skin. (c) Galvanostatic charging and discharging curves and transmittance spectra. (d) Photographs of ultrathin battery on a cylinder during discharging. Reprinted with permission from Wiley.[64]

**Table 9.1.** Summary of advantages and limitations of hydrogel electrolytes in batteries.

|  | Advantage | Limitation |
|---|---|---|
| Mechanical property | • Designable networks | • Sluggish ion transport |
|  | • Flexibility and stretchability | • Interface compatibility |
|  | • Shape diversity | • Interfacial resistance |
| Temperature adaptation | • Freezing resistance | • Surface stability |
|  | • Non-flammability | • Encapsulation technique |
| Chemical stability | • Ion flux homogenization | • Rate capability |
|  | • Dendrite suppression | • Water electrolysis |
|  | • Active species retention | • Pore wettability |

# 9.4 Perspectives

Hydrogels provide a variety of functions for smart batteries. When the temperature adaptability and mechanical properties of hydrogels are properly tuned, batteries have been initially applied to wearable electronics under different environments (Table 9.1). It is expected that smart batteries will evolve into powerful energy cornerstones to be part of the Internet of Things (Figure 9.10). Our perspective for the future is as follows.

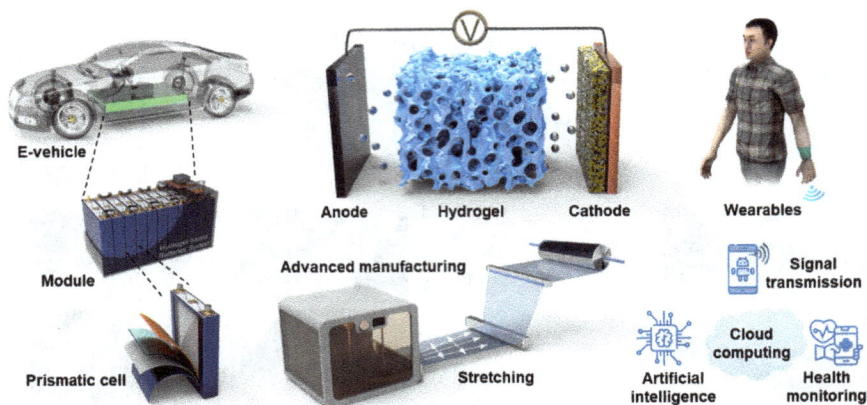

**Figure 9.10.** Perspective of hydrogels for future smart batteries.

## 9.4.1 *Advanced Manufacturing*

It is worth noting that the above-mentioned advantages may not co-exist if only one type of hydrogel electrolyte is used, and neither do the limitations all present on one hydrogel. However, it is likely that the functions of hydrogels can be enriched by composition engineering and advanced manufacturing technologies. Ultrathin hydrogel films endow them with excellent mechanical properties and rapid stimuli response. Surface-initiated polymerizations including atom transfer radical polymerization and electrochemical deposition have distinct advantages,[65,66] through which a precise control of polymer composition, topology, and site-specific functionality is achievable. In contrast, additive manufacturing is a suitable technology for complex geometries and large-scale production, which can be compatible with structural batteries.[67] Recently, 4D-printed hydrogels have spurred burgeoning interests.[68] The so-called 4D-printed hydrogels are 3D-printed jelly objects that can change their properties and functions when triggered by external electrical, mechanical, or thermal stimuli. More investigations on their responsiveness to stimuli could be the future direction of smart batteries. The advanced manufacturing technologies should be ideally compatible with the current battery industry, so that smart hydrogels can be applied to first cells and modules, ultimately serving battery packs.

## 9.4.2 *Beyond Room Temperature*

Temperature adaptability of rechargeable batteries determines whether they can be operable in intercontinental regions.[69] As a typical example, hydrogel electrolytes in Zn-MnO$_2$ batteries can be stable over a wide temperature range from −40 to 60°C.[70] At low temperatures, the electrolyte needs to have prominent conductivity to keep the battery running. In terms of anti-freezing thermodynamics and ion diffusion kinetics, double networks, functional groups, and additives that interact with water can be introduced in the hydrogels to restrain the formation of hydrogen bonds between water molecules.[71–73] In contrast, at excessive temperatures, the battery should preferably be in a shutdown state to ensure safety. The thermal self-protection function is achievable by employing thermosensitive groups to control ion mobility. Research may be directed to engineering of polymer chains by appropriately designing polyanionic, water-interactive and thermal-responsive groups in one hydrogel electrolyte. Hence, high-performance and stable batteries for all seasons can be realized in the near future.

## 9.4.3 *Wearable Design*

Wearables directly on the human body need to consider biocompatibility of materials and mechanical adaptability of devices. Batteries with all-hydrogel components present great potential on comfort and implantability.[74] The design principle of each component requires special attention, in which electrolyte and electrode should be ionically and electrically conductive, respectively. Energy autonomy is another concern to the next generation of portable and wearable devices. In this context, sweat-activated battery offers a promising choice.[75] These batteries can be assembled with salt-free hydrogels for storage and activated by sweat for operation.[76] The sweat can further act as a health monitoring parameter according to the battery performance, eventually to realize smart wearables. Multifunctionality rendered by various hydrogels should not result in significant sacrifice in energy density, power density, and lifespan.

## 9.5 Conclusion

In conclusion, functional hydrogel is a wonderful gift to emergent batteries. Despite fundamental progress made in research labs, hydrogel-based smart batteries are still in their infancy and far from industrial production. Clear understandings of water and ion transport in hydrogels are still lacking, and more study is required to elucidate the electrochemical behavior and interfacial chemistry. It is likely that hydrogels will be a key material for fabricating energy storage devices with a market share in widely adaptable wearable electronics. With ongoing progress in materials, electronics and manufacturing technologies, the potential of smart batteries will continue to grow, ranging from human health, e-vehicles, and power regulation. Finally, smart communities will be constructed by integrating cloud computing and the Internet of Things (Figure 9.9), so that human beings can live in a more refined and intelligent way.

## Acknowledgments

H.J.F. acknowledges financial support from the Singapore Ministry of Education by Academic Research Fund Tier 1 (RG 85/20) and Tier 2 (MOE-T2EP50121-0006). P.Y. acknowledges Wuhan University for the startup support and the National Natural Science Foundation of China (22209124). J.-L.Y. is thankful for the financial support by the China Scholarship Council (No. 202006210070).

## References

[1]  A. Sumboja, J. Liu, W. G. Zheng, Y. Zong, H. Zhang, Z. Liu, *Chem. Soc. Rev.* **2018**, 47, 5919.

[2]  Q. Zhao, S. Stalin, C.-Z. Zhao, L. A. Archer, *Nat. Rev. Mater.* **2020**, 5, 229.

[3]  D. Chao, W. Zhou, F. Xie, C. Ye, H. Li, M. Jaroniec, S.-Z. Qiao, *Sci. Adv.* **2020**, 6, eaba4098.

[4]  P. Ruan, S. Liang, B. Lu, H. J. Fan, J. Zhou, *Angew. Chem. Int. Ed.* **2022**, 61, e202200598.

[5]  M. Chen, Y. Zhang, G. Xing, S.-L. Chou, Y. Tang, *Energy Environ. Sci.* **2021**, 14, 3323.

[6]  C. Yang, Z. Suo, *Nat. Rev. Mater.* **2018**, 3, 125.

[7]  Y. Guo, J. Bae, F. Zhao, G. Yu, *Trends Chem.* **2019**, 1, 335.

[8]  J. Duan, W. Xie, P. Yang, J. Li, G. Xue, Q. Chen, B. Yu, R. Liu, J. Zhou, *Nano Energy* **2018**, 48, 569.

[9]  Y. Guo, J. Bae, Z. Fang, P. Li, F. Zhao, G. Yu, *Chem. Rev.* **2020**, 120, 7642.

[10]  F. Fu, J. Wang, H. Zeng, J. Yu, *ACS Mater. Lett.* **2020**, 2, 1287.

[11]  Y. S. Zhang, A. Khademhosseini, *Science* **2017**, 356, eaaf3627.

[12]  T. Nonoyama, J. P. Gong, *Annu. Rev. Chem. Biomol. Eng.* **2021**, 12, 393.

[13]  T. Yasui, E. Kamio, H. Matsuyama, *Langmuir* **2018**, 34, 10622.

[14] Q.-F. Guan, H.-B. Yang, Z.-M. Han, Z.-C. Ling, C.-H. Yin, K.-P. Yang, Y.-X. Zhao, S.-H. Yu, *ACS Nano* **2021**, 15, 7889.

[15] J. Li, C. Wu, P. K. Chu, M. Gelinsky, *Mater. Sci. Eng. R* **2020**, 140, 100543.

[16] V. M. Gun'ko, I. N. Savina, S. V. Mikhalovsky, *Gels* **2017**, 3, 37.

[17] C. Yan, P. L. Kramer, R. Yuan, M. D. Fayer, *J. Am. Chem. Soc.* **2018**, 140, 9466.

[18] X. Liu, W. Wei, M. Wu, K. Li, S. Li, *Mol. Phys.* **2019**, 117, 3852.

[19] Y. Liu, X. Liu, B. Duan, Z. Yu, T. Cheng, L. Yu, L. Liu, K. Liu, *J. Phys. Chem. Lett.* **2021**, 12, 2587.

[20] S. A. Roget, Z. A. Piskulich, W. H. Thompson, M. D. Fayer, *J. Am. Chem. Soc.* **2021**, 143, 14855.

[21] X. Peng, H. Liu, Q. Yin, J. Wu, P. Chen, G. Zhang, G. Liu, C. Wu, Y. Xie, *Nat. Commun.* **2016**, 7, 11782.

[22] F. Mo, Z. Chen, G. Liang, D. Wang, Y. Zhao, H. Li, B. Dong, C. Zhi, *Adv. Energy Mater.* **2020**, 10, 2000035.

[23] K. Leng, G. Li, J. Guo, X. Zhang, A. Wang, X. Liu, J. Luo, *Adv. Funct. Mater.* **2020**, 30, 2001317.

[24] H. Yuk, B. Lu, X. Zhao, *Chem. Soc. Rev.* **2019**, 48, 1642.

[25] C.-J. Lee, H. Wu, Y. Hu, M. Young, H. Wang, D. Lynch, F. Xu, H. Cong, G. Cheng, *ACS Appl. Mater. Interfaces* **2018**, 10, 5845.

[26] B. D. Paulsen, S. Fabiano, J. Rivnay, *Annu. Rev. Mater. Res.* **2021**, 51, 73.

[27] T. Wang, C. Li, X. Xie, B. Lu, Z. He, S. Liang, J. Zhou, *ACS Nano* **2020**, 14, 16321.

[28] Y. Zuo, K. Wang, P. Pei, M. Wei, X. Liu, Y. Xiao, P. Zhang, *Mater. Today Energy* **2021**, 20, 100692.

[29] P. Yang, C. Feng, Y. Liu, T. Cheng, X. Yang, H. Liu, K. Liu, H. J. Fan, *Adv. Energy Mater.* **2020**, 10, 2002898.

[30] D.-S. Kwon, H. J. Kim, J. Shim, *Macromol. Rapid Commun.* **2021**, 42, 2100279.

[31] C. Monroe, J. Newman, *J. Electrochem. Soc.* **2005**, 152, A396.

[32] X. Kong, P. E. Rudnicki, S. Choudhury, Z. Bao, J. Qin, *Adv. Funct. Mater.* **2020**, 30, 1910138.

[33] S. Jin, J. Yin, X. Gao, A. Sharma, P. Chen, S. Hong, Q. Zhao, J. Zheng, Y. Deng, Y. L. Joo, L. A. Archer, *Nat. Commun.* **2022**, 13, 2283.

[34] Y. Tang, C. Liu, H. Zhu, X. Xie, J. Gao, C. Deng, M. Han, S. Liang, J. Zhou, *Energy Storage Mater.* **2020**, 27, 109.

[35] B. Zhang, L. Qin, Y. Fang, Y. Chai, X. Xie, B. Lu, S. Liang, J. Zhou, *Sci. Bull.* **2022**, 67, 955.

[36] Y. Cui, Q. Zhao, X. Wu, X. Chen, J. Yang, Y. Wang, R. Qin, S. Ding, Y. Song, J. Wu, K. Yang, Z. Wang, Z. Mei, Z. Song, H. Wu, Z. Jiang, G. Qian, L. Yang, F. Pan, *Angew. Chem. Int. Ed.* **2020**, 59, 16594.

[37] J. Cong, X. Shen, Z. Wen, X. Wang, L. Peng, J. Zeng, J. Zhao, *Energy Storage Mater.* **2021**, 35, 586.

[38] H. Fan, M. Wang, Y. Yin, Q. Liu, B. Tang, G. Sun, E. Wang, X. Li, *Energy Storage Mater.* **2022**, 49, 380.

[39] S. Wang, Y. Jiang, X. Hu, *Adv. Mater.* **2022**, 34, 2200945.

[40] J.-L. Yang, J. Li, J.-W. Zhao, K. Liu, P. Yang, H. J. Fan, *Adv. Mater.* **2022**, 34, 2202382.

[41] R. Chen, Q. Liu, L. Xu, X. Zuo, F. Liu, J. Zhang, X. Zhou, L. Mai, *ACS Energy Lett.* **2022**, 7, 1719.

[42] H. Dong, J. Li, S. Zhao, Y. Jiao, J. Chen, Y. Tan, D. J. L. Brett, G. He, I. P. Parkin, *ACS Appl. Mater. Interfaces* **2021**, 13, 745.

[43] C.-Y. Wang, G. Zhang, S. Ge, T. Xu, Y. Ji, X.-G. Yang, Y. Leng, *Nature* **2016**, 529, 515.

[44] S. Liu, R. Zhang, J. Mao, Y. Zhao, Q. Cai, Z. Guo, *Sci. Adv.* **2022**, 8, eabn5097.

[45] Y. Jian, S. Handschuh-Wang, J. Zhang, W. Lu, X. Zhou, T. Chen, *Mater. Horiz.* **2021**, 8, 351.

[46] F. Mo, G. Liang, Q. Meng, Z. Liu, H. Li, J. Fan, C. Zhi, *Energy Environ. Sci.* **2019**, 12, 706.

[47] K. Shang, W. Liao, J. Wang, Y.-T. Wang, Y.-Z. Wang, D. A. Schiraldi, *ACS Appl. Mater. Interfaces* **2016**, 8, 643.

[48] H. Zhang, Z. Liu, J. Mai, N. Wang, H. Liu, J. Zhong, X. Mai, *Adv. Sci.* **2021**, 8, 2100320.

[49] Z. Chen, P.-C. Hsu, J. Lopez, Y. Li, J. W. F. To, N. Liu, C. Wang, Sean C. Andrews, J. Liu, Y. Cui, Z. Bao, *Nat. Energy* **2016**, 1, 15009.

[50] F. Mo, H. Li, Z. Pei, G. Liang, L. Ma, Q. Yang, D. Wang, Y. Huang, C. Zhi, *Sci. Bull.* **2018**, 63, 1077.

[51] J. Zhu, M. Yao, S. Huang, J. Tian, Z. Niu, *Angew. Chem. Int. Ed.* **2020**, 59, 16480.

[52] Z. Pan, J. Yang, J. Jiang, Y. Qiu, J. Wang, *Mater. Today Energy* **2020**, 18, 100523.

[53] D. G. Mackanic, M. Kao, Z. Bao, *Adv. Energy Mater.* **2020**, 10, 2001424.

[54] W.-J. Song, S. Lee, G. Song, S. Park, *ACS Energy Lett.* **2019**, 4, 177.

[55] L. Ma, S. Chen, D. Wang, Q. Yang, F. Mo, G. Liang, N. Li, H. Zhang, J. A. Zapien, C. Zhi, *Adv. Energy Mater.* **2019**, 9, 1803046.

[56] S. I. Rich, S. Lee, K. Fukuda, T. Someya, *Adv. Mater.* **2022**, 34, 2106683.

[57] M. Yao, Z. Yuan, S. Li, T. He, R. Wang, M. Yuan, Z. Niu, *Adv. Mater.* **2021**, 33, 2008140.

[58] C. Wang, K. He, J. Li, X. Chen, *SmartMat* **2021**, 2, 252.

[59] M. Hu, J. Zhang, Y. Liu, X. Zheng, X. Li, X. Li, H. Yang, *Macromol. Rapid Commun.* **2022**, 43, 2200047.

[60] M. Wang, D. Vecchio, C. Wang, A. Emre, X. Xiao, Z. Jiang, P. Bogdan, Y. Huang, N. A. Kotov, *Sci. Robot.* **2020**, 5, eaba1912.

[61] H. Li, C. Han, Y. Huang, Y. Huang, M. Zhu, Z. Pei, Q. Xue, Z. Wang, Z. Liu, Z. Tang, Y. Wang, F. Kang, B. Li, C. Zhi, *Energy Environ. Sci.* **2018**, 11, 941.

[62] C. Liu, W. Xu, C. Mei, M. L., W. Chen, S. Hong, W.-Y. Kim, S.-y. Lee, Q. Wu, *Adv. Energy Mater.* **2021**, 11, 2003302.

[63] P. Yang, J. Li, S. W. Lee, H. J. Fan, *Adv. Sci.* **2022**, 9, 2103894.

[64] J. Li, P. Yang, X. Li, C. Jiang, J. Yun, W. Yan, K. Liu, H. J. Fan, S. W. Lee, *ACS Energy Lett.* **2023**, 8, 1.

[65] K. Matyjaszewski, *Adv. Mater.* **2018**, 30, 1706441.

[66] E. R. Cross, *SN Appl. Sci.* **2020**, 2, 397.

[67] Z. Zhu, D. W. H. Ng, H. S. Park, M. C. McAlpine, *Nat. Rev. Mater.* **2021**, 6, 27.

[68] Y. Dong, S. Wang, Y. Ke, L. Ding, X. Zeng, S. Magdassi, Y. Long, *Adv. Mater. Technol.* **2020**, 5, 2000034.

[69] H. Wang, Z. Chen, Z. Ji, P. Wang, J. Wang, W. Ling, Y. Huang, *Mater. Today Energy* **2021**, 19, 100577.

[70] M. Chen, J. Chen, W. Zhou, X. Han, Y. Yao, C.-P. Wong, *Adv. Mater.* **2021**, 33, 2007559.

[71] D. Zhang, Y. Liu, Y. Liu, Y. Feng, Y. Tang, L. Xiong, X. Gong, J. Zheng, *Adv. Mater.* **2021**, 33, 2104006.

[72] S. He, Q. Cheng, Y. Liu, Q. Reng, M. Liu, *Sci. China Mater.* **2022**, 65, 1980.

[73] L. Jiang, D. Dong, Y.-C. Lu, *Nano Res. Energy* **2022**, 1, e9120003.

[74] T. Ye, J. Wang, Y. Jiao, L. Li, E. He, L. Wang, Y. Li, Y. Yun, D. Li, J. Lu, H. Chen, Q. Li, F. Li, R. Gao, H. Peng, Y. Zhang, *Adv. Mater.* **2022**, 34, 2105120.

[75] L. Manjakkal, L. Yin, A. Nathan, J. Wang, R. Dahiya, *Adv. Mater.* **2021**, 33, 2100899.

[76] J. Lv, G. Thangavel, Y. Li, J. Xiong, D. Gao, J. Ciou, M. W. M. Tan, I. Aziz, S. Chen, J. Chen, X. Zhou, W. C. Poh, P. S. Lee, *Sci. Adv.* **2021**, 7, eabg8433.

© 2024 World Scientific Publishing Company
https://doi.org/10.1142/9789811278327_0010

Chapter 10

# The Rise of Chalcogens for High-energy Zinc Batteries

Ze Chen, Chunyi Zhi*

*Department of Materials Science and Engineering,
City University of Hong Kong, 83 Tat Chee Avenue,
Kowloon, Hong Kong 999077, China*

Aqueous zinc batteries (AZBs) are regarded as strong contenders for next-generation energy storage, thanks to the abundance and high theoretical capacity of zinc metal. However, conventional cathode materials are mainly based on ion-insertion electrochemistry, which can only deliver limited capacity. The conversion-type aqueous zinc∥chalcogen batteries combine the strengths of chalcogen cathodes (S, Se, Te, and interchalcogens) with zinc anodes to significantly boost capacity and energy output. While research on zinc∥chalcogen batteries has been growing, it is still in its early stages. The selection and regulation of cathode material systems need more comprehensive and systematic approaches, and deeper investigation into the mechanisms is required. In this overview, we provide a detailed account of the recent progress in zinc∥chalcogen batteries and offer comprehensive guidelines for future research. We summarize the fundamental conversion mechanisms of chalcogens in AZBs and explore various optimization strategies to improve the performance of chalcogen-based batteries. These strategies include electrode engineering, electrolyte engineering, decoupling charge carriers, and altering the redox pathways of chalcogens. Finally, we discuss the challenges and prospects in developing chalcogen-based AZBs.

## 10.1 Introduction

Because of frequent safety accidents caused by non-aqueous lithium-ion batteries (LIBs), aqueous batteries, as an alternative technology employing water molecules as

---

*Corresponding author: cy.zhi@cityu.edu.hk

the sole or main solvent in the electrolyte, are attracting widespread attention in terms of energy storage.[1–7] Compared with traditional non-aqueous LIBs, aqueous batteries more effectively address the safety concern arising from the use of flammable electrolytes in LIBs, and they have the additional advantages of low cost, ambient assembly, and environmental friendliness.[8–11] In recent years, diverse types of aqueous battery systems have been developed, including LIBs,[12] sodium-ion batteries,[13] aluminum-ion batteries,[14] and zinc batteries.[15] Aqueous zinc batteries (AZBs) are considered one of the most promising battery systems.[16, 17] AZBs exhibit a high energy density owing to the use of high-capacity zinc metal as the anode (820 mAh g$^{-1}$).[18] This property enables them to exclude electrochemically inactive components at the anode side, including the current collector, binder, and conductive additive, enhancing the energy density from the perspective of atomic economy. Moreover, the Zn metal anode demonstrates a low redox potential ($Zn^{2+}/Zn$, –0.76 V vs. standard hydrogen electrode (SHE)), approaching the limit of water decomposition. Thus, AZBs, with their intrinsic capacity and redox potential advantages at the anode side, have superior capability for enhancing the energy density of aqueous batteries.[19, 20]

In recent years, rapid advancements in cathode materials have led to enhancements in the energy density of AZBs. The continuous enrichment of cathode materials (e.g., manganese oxides,[21] vanadium oxides,[22] Prussian blue analogs,[23] and organic compounds[24]) and the accompanying optimization of the cathode structure (e.g., interlayer spacing mediation and oxygen vacancy construction) have increased the specific capacity of cathodes and thus enhanced the energy density of AZBs to some degree.[16] However, the energy density of AZBs is still unsatisfactory, which still lags behind the energy requirement of practical applications such as wearable electronics. [25] Although the cathode specific capacity of AZBs has been progressively enhanced, their limitation related to single-electron transfer reactions has not yet been overcome (e.g., $MnO_2$ cathode with only 308 mAh g$^{-1}$ capacity).[26] The discharge potential of the current cathode species versus that of the Zn anode is another problem limiting the energy density of AZBs; this problem results from the failure in utilizing the relatively narrow voltage window of aqueous electrolytes (e.g., <1 V for vanadium oxides).[27] Moreover, the energy density of AZBs is associated with the limitation that, on the one hand, the volumetric capacity of cathode materials for AZBs is considerably lower than that of traditional cathode materials, such as nickel cobalt manganese oxides;[28] on the other hand, the specific capacity of cathode materials sharply declines when the loading mass or thickness is increased. Thus, new approaches to enhance the energy density of AZBs in terms of cathode specific capacity and discharge potential should be urgently developed.

Chalcogens (Group VIA elements), including sulfur (S), selenium (Se), and tellurium (Te), have attracted considerable research attention in the extensive study of metal-chalcogen batteries.[29–33] Because of the two-electron conversion reaction occurring between the chalcogen cathode and metal anode, chalcogen-based electrodes exhibit a substantially high gravimetric capacity (S: 1672 mAh g$^{-1}$, Se: 675 mAh g$^{-1}$, and Te: 419 mAh g$^{-1}$).[34] Chalcogen cathodes can easily host additional metal ions through the multielectron transfer process, making them promising materials for realizing the development of next-generation high-energy batteries. Moreover, from the economic perspective, S exerts little impact on the environment and human health and is among the most abundant elements on the earth (Figure 10.1a).[31] Because of their high electronic conductivity, Se and Te cathodes are suitable for realizing the development of high-energy rechargeable metal-ion batteries.[35] Moreover, chalcogen-based electrodes demonstrate a high volumetric capacity (S: 3467 mAh cm$^{-3}$ at a density of 2.07 g cm$^{-3}$, Se: 3253 mAh cm$^{-3}$ at a density of 4.82 g cm$^{-3}$, and Te: 2621 mAh cm$^{-3}$ at a density

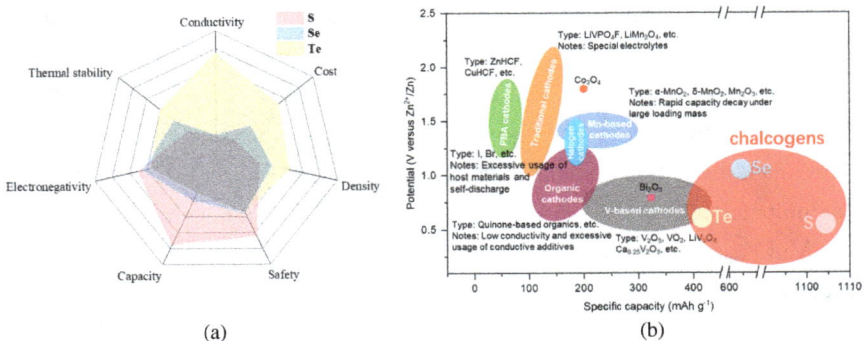

(a)                                    (b)

**Figure 10.1.** (a) Comparison of S, Se, and Te on the basis of various properties; (b) Energy density and plateau voltage of Zn‖chalcogen batteries compared with other classical AZBs. Reproduced with permission from Refs. [16, 35].

of 6.24 g cm$^{-3}$) and thus a substantially high volumetric energy density output. Metal-chalcogen batteries are beneficial for volume-constrained applications.

Although chalcogens have been employed as high-energy cathode materials in various metal-chalcogen batteries, the development of Zn||chalcogen batteries is still in the nascent stage.[36] In 1998, Licht patented the first Zn||S battery based on the conversion process of S/HS$^-$/ZnS; however, the Zn||S battery is a primary cell without reversibility.[37] Furthermore, in 2001, Licht and coworkers explored the possibility of using a Zn||S fuel cell in alkaline electrolytes with CoS as the electrocatalytic additive in the cathode.[38] In the following years, the development of Zn||S cell almost stalled. In 2020, Zhi and coworkers developed a reversible aqueous Zn||S battery with polysulfide as the cathode.[39] In the same year, Jiang and coworkers successfully developed an aqueous Zn||S battery with S/carbon nanotubes (CNTs) as the cathode;[40] this can be regarded as the first true Zn||S battery with high reversibility. Subsequently, many studies on Zn||S batteries have been conducted, highlighting the increasing research interest on this topic. Similar to Zn||S batteries, substantial progress has been made in research on Zn||Se and Zn||Te batteries in the past two years. In 2021, Zhi and coworkers developed Zn||Te batteries with Te nanosheets as the cathode for the first time.[41] In 2021, the same research group reported the development of the first Zn||Se batteries, which is highly compatible with both organic and aqueous electrolytes.[42] Although Zn||chalcogen batteries have been well researched recently, chalcogen-based electrodes warrant more attention considering their competitive gravimetric or volumetric capacity and high stability in batteries, which are highly promising for the development of high-energy AZBs (Figure 10.1b).

In this chapter, we describe the progress of Zn||chalcogen batteries, including battery reactions and mechanisms. Moreover, we discuss promising strategies for improving the electrochemical performance of Zn||chalcogen batteries from the perspective of cathode and electrolyte engineering. Finally, we highlight the potential of chalcogens as a new approach to enhance the energy density of AZBs and present the possible challenges to guide the rapid development of high-energy-density Zn||chalcogens batteries in the near future.

## 10.2 Electrochemistry of Chalcogen Cathodes Versus Zn

The high theoretical energy output of rechargeable Zn||chalcogen batteries typically originates from the two-electron redox reaction between the Zn metal and chalcogens, which can be expressed as follows:

$$C + 2e^- + Zn^{2+} \leftrightarrow ZnC$$

where C denotes the chalcogen element (S, Se, or Te).

The zero-valent metal Zn and Zn chalcogenide compounds are generated through the conversion reaction. However, the conversion electrochemistry of the chalcogen cathode versus Zn shows its own features. Chalcogen elements exist as different crystal structures, and the S crystal comprises $S_8$ rings. Se crystals contain either $Se_8$ rings or Se chains.[43] The Te crystal has only chain-like structures, which exhibit an infinite helix structure (Figure 10.2).[44] The electrochemical reactions of chalcogens with Zn typically proceed in different steps and even involve the generation of intermediate products.[45, 46] In this section, we first review the entire development process of chalcogen-based electrodes in AZBs by summarizing all published Zn||chalcogen batteries (Table 10.1). Subsequently,

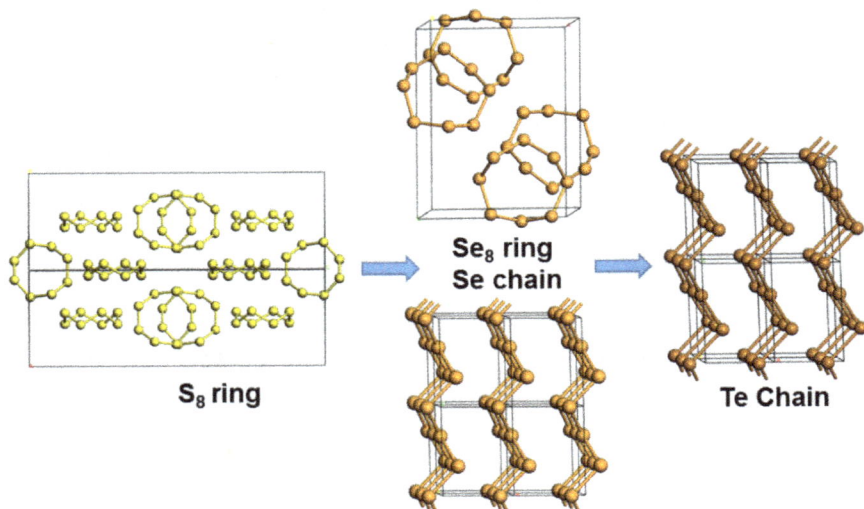

**Figure 10.2.** Crystal structure transformations of S, Se, and Te. The S crystal is constructed by $S_8$ rings. Se crystals contain either $Se_8$ rings or Se chains. The Te crystal only has chain-like structures. The difference in the structure indicates that three congener elements exhibit distinct structural changes during the charge/discharge process. Reproduced with permission from Ref. [35].

**Table 10.1.** Summary of various Zn‖chalcogen batteries reported in the literature.

| Cathode | Electrolyte[a] | Capacity (mAh g$^{-1}$)@ C rate | Plateau voltage (V) | Cycling life[b] | Ref. |
|---|---|---|---|---|---|
| S@CNTs | 1 M Zn(CH$_3$COO)$_2$ with 0.05 wt% I$_2$ | 1105 (0.1 A g$^{-1}$) | 0.51 | 85% (50)@2 A g$^{-1}$ | 40 |
| Li$_2$S$_6$ | 1 M Zn(TFSI)$_2$ | 1148 (0.3 A g$^{-1}$) | 1.21 | 49.3% (700)@1 A g$^{-1}$ | 39 |
| S@kejent black | 1 M ZnSO$_4$ with thiourea (2 g L$^{-1}$) | 1785 (1 A g$^{-1}$) | 0.45 | 67% (300)@5 A g$^{-1}$ | 47 |
| ZnS@carbon fibers | 3 M ZnSO$_4$ with 1 wt% (thiourea$^{2+}$)(I$^-$)$_2$ | 465 (0.1 A g$^{-1}$) | 0.61 | 65% (300)@2 A g$^{-1}$ | 48 |
| S@CNTs | 1 M Zn(CH$_3$COO)$_2$ with PEG | 1013 (0.1 A g$^{-1}$) | 0.76 | 82% (200)@1 A g$^{-1}$ | 49 |
| S@carbon black | 0.5 M ZnCl$_2$ + 0.5 M LiCl in urea/choline chloride + 10% acetonitrile | 846 (0.5 A g$^{-1}$) | 0.57 | 25% (400)@1 A g$^{-1}$ | 50 |
| S@Fe -polyaniline | 2 M ZnSO$_4$ | 1205 (0.2 A g$^{-1}$) | 0.58 | 54% (200) | 51 |
| Carbon nanofiber-S | Zn(CH$_3$COO)$_2$ with ethylene glycol + poly(vinyl alcohol) + I$_2$ | 667 (0.5 A g$^{-1}$) | 0.34 | 39% (300)@1 A g$^{-1}$ | 52 |
| S@FeNC/ NC/CC | ZnSO$_4$/PVA hydrogel | 1143 (0.2 A g$^{-1}$) | 0.75 | 58% (300)@0.5 A g$^{-1}$ | 53 |
| S@C | Gelation/ZnSO$_4$ gel | 2063 (0.1 A g$^{-1}$) | 1.15 | 78% (100)@0.5 A g$^{-1}$ | 54 |
| S@C | CuSO$_4$ gel/ZnSO$_4$ gel | 48 mAh cm$^{-2}$ (1 mA cm$^{-2}$) | 1.15 | ~100% (100)@1 mA cm$^{-2}$ | 55 |
| S@carbon spheres | 2 M Zn(OTf)$_2$/ tetraglyme/I$_2$ | 775 (2 A g$^{-1}$) | 0.45 | 70% (600)@4 A g$^{-1}$ | 56 |
| CMK-3@S | 3 M Zn(OTf)$_2$ | 788 (0.2 A g$^{-1}$) | 0.35 | 65% (100)@1 A g$^{-1}$ | 57 |
| Se@CMK-3 | 2 M Zn(TFSI)$_2$ in PEG/ water | 611 (0.1 A g$^{-1}$) | 1.23 | 80.3% (1000)@1 A g$^{-1}$ | 42 |
| Se-in-Cu [Co(CN)$_6$] | 4 M Zn(OTf)$_2$ with 20% PEG | 664 (0.2 A g$^{-1}$) | 1.21 | 90.6% (6000)@10 A g$^{-1}$ | 58 |

*(Continued)*

**Table 10.1.** (*Continued*)

| Cathode | Electrolyte[a] | Capacity (mAh g$^{-1}$)@ C rate | Plateau voltage (V) | Cycling life[b] | Ref. |
|---|---|---|---|---|---|
| Triphenyl-phosphine selenide | 1 M Zn(OTf)$_2$ in acetonitrile/water (v/v = 85:15) | 72.9 (0.5 A g$^{-1}$) | 1.96 | 85.3% (4300)@2 A g$^{-1}$ | 59 |
| Se@C | 0.5 M CuSO$_4$/0.5 M ZnSO$_4$ | 1263 (0.5 A g$^{-1}$) | 1.2 | 95% (300)@5 A g$^{-1}$ | 60 |
| Te nanosheets | 1 M ZnSO$_4$ | 419 (0.05 A g$^{-1}$) | 0.59 | 83% (500)@1 A g$^{-1}$ | 41 |
| Te powder | 0.8 M Zn(OTf)$_2$ in water/pentanediol | 386 (0.2 A g$^{-1}$) | 0.53 | 56% (100)@0.8 A g$^{-1}$ | 61 |
| Te powder | 1 M ZnSO$_4$ | 460 (0.05 A g$^{-1}$) | 0.62 | 60% (50)@0.05 A g$^{-1}$ | 62 |
| SeS$_{5.76}$@3D-NPCF | 3 M ZnSO$_4$ with 0.1 wt% I$_2$ | 1222 (0.2 A g$^{-1}$) | 0.71 | 75% (500)@4 A g$^{-1}$ | 63 |
| SeS$_2$@PCS | 1 M ZnSO$_4$ with 0.1 wt% I$_2$ | 1107 (0.05 A g$^{-1}$) | 0.74 | 85% (1000)@5 A g$^{-1}$ | 64 |

[a] All electrolytes are aqueous unless otherwise specified in the corresponding positions.

[b] The format of the cycling life column is referred as capacity retention (cycle numbers)@current density during cycling.

*CNT: carbon nanotube; CMK: ordered mesoporous carbon; TFSI: bis(trifluoromethanesulfonyl)imide; PEG: polyethylene glycol; PEO: polyethylene oxide; PVA: polyvinyl alcohol; NPCF: nitrogen, phosphorus co-doped carbon foam; PCS: porous carbon sphere.

we present the fundamental electrochemistry of different chalcogen element cathodes versus Zn.

## 10.2.1 *Electrochemistry of the S Cathode Versus Zn*

A typical Zn||S cell consists of an S cathode, a Zn metal anode, an electrolyte, a separator, and other components (Figure 10.3a). During the discharge process of the cell, S and Zn undergo a two-electron conversion reaction (S + Zn$^{2+}$ + 2e$^-$ → ZnS) to generate zinc sulfide. By contrast, during charging, ZnS is decomposed into S and Zn. Because of the strong catenation tendency of the S atom, S has many allotropes. The most stable allotrope is cyclo-S$_8$, which has an octatomic ring-like configuration and crystallizes into orthorhombic $\alpha$-S$_8$ at room temperature.[65, 66] In the conventional cathode structure, cyclo-S$_8$ is attached to conductive support. During the discharge process,

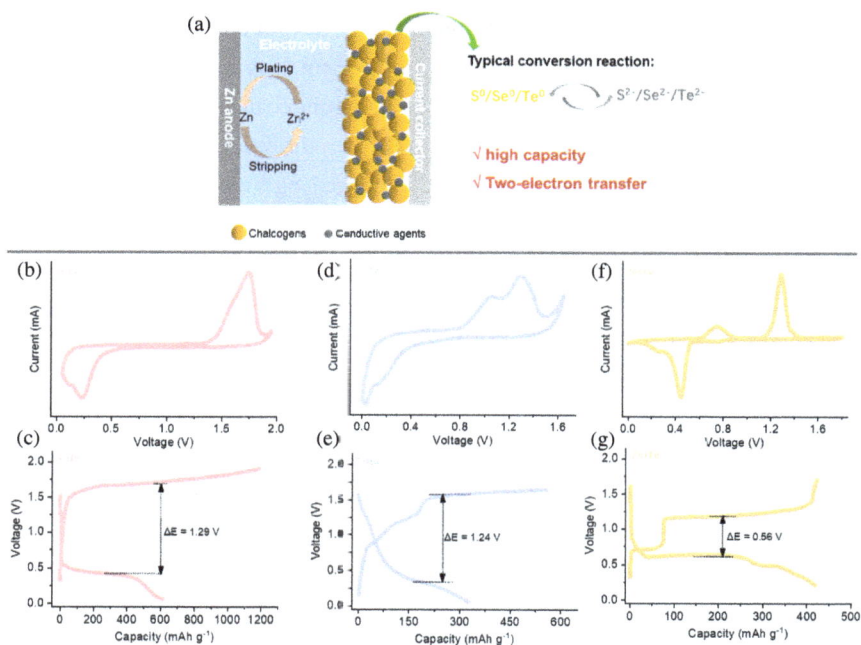

**Figure 10.3.** (a) Schematic model of Zn |chalcogen batteries based on the two-electron conversion type reaction mechanism. Typical cyclic voltammetry and galvanostatic charge-discharge (GCD) curves of Zn||chalcogens batteries with 1 M ZnSO$_4$ electrolytes: (b) and (c) Zn||S cell, (d) and (e) Zn||Se cell, (f) and (g) Zn||Te cell.

the reduction of cyclo-S$_8$ by Zn superficially involves one step, and the discharge profile of the Zn||S cell exhibits a long and flat plateau with a voltage of approximately 0.5 V (vs. Zn$^{2+}$/Zn; Figure 10.3b and 10.3c). In addition, cyclo-S$_8$ undergoes the conventional redox process; that is, the S$_8$ ring is opened and ZnS is formed during the discharge process, and ZnS is oxidized to S$_8$ chains during the charge process. Although S has a high theoretical capacity (1675 mAh g$^{-1}$) based on the two-electron conversion reaction, the Zn||S cell typically delivers approximately 1000 mAh g$^{-1}$ discharge capacity, which is mainly attributed to the low electrical conductivity of S (1 $\times$ 10$^{-15}$ S m$^{-1}$).[67] The Zn||S conversion reaction results in the high gravimetric specific capacities of both electrodes; however, it can lead to many adverse outcomes. S and its discharge products are insulated, which increase the voltage hysteresis of the cell and decrease efficiency.[34, 68] Moreover, the accumulation of poorly conductive insoluble compounds (mainly ZnS$_2$ and ZnS) on the surface of S particles prevents the further reduction of the interior S, leading to the low utilization of active S and the rapid decay of the discharge capacity.[40] In addition, because the density of ZnS (4.09 g cm$^{-3}$) is higher than that of S (2.07 g cm$^{-3}$), the S cathode undergoes substantial volume changes upon cycling, which reduces the electrical contact of active S with the current collector.[69] Because of the amorphous nature of the generated ZnS, the volume changes are more prominent and severely impair the cycling stability of the cathode.[49, 70]

In contrast to the polysulfide shuttling problem encountered in the Li||S cell, no polysulfides are generated during the cycling of the Zn||S cell; this phenomenon indicates the potential of the Zn||S cell for achieving S-based redox chemistry with high stability.[38, 56] Thus, increasing the utilization rate of S active materials and constructing stable cathode materials are vital for achieving a high-performance Zn||S cell. The performance of Zn||S cells can be improved by using strategies applied for Li||S cells as well as based on the advances in the Li||S cell in the recent decade. An S cathode with an increased S utilization rate and enhanced structural stability can be achieved by developing a carbon host with a high specific surface area, high conductivity, and favorable wettability for the electrolyte.[71, 72] Various carbon hosts including CNTs, carbon fibers, carbon cloth, carbon spheres, and polyaniline have been introduced into Zn||S cells. In addition, introducing electrocatalysts, decoupling charge carriers, optimizing S redox pathways, and regulating electrolytes can improve the electrochemical performance of the battery.

## 10.2.2 *Electrochemistry of Se and Te Cathodes Versus Zn*

As chalcogen elements, Se and Te are similar to S; however, their conversion electrochemistry versus Zn is provided with their own features.[73–75] In aqueous electrolytes, the electrochemical reactions of Zn||Se batteries are similar to conversions occurring in the Zn||S battery. The involved two-electron conversion reaction can be summarized as follows: Se + Zn$^{2+}$ + 2e$^-$ → ZnSe.[42] However, the low discharge capacity (<300 mAh g$^{-1}$) and low plateau voltage (~0.4 V vs. Zn$^{2+}$/Zn) of the common Zn||Se cell can be ascribed to poor compatibility between the Se cathode and electrolytes (Figure 10.3d and 10.3e). In addition, Se chains are more thermodynamically favorable than cyclo-S$_8$. Compared with the ring structure, the chain structure of Se molecules leads to higher electronic conductivity and electrochemical activity as well as stronger chemical interaction with the C substrate.[31] Thus, Zn||Se batteries based on optimized electrolytes may not only maintain the high mass loading of active Se on the cathode but also exhibit reaction kinetics superior to that of S. However, additional studies on this phenomenon are warranted.[58]

Similar to S and Se, Te, as a cathode material, involves the two-electron conversion reaction between Te and Zn. In contrast to the single plateau reaction observed in Zn||S and Zn||Se cells, two plateaus appear in the discharge-charge profiles of the two-step conversion reaction in the Zn||Te cell, indicating the formation of intermediate products. [28] The Zn||Te cell undergoes conversion from Te to ZnTe$_2$ to ZnTe. Zhi and coworkers reported a Zn||Te cell prepared using Te nanosheets as the cathode material,[41] which delivered an ultrahigh volumetric capacity of 2619 mAh cm$^{-3}$ (419 mAh g$^{-1}$), 74.1% of which was derived from the first conversion (Te to ZnTe$_2$), with an ultraflat discharge plateau (Figure 10.3f and 10.3g). Although this is the first reported Zn||Te battery with a challenging aqueous environment, it demonstrated an excellent retention capacity of >82.8% after 500 cycles. Similar to the Zn||S cell, the formation of polytelluride intermediates during the conversion reaction of the Te cathode was not detected.[76]

## 10.2.3 *Electrochemistry of the Interchalcogen Cathode Versus Zn*

Chalcogen elements can form a series of miscible solid solutions (interchalcogen compounds).[77] For example, as two adjacent Group VIA elements, both S and Se have solid-state structures comprising eight-membered ring units, and the difference between their atomic radii is small ($r_S$ = 102 pm; $r_{Se}$ = 119 pm).[78] Therefore, Se and S can form a series of selenium sulfides (Se$_x$S$_y$), which can be prepared in a wide range of proportions from Se$_5$S to SeS$_{=0}$ (Figure 10.4a).[79, 80] Thus, Se and S are infinitely miscible, and S-Se compounds with a wide range of proportions can be prepared. Se is substantially less abundant in the earth's crust and is more costly than S.[81, 82] As cathode materials, Se$_x$S$_y$ compounds exhibit considerably high gravimetric specific capacities (675 mAh g$^{-1}$ for pure Se and 1675 mAh g$^{-1}$ for pure S, and these values are adjustable according to the Se$_x$S$_y$ composition), which is beneficial for developing high-energy rechargeable AZBs. Because of the semiconductor nature of Se, the electron density is redistributed at the S/Se sites of Se$_x$S$_y$, and the altered electron structure may considerably affect the conversion electrochemistry of the interchalcogen compound cathode (Figure 10.4b).[63] Moreover, the different compatibilities of

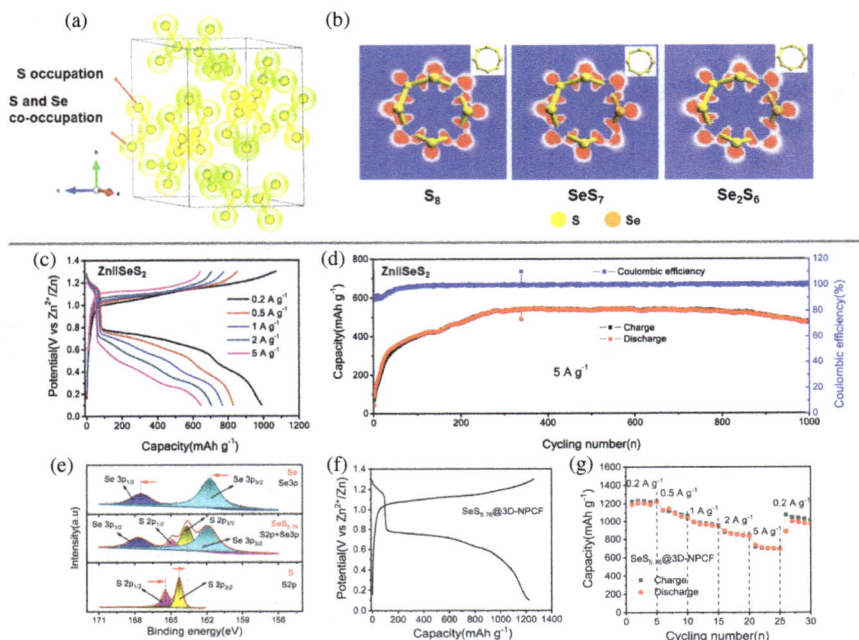

**Figure 10.4.** (a) Crystal structure of $Se_xS_y$; (b) Electron density differences for $S_8$, $SeS_7$, and $Se_2S_6$ compounds; (c) Charge and discharge profiles of the $SeS_2$ electrode at different current densities; (d) Cycling performance of the $SeS_2$ electrode at 5 A g$^{-1}$; (e) X-ray photoelectron spectra of Se3p in pure Se, S2p + Se3p in $SeS_{5.76}$, and S2p in pure S; (f) Charge and discharge profiles of the $SeS_{5.76}$ electrode, and (g) the corresponding rate performance. Reproduced with permission from Refs. [63, 64].

S/Se with the electrolyte may effectively improve the interface kinetics of the $Se_xS_y$ electrode in the cell.[34, 83]

Wang and coworkers investigated the electrochemical performance of phosphorus-doped carbon sheet-encapsulated $SeS_2$ in rechargeable AZBs.[64] The as-developed Zn||$SeS_2$ cell exhibited a reversible capacity of 1107 mAh g$^{-1}$, with a flat discharge potential of 0.74 V (vs. Zn$^{2+}$/Zn), and high cycling performance (Figure 10.4c and 4d). In terms of the reaction mechanism, a reversible conversion of $SeS_2$ to ZnSe and ZnS was observed. Subsequently, the same group exploited a series of selenium-sulfur solid solutions and their composites ($SeS_{14}$, $SeS_{5.76}$, and $SeS_{2.4}$).[63] Because of the introduction of Se and its synergistic effect with S, the physical and electrochemical properties of Se/S interchalcogens were manipulated (Figure 10.4e). In particular, by optimizing the Se content in these composites, $SeS_{5.76}$ exhibited a capacity of 1222 mAh g$^{-1}$ and a flat plateau of 0.71 V at 0.2 A g$^{-1}$ (Figure 10.4f and 10.4g). The Zn storage kinetics was determined on the basis of the discharge process, during which $SeS_{5.76}$ was converted into ZnSe and ZnS. The results revealed that Se led to differences in the electron density and reaction energy of S, thus increasing its conductivity and reactivity to facilitate the electrochemical reaction with Zn.[80, 84]

Theoretically, Te can form interchalcogen compounds with S.[85] However, Te has only one crystalline form, and the atomic radius of Te ($r_{Te}$ = 140 pm) is significantly larger than that of S, and the melting point of Te (452°C) is higher than the boiling point of S (445°C). Therefore, the synthesis of S-Te interchalcogens compounds under mild conditions or in an open system is difficult.[86] However, under specific conditions, a limited number of S–Te compounds can be prepared. For example, $TeS_7$ and $Te_2S_6$ are two known Te–S interchalcogen compounds, and further increasing the Te/S atomic ratio has been reported to be difficult.[87] With a higher p orbital of Te, the introduction of Te atoms into the S crystal introduces more energy states in electronic structures and weakens the electron localization of S atoms around incorporated Te sites.[88] Although the compounds exhibit improved electrical conductivity and reaction kinetics for electrochemical conversion in alkaline metal batteries, the application of S–Te compounds in AZBs has never been reported.[35]

Investigating interchalcogen compounds can provide insights into the optimal design of chalcogen cathodes for developing high-energy rechargeable AZBs in the near future. Compared with S, Se and Te exhibit significantly higher electronic conductivity and more favorable compatibility with electrolytes. By forming interchalcogen compounds with S, the electrochemical activity and cycling stability of the cathode can be improved. Thus, the cathode may exhibit a high specific capacity, a long cycle life, and improved rate performance in rechargeable AZBs.[34] S overcomes the drawbacks associated with low resource abundance and high Se/Te cost, and its use increases the upper limit of the gravimetric specific capacity of the interchalcogen compound cathode (although it exerts less impact on the volumetric capacity density of the cathode). Because of the solid nature of the solution, the elemental composition of the interchalcogen compound cathode can be precisely adjusted to meet the requirements of target use. For high-power applications, the content of Se/Te in the compound can be increased. Increasing the mass content of S can be beneficial if the emphasis is on the specific energy and cost of the battery. To improve the conversion electrochemistry of the interchalcogen compound cathode, new electrolytes and battery configurations should be developed.

## 10.3 Optimization Principle for Zn ‖ Chalcogen Batteries

Chalcogens exhibit a high specific capacity because of their reduction; however, a high energy barrier needs to be overcome for their reverse oxidation to chalcogens, leading to a large overpotential and poor reversibility for Zn‖chalcogen batteries.[52, 89] In addition, the reduction of chalcogens usually occurs at a low electrochemical potential (less than –0.7 V vs. SHE) and results in the low operation voltage of Zn‖chalcogen batteries.[16, 90] Therefore, limited reversibility, low electrochemical potential, and poor energy efficiency are the main obstacles for the practical use of Zn‖chalcogen batteries. In this section, we discuss promising strategies for improving the electrochemical performance of Zn‖chalcogen batteries in terms of reaction kinetics, cycling stability, energy density, and energy efficiency.

### 10.3.1 Cathode Engineering Through Redox Catalysis

To date, various structural and compositional designs have been derived to fabricate stable and high-capacity chalcogen cathodes.[34, 91] In particular, a porous carbon host is widely used for developing high-performance chalcogen-based batteries, which overcomes the inferior electronic conductivity of S and its discharge products and alleviates volumetric changes in S species.[92] However, despite the use of a high-performance carbon host, intrinsic problems are prevalent for Zn-chalcogen batteries, including sluggish chalcogen conversion kinetics and a high reactivation energy barrier for zinc chalcogenide oxidation; these problems severely affect the electrochemical performance of Zn‖chalcogen batteries. Thus, in addition to the introduction of a carbon host, more advanced approaches should be developed. In this section, we discuss promising strategies for solving the aforementioned problems from the perspective of cathode engineering.

In recent years, the concept of cathode engineering through catalysis has been introduced to investigate the electrocatalytic effects on typical Li‖S batteries.[93] Catalysis is a process that increases the reaction rate without modifying the overall standard Gibbs energy change in the reaction.[94] In Li‖S batteries, catalysis enhances redox kinetics and improves the reaction efficiency of Li||S chemistry (even at a high loading mass of active materials). Thus, catalytic materials in batteries comprise materials that can significantly enhance the electrochemical conversion process of active material species. With advances in electrocatalysis in Li||S batteries, electrocatalytic components have been utilized to regulate the related chalcogen conversion kinetics in Zn||chalcogen batteries for improving electrochemical performance (Figure 10.5a).

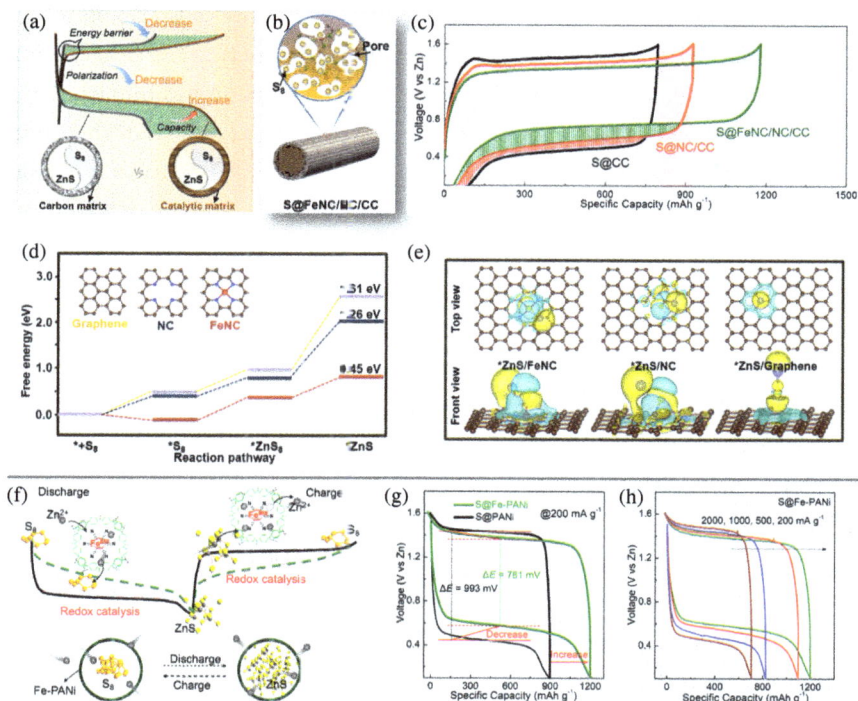

**Figure 10.5.** (a) Schematic diagram of the working mechanism and comparison of typical galvanostatic charge/discharge potential profiles of the carbon- and catalytic matrix-supported sulfur cathodes; (b) Schematic illustration for the fabrication of the FeNC-supported sulfur cathode; (c) Voltage profiles of different cathodes at a current density of 0.2 A g$^{-1}$; (d) Energy profiles for the reduction of sulfur species on FeNC, NC, and graphene on the basis of density functional theory results; (e) Top and front view of the band-decomposed charge density distribution of the FeNC-, NC-, and graphene-based *ZnS adsorption system. (f) Schematic for sulfur redox in Zn||S cells with redox catalysis facilitated an efficient cation transport pathway and sulfur reversible conversion based on the Fe-PANi system; (g) Comparison of charge-discharge voltage profiles of S@FePANi and S@PANi at a current density of 0.2 A g$^{-1}$; (h) rate performance of the Zn||S cell assembled with the S@Fe-PANi cathode at current densities ranging from 0.2 to 2 A g$^{-1}$. Reproduced with permission from Refs. [51, 53].

Lu and coworkers developed atomically dispersed Fe sites with Fe-N4 coordination as bidirectional electrocatalytic hotspots to simultaneously manipulate the complete conversion of sulfur and minimize the energy barrier of ZnS decomposition (Figure 10.5b).[53] Fe sites are favorable for strong sulfur and possible zinc polysulfide intermediate adsorption and for ensuring the nearly complete conversion of sulfur to ZnS during the discharge process. For the following recharge process, electrodeposited ZnS is reversibly converted into S without a noticeable activation overpotential for Fe-N4 moieties compared with pure carbon matrices. The developed sulfur cathode exhibits a high specific capacity of 1143 mAh g$^{-1}$ and a lower voltage hysteresis of 0.61 V (Figure 10.5c–e). Similarly, a high-energy Zn||S system was fabricated through the *in situ* interfacial polymerization of Fe(CN)$_6$$^{4-}$-doped polyaniline within the sulfur nanoparticle (Figure 10.5f).[51] Compared with sulfur, Fe$^{II/III}$(CN)$_6$$^{4/3-}$ redox mediators exhibited substantially faster cation (de)insertion kinetics. The higher cathodic potential (~0.8 V for Fe$^{II}$(CN)$_6$$^{4-}$/Fe$^{III}$(CN)$_6$$^{3-}$ vs. ~0.4 V for S/S$^{2-}$) spontaneously catalyzed the full reduction of sulfur during battery discharge (S$_8$ + Zn$_2$Fe$^{II}$(CN)$_6$ $\leftrightarrow$ ZnS + Zn$_{1.5}$Fe$^{III}$(CN)$_6$). The open iron redox species resulted in a lower energy barrier to ZnS activation during the reverse charging process, and the facile Zn$^{2+}$ intercalative transport facilitated highly reversible conversion between S and ZnS. The cathode with a yolk-shell structure and 70 wt% sulfur displayed a reversible capacity of 1205 mAh g$^{-1}$, with a flat operation voltage of 0.58 V and a fade rate of 0.23%/cycle over 200 cycles (Figure 10.5g and 10.5h).

In addition to electrocatalysis in the Zn||S cell, Zhi and coworkers proposed an electrocatalytic Se reduction/oxidation reaction strategy to realize high-Se-loading Zn||Se batteries with fast kinetics and high Se utilization.[58] The synergetic effects of Cu and Co transition metal species in the channel structure of the host can effectively immobilize and catalytically convert Se during cycling, thus facilitating Se utilization and conversion kinetics. Accordingly, the Zn battery employing a Se-in-Cu[Co(CN)$_6$] cathode exhibited a capacity of 664.7 mAh g$^{-1}$ at 0.2 A g$^{-1}$, and an excellent rate capacity of 430.6 mAh g$^{-1}$ was achieved even at 10 A g$^{-1}$. Furthermore, an A-h-level (~1350 mAh) Zn||Se pouch-type battery with high Se loading (~12.3 mg$_{(Se)}$ cm$^{-2}$) exhibited a high Se utilization of 83.3% and excellent cyclic stability, with 89.4% initial capacity retained after 400 cycles at a Coulombic efficiency of >98%.

These findings highlight the feasibility of using catalysts for manipulating the chalcogen redox kinetics, which can realize the practical application of Zn||chalcogen batteries. Despite considerable progress in this field, an in-depth understanding of these catalytic mechanisms is required. Information on complete reaction processes is crucial to establish fundamental theory for guiding the exploration and design of catalytic materials. In addition, lightweight catalytic materials with high efficiency, low cost, and controllable structures are highly recommended, and they can be beneficial for developing Zn||chalcogen batteries.[95] In this regard, the configuration of conductive carbon substrates and metal compounds (nitrides, metal oxides, and sulfides) is highly preferable.[96]

## 10.3.2 *Electrolyte Engineering Through a Redox Mediator*

When an electrocatalyst is integrated with a cathode, the electrocatalytic performance usually declines due to the accumulation of insulating metal chalcogenide deposits on electrocatalytic active sites. Thus, in contrast to the adsorption reaction mechanism occurring on the surface of electrocatalysts, soluble redox mediators (RMs) with suitable redox potentials have been reported to be effective for promoting conversion kinetics in Li||S batteries through a solution reaction circulation bypass.[97] For Zn||chalcogen batteries with poor conversion kinetics, exploiting RMs to accelerate the conversion of chalcogens into zinc chalcogenides is promising for constructing high-performance AZBs.

Jiang and coworkers employed the $I_2$ additive as the RM in 1 M $Zn(CH_3COO)_2$ aqueous electrolytes, where $I_2$ served as the medium of $Zn^{2+}$ ions for reducing the voltage hysteresis of the S cathode, improving conversion kinetics, and stabilizing Zn stripping/plating (Figure 10.6a–c).[40] The as-developed Zn||S battery exhibited a high capacity of 1105 mAh g$^{-1}$, with a flat discharge voltage of 0.5 V, realizing an energy density of 502 Wh kg$^{-1}$. A recent study explored a novel bifunctional RM, namely thiourea (TU), for improving the sluggish reaction kinetics for the conversion of ZnS to S and the formation of the irreversible byproduct of $SO_4^{2-}$ during charging in the Zn||S battery (Figure 10.6d and 10.6e).[47] TU can undergo reversible redox reactions during cycles, which not only enhances the reversibility between ZnS and S but also contributes to additional capacities. The negative and positive centers of the intermediates of TU can interact with ZnS to weaken the Zn||S bond for improving electrochemical dynamics. Simultaneously, carbonium ions in the intermediate of TU

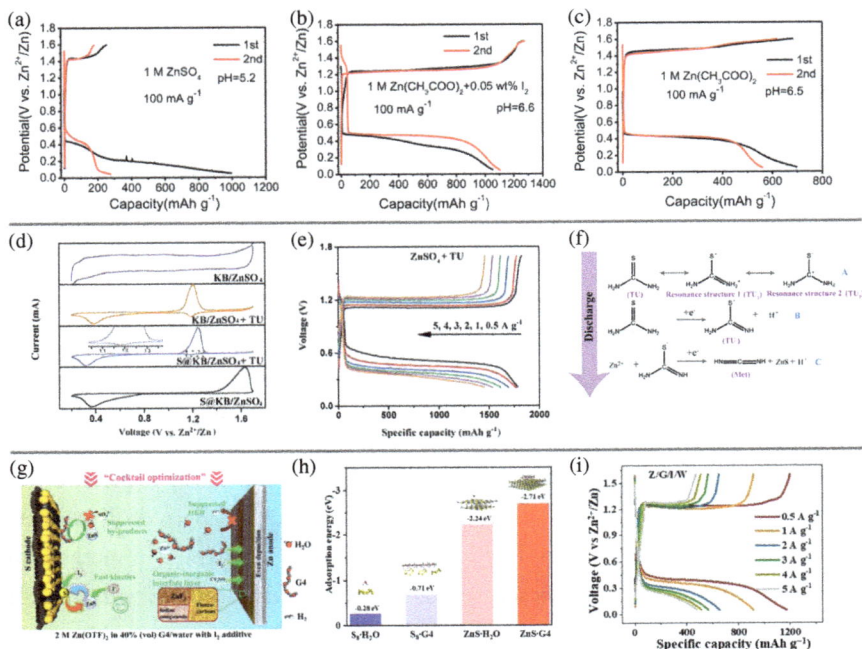

**Figure 10.6.** Charge and discharge curves of the S@CNTs cathode in (a) 1 M ZnSO$_4$, (b) 1 M Zn(CH$_3$COO)$_2$, and (c) 1 M Zn(CH$_3$COO)$_2$ with 0.05 wt% I$_2$ as additive; (d) Cyclic voltammetry curves of the S electrode in different electrolytes at 0.2 mV s$^{-1}$; (e) GCD curves of the Zn||S cell with TU additive at different current densities; (f) The discharge and charge mechanism of additive TU and the electrochemical reactions of TU during discharging; (g) Schematic illustration of aqueous Zn||S batteries using (up) 2 M Zn(OTF)$_2$ in 40% (vol.) G4/water with the I$_2$ additive; (h) Calculated adsorption energy values of S$_8$·H$_2$O, S$_8$·G4, ZnS·H$_2$O, and ZnS·G4 ligands; (i) GCD curves of the Zn||S battery in Z/G/I/W at different current densities. Reproduced with permission from Refs. [40, 48, 56].

exhibit strong reactivity for ZnS, resulting in the inhibition of the formation of SO$_4^{2-}$ (Figure 10.6f). The S electrode exhibited excellent cyclic performance at a capacity of 763.7 mA h g$^{-1}$ after 300 cycles at 5 A g$^{-1}$, corresponding to a low decay rate of only 0.11% per cycle. Furthermore, Huang and coworkers proposed nanoscale ZnS with carbon as the cathode and iodinated TU as an RM, in which the activation barrier of ZnS was decreased to 1.26 V.[48] The ZnS cathode exhibited a high capacity of 465 mAh g$_{ZnS}^{-1}$, a high specific energy density of 274 Wh kg$_{ZnS}^{-1}$ (832 Wh kg$_S^{-1}$), and excellent rate performance (197 mAh g$_{ZnS}^{-1}$ at 9.04 C). However, the commonly used nonpolar $I_2$ catalyst weakly interacts with sulfur and usually leads to limited improvement. Moreover, the added $I_2$ catalyst accelerates the corrosion of zinc metal, further exacerbating battery performance. Zhu and coworkers developed a "cocktail optimized" electrolyte containing tetraglyme (G4) and water as cosolvents and $I_2$ as an additive (Figure 10.6g).[56] G4–$I_2$ synergy activated efficient polar $I_3^-/I^-$ catalyst coupling and shielded the cathode from water, thus facilitating the conversion kinetics of S and suppressing interfacial side reactions. Simultaneously, it stabilized the Zn anode by forming an organic-inorganic interphase upon cycling. With the enhanced reversibility of electrodes, the Zn||S cell displayed a high capacity of 775 mAh $g^{-1}$ at 2 A $g^{-1}$ and retained more than 70% capacity after 600 cycles at 4 A $g^{-1}$ (Figure 10.6h and 10.6i).

RMs can considerably enhance electron transfer from the active material to the electrode, without the need for physical contact between the electroactive surface and active material.[98] However, RMs should have a reversible potential that is higher than the oxidation potential of active materials. In addition, the mediation process of RMs considerably depends on the conductive surface. Active materials are easily covered by insulating discharging zinc chalcogenide products and consequently lose their functions, especially at a higher depth of discharge. Moreover, modulating the solvation structure of RMs can relieve the dependence of the regulation strategy on conductive surfaces. To improve the electrochemical properties of chalcogen-based batteries with the RM additive, more heterogeneous mediators should be developed, including organic chalcogenides and nitrides, accompanied with tuning the composition and concentration of the electrolyte solvent and supporting salts.

## 10.3.3 Decoupling Charge Carriers

Zn||chalcogen batteries exhibit sluggish conversion kinetics between chalcogens and zinc chalcogenides, leading to poor rate capability and decayed reversible capacity.[40] The inability to enhance conversion kinetics can be attributed to the intrinsic insulating property of chalcogens and zinc chalcogenides, especially S and ZnS. For example, sulfur typically displays a two-electron reaction (S → ZnS) and low redox potential (approximately −0.26 V vs. SHE) when $Zn^{2+}$ ions are the charge carriers. Thus, decoupling charge carriers at the cathode or anode side (i.e., chalcogens bind to other metal ions rather than $Zn^{2+}$ as a charge carrier for forming high-electroactivity metal chalcogenide products) is promising for enhancing the conversion kinetics of chalcogen electrodes.[99] The known shuttle effect in chalcogen electrodes mainly results from the high solubility of metal chalcogenides, which leads to the rapid decay of capacity.[100] However, the low solubility of metal chalcogenides impairs the compatibility of active materials and electrolytes, leading to decreased interfacial electrochemical activity.[99] A study determined the solubility product constant ($K_{sp}$) of various metal chalcogenides ($pK_{sp}$ is the negative logarithm of $K_{sp}$) and indicated that $Cu^{2+}$, $Fe^{2+}$, and $Ni^{2+}$ can theoretically serve as cathode side charge carriers (Figure 10.6a). The standard electrode potential, including the electromotive force (EMF), cathode potential, and anode potential, is another crucial factor determining the applicability of charge carriers for chalcogens. Aqueous electrolytes typically exhibit a narrow window of electrochemical stability.[101] As presented in Figure 10.6b, the range from 1.229 to −0.828 V (vs. SHE) is the electrochemical stability window of aqueous batteries, and the reasonable values of the EMF and electrode potential must be within the range. Thus, aqueous batteries based on Ni-, Fe-, and Cu-based charge carriers are suitable for constructing decoupled Zn∥chalcogens batteries. Moreover, Cu, Ni, and Fe chalcogenides can be useful for relieving the volume expansion of electrode materials.

Inspired by the four-electron electrode reaction with S ($S + 2Cu^{2+} + 4 e^- \rightarrow Cu_2S$) that has a high redox potential (0.5 V vs. SHE), Qu and coworkers proposed a strategy for decoupling charge carriers.[54] They used $Zn^{2+}$ ions in the anolyte and $Cu^{2+}$ ions in the catholyte to fully tap the potential of Zn and S. The movement of $SO_4^{2-}$ ions between the anolyte and catholyte connected the cathode and anode, facilitating charge exchange between them. The spontaneous reaction between $Cu^{2+}$ and the Zn anode was inhibited by the introduction of an anion exchange membrane that hindered the diffusion of cations between the anolyte and catholyte. The fabricated aqueous Zn∥S hybrid battery exhibited a high reversible specific capacity of 2063 mAh $g_S^{-1}$. In addition, the extremely low voltage hysteresis (~0.2 V) and excellent rate capability (938 mAh $g_S^{-1}$ at 2 A $g_S^{-1}$) were observed in the battery, surpassing all reports of other Zn∥S batteries. Furthermore, as a variable-valence charge carrier, the valence state of $Cu^{2+}$ could change during the charge/discharge process, contributing additional capacity for the battery. Thus, Qu and coworkers developed a cascade battery in which solid S was used as the cathode in the first discharge step and the generated $Cu_2S$ catalyzed the reduction of $Cu^{2+}$ to $Cu/Cu_2O$ in the second discharge step (Figure 10.6c and 10.6d).[55] An ultrahigh areal capacity of 48 mAh cm$^{-2}$ was achieved even at a low solid cathode loading (9.6 mg cm$^{-2}$). The cascade battery design deviates from the conventional battery configuration, providing a paradigm for constructing two-in-one batteries. Similarly, Qu and coworkers adopted the charge carrier decoupling strategy to circumvent the poor rate capability of Se-based cathodes (Figure 10.6e).[60] The as-developed laboratory-scale Zn||Se@C cell displayed a discharge voltage of approximately 1.2 V at 0.5 A g$^{-1}$ and an initial discharge capacity of 1263 mAh g$_{Se}^{-1}$. When a specific charging current of 6 A g$^{-1}$ was applied, the Zn||Se@C cell exhibited a stable discharge capacity of approximately 900 mAh g$_{Se}^{-1}$, which was independent from the discharge rate, suggesting excellent fast-charging performance (Figure 10.6f).

For Zn||chalcogen batteries, the unique charge carrier decoupling strategy facilitated faster reaction kinetics, higher output voltage, higher discharge capacity, and longer cycle life than those obtained using a single Zn$^{2+}$ cation charge carrier. However, the employed ion exchange membranes leading to decreased cost efficiency and the tedious assembly process of the batteries cannot be neglected. In addition, compared with Cu$^{2+}$, Ni$^{2+}$ (as a decoupling charge carrier) demonstrated more potential for improving the conversion kinetics and increasing the output voltage of Zn||chalcogen batteries, which warrants more attention.

### 10.3.4 Optimizing Redox Pathways of Chalcogens

Current Zn||chalcogen batteries depend on the negative valence conversion of chalcogens from –2 to 0, with the disadvantages of limited reversibility, low electrochemical potential, and poor energy efficiency (Figure 10.8a).[41, 102] Considering the multivalent nature of chalcogens (–2, 0, +2, +4, +6), altering the chalcogen redox pathway to achieve chalcogen oxidation is a promising strategy because it can overcome the intrinsic low-voltage shortcoming of chalcogen reduction (Figure 10.8b).[59, 89] Tapping into the innovative conversion chemistry in Zn||chalcogen batteries can promote their electrochemical performance and thus their voltage and capacity. The positive valence conversion for chalcogens with high reversibility is difficult to achieve in batteries.[84, 103] Given the high surface negative charge density of chalcogens, especially S and Se, chalcogens are natural electron acceptors, difficult to oxidize, and stabilized in AZBs.[104]

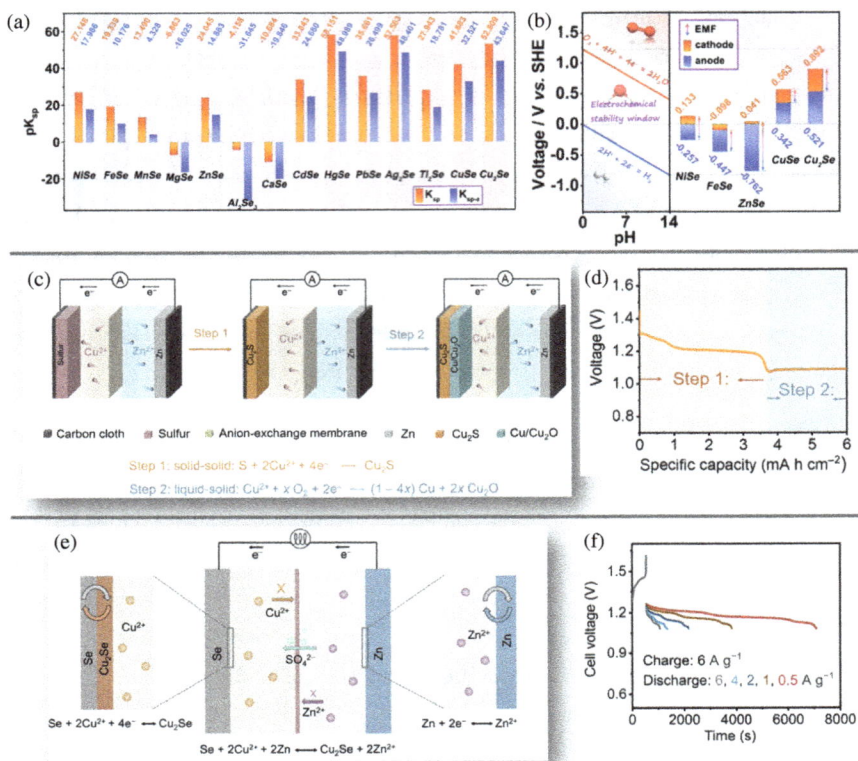

**Figure 10.7.** (a) The solubility product constants of typical metal selenides; (b) Electrochemical working window for several typical aqueous metal-Se batteries. Concept and merits of the Zn||S cascade battery: (c) Discharging process of the battery: the first and second discharge steps of the battery are based on solid-solid and liquid-solid conversion working mechanisms, respectively; (d) The typical discharge curve. (e) Schematic diagram of the aqueous Zn||Se full cell and (f) the fast-charging rate performance of the aqueous Zn||Se cell: charging at $6\,A\,g^{-1}$, discharging at different currents. Reproduced with permission from Refs. [55, 60, 99].

Recently, some significant achievements have been made regarding the multivalence of chalcogens in AZBs. Zhi and coworkers invented a conversion-type Zn||Se battery that delivers superior performance in both organic and aqueous electrolytes, benefiting from a highly reversible six-electron conversion reaction between $Se^{2-}$ and $Se^{4+}$ (Figure 10.8c and 10.8d).[42] They achieved excellent capacities for the organic system ($551\,mAh\,g^{-1}$) and aqueous system ($611\,mAh\,g^{-1}$), accompanied with a remarkable rate performance and cycling performance in both the systems. In addition, very low voltage

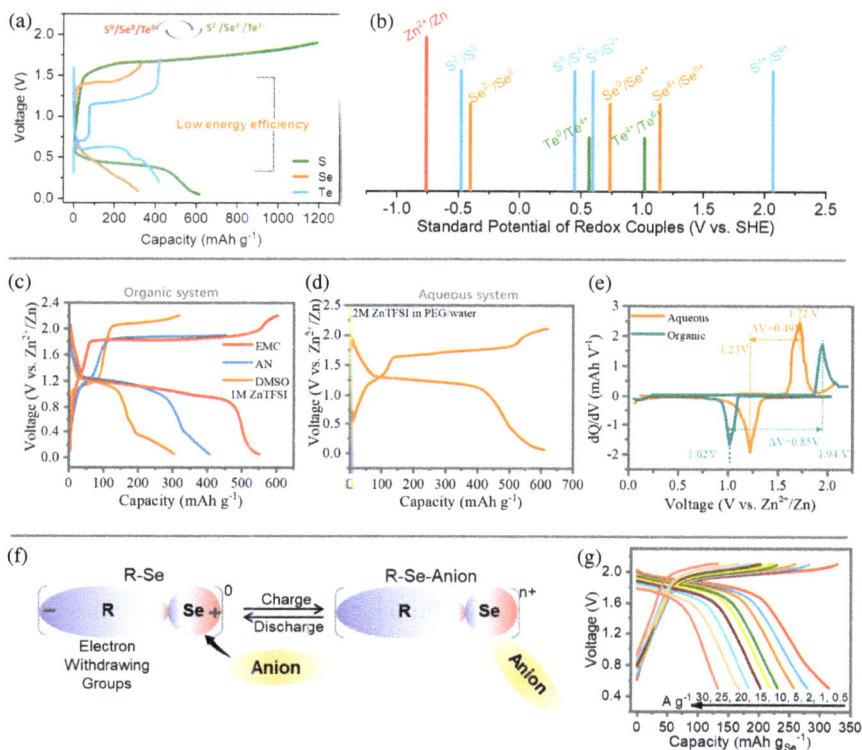

**Figure 10.8.** (a) GCD curves of conventional aqueous Zn||chalcogen batteries with 1 M ZnSO$_4$ electrolytes; (b) Standard potential of redox couples from chalcogen derivates. High-voltage Zn||Se battery: (c) GCD curves in the organic system and (d) GCD curves in the aqueous system; (e) dQ/dV of the discharge-charge profile of Zn||Se batteries at 0.1 A g$^{-1}$. Zn||organic Se battery: (f) Schematic illustration of the strategy for achieving positive valence conversion of chalcogen-based functional groups (both electron-withdrawing group and anions facilitate the positive valence conversion); (g) GCD curves at different current densities. Reproduced with permission from Refs. [42, 59].

plateau slopes, 0.94 V/(Ah g$^{-1}$) and 0.61 V/(Ah g$^{-1}$), were obtained for the organic and aqueous systems, respectively, due to the advanced conversion mechanism (Figure 10.8e). These unique features provide Zn||Se batteries with unprecedent energy densities up to 581 Wh kg$_{Se}^{-1}$ for the organic system and 751 Wh kg$_{Se}^{-1}$ for the aqueous system. Furthermore, by applying a molecular-level design strategy, Zhi and coworkers constructed organic Se-based molecules through the attachment of associated Se atoms with strong electron-withdrawing groups, aiming to undermine the surface charge density of Se to promote the positive valence conversion of Se (Figure 10.8f and 10.8g).[59] A high-potential triphenylphosphine selenide organic cathode (TP-Se) was developed with (TP-Se)$^-$ to (TP-Se)$^0$ to (TP-Se)$^+$ conversion. The Zn||TP-Se batteries exhibited a flat discharge plateau at 1.96 V and a superior discharge capacity. With stable TP molecular structures and optimized hybrid electrolytes, excellent cycling performance was achieved (up to 85.3% capacity retention after 4,300 cycles). The system is promising due to its high discharge voltage, which is higher than that ever reported for the organic cathodes of zinc batteries.

Promoting the unconventional positive valence conversion of chalcogens can enhance the electrochemical performance of Zn||chalcogen batteries in terms of output voltage, discharge capacity, and conversion kinetics. Such a strategy is still in the nascent stage and has not been applied to other Zn||chalcogen systems. However, it is difficult to achieve the positive valence conversion of chalcogens, and the key is promoting the oxidation of chalcogens as well as stabilizing and suppressing the dissolution of positive valence products. This involves the regulation of electrodes and electrolytes, including electrode structure optimization, catalyst introduction, appropriate anion selection, electrolyte viscosity tuning, and separator modification.

## 10.4 Perspectives

Although the history of Zn||chalcogen batteries as a new member of AZB systems is limited, it has shown remarkable energy storage performance and can compete with other AZB counterparts. Zn||chalcogen batteries have many advantages, including high discharge capacity, stable energy output, and low cost. In this chapter, we systematically discussed the conversion electrochemistry and stabilization strategies of different chalcogen elements and interchalcogen compound cathodes. Because the new type of Zn||chalcogen batteries are still in the nascent stage, there is high potential for their development and improvement. On the basis of the results of this new battery chemistry, the following issues should be addressed for further improvement:

i)   Research should be conducted to reveal the mechanism of relevant reactions, chemical composition, and structural evolution of chalcogen and interchalcogen compounds during the conversion reaction and the unconventional electrochemistry of S, Se, and Te allotropes and their combinations. Although Zn||chalcogen batteries undergo stable solid-to-solid conversion without the apparent formation of soluble products during cycling, the reported intermediates, such as $ZnS_2$ and $ZnTe_2$, indicate that the underlying mechanism of Zn||chalcogen batteries is complicated. Similarly, for the unconventional positive valence conversion of chalcogens, the involuted reaction mechanism may involve a spontaneous multielectron conversion process. More efforts on characterization techniques are needed to clarify the new chemistry.

ii) The interchalcogen compound cathode, especially S-rich compounds, can be the optimal choice because it features the advantages of low material cost, high gravimetric specific capacity, and high conductivity. The precise control of compound composition and the development of suitable electrolytes can provide more opportunities in developing high-energy batteries. Based on the current results, S, Se, and Te have their own advantages and disadvantages. S provides a high discharge capacity but is associated with electronic insulating problems. Se and Te exhibit distinct electronic conductivity, but the capacity of Te is less than that of S. Thus, for developing promising Zn||chalcogen batteries, the interchalcogen compound is the most favorable option for achieving high-energy AZBs.

iii) The critical performance metrics of zinc metal anodes are crucial for the practical realization of rechargeable AZBs, including the Coulombic efficiency, areal capacity, N/P ratio, critical current density (especially in quasi/solid-state batteries), shorting/failure time, compact density, and volume variation (during electrochemical stripping/plating). The Zn anode typically exhibits the problems of dendrite growth, hydrogen generation, and Zn metal corrosion and passivation. To build long-lasting, safe Zn||chalcogen batteries, polymer-based electrolytes and quasi-solid-state batteries should be developed. For the unconventional positive valence conversion of chalcogens with the inevitable dissolution of positive valence intermediates and final products, quasi-solid-state or solid-state electrolytes are promising for relieving or even eliminating dissolution problems.

iv) For these new battery architectures, studies should focus on improving the thermodynamic stability and kinetics of mass/charge transfer at heterointerfaces as well as in the bulk of the electrode/electrolyte, the chemical evolution and parasitic reaction of key materials/components, and the performance fade/failure mechanism of the battery. Solving these problems will require joint efforts from interdisciplinary fields and can facilitate the development of sustainable storage technology and economics in the future.

v) Several studies have verified the possibility of using chalcogens as anodes, but they have not received adequate research attention. For current energy storage devices, the anode has always been a huge obstacle to the practical application of high-performance batteries. Therefore, an anode with a suitable redox potential, high conductivity, high capacity, and excellent stability is urgently required. On the basis of the reliable conversion reaction mechanism, chalcogens can deliver stable ion storage performance, output competitive voltage and capacity, maintain distinct cycling performance, and exhibit high safety. When employed in aqueous batteries, the cycling life can be further enhanced ("shuttle free"), which can provide a favorable and reliable solution for high-performance energy storage devices.

# References

[1] Y. Liang, Y. Yao, *Nat. Rev. Mater.* **2023**, 8, 109–122.

[2] D. Chao, W. Zhou, F. Xie, C. Ye, H. Li, M. Jaroniec, S.-Z. Qiao, *Sci. Adv.* **2020**, 6, eaba4098.

[3] J. O. G. Posada, A. J. R. Rennie, S. P. Villar, V. L. Martins, J. Marinaccio, A. Barnes, C. F. Glover, D. A. Worsley, P. J. Hall, *Renewable Sustainable Energy Rev.* **2017**, 68, 1174–1182.

[4] Z. Ju, Q. Zhao, D. Chao, Y. Hou, H. Pan, W. Sun, Z. Yuan, H. Li, T. Ma, D. Su, B. Jia, *Adv. Energy Mater.* **2022**, 12, 2201074.

[5] Z. Liu, Y. Huang, Y. Huang, Q. Yang, X. Li, Z. Huang, C. Zhi, *Chem. Soc. Rev.* **2020**, 49, 180–232.

[6] J. M. Tarascon, M. Armand, *Nature* **2001**, 414, 359.

[7] P. Ruan, S. Liang, B. Lu, H. J. Fan, J. Zhou, *Angew. Chem. Int. Ed.* **2022**, 61, e202200598.

[8] G. Fang, J. Zhou, A. Pan, S. Liang, *ACS Energy Lett.* **2018**, 3, 2480–2501.

[9] Y. Zhao, Z. Chen, F. Mo, D. Wang, Y. Guo, Z. Liu, X. Li, Q. Li, G. Liang, C. Zhi, *Adv. Sci.* **2021**, 8, 2002590.

[10] Z. Tie, Z. Niu, *Angew. Chem. Int. Ed.* **2020**, 59, 21293–21303.

[11] L. Kang, M. Cui, Z. Zhang, F. Jiang, *Batteries Supercaps* **2020**, 3, 966–1005.

[12] A. von Wald Cresce, K. Xu, *Carbon Energy,* **2021**, 3, 721–751.

[13] M. Liu, H. Ao, Y. Jin, Z. Hou, X. Zhang, Y. Zhu, Y. Qian, *Mater. Today Energy* **2020**, 17, 100432.

[14] B.-E. Jia, A. Q. Thang, C. Yan, C. Liu, C. Lv, Q. Zhu, J. Xu, J. Chen, H. Pan, Q. Yan, *Small* **2022**, 18, 2107773.

[15] H. Li, L. Ma, C. Han, Z. Wang, Z. Liu, Z. Tang, C. Zhi, *Nano Energy* **2019**, 62, 550–587.

[16] Q. Yang, X. Li, Z. Chen, Z. Huang, C. Zhi, *Acc. Mater. Res.* **2022**, 3, 78–88.

[17] J. Shin, J. Lee, Y. Park, J. W. Choi, *Chem. Sci.* **2020**, 11, 2028–2044.

[18] C. Xu, B. Li, H. Du, F. Kang, *Angew. Chem. Int. Ed.* **2012**, 51, 933–935.

[19] S. Huang, J. Zhu, J. Tian, Z. Niu, *Chem. Eur. J.* **2019**, 25, 14480–14494.

[20] N. Ma, P. Wu, Y. Wu, D. Jiang, G. Lei, *Funct. Mater. Lett.* **2019**, 12, 1930003.

[21] F. Mo, Z. Chen, G. Liang, D. Wang, Y. Zhao, H. Li, B. Dong, C. Zhi, *Adv. Energy Mater.* **2020**, 10, 2000035.

[22] Z. Chen, T. Wang, Y. Hou, Y. Wang, Z. Huang, H. Cui, J. Fan, Z. Pei, C. Zhi, *Adv. Mater.* **2022**, 34, 2207682.

[23] Z. Chen, X. L. Li, D. H. Wang, Q. Yang, L. T. Ma, Z. D. Huang, G. J. Liang, A. Chen, Y. Guo, B. B. Dong, X. Y. Huang, C. Yang, C. Y. Zhi, *Energy Environ. Sci.* **2021**, 14, 3492–3501.

[24] H. Cui, L. Ma, Z. Huang, Z. Chen, C. Zhi, *SmartMat* **2022**, 3, 565–581.

[25] D. Wang, C. Han, F. Mo, Q. Yang, Y. Zhao, Q. Li, G. Liang, B. Dong, C. Zhi, *Energy Storage Mater.* **2020**, 28, 264–292.

[26] T. Xue, H. J. Fan, *J. Energy Chem.* **2021**, 54, 194–201.

[27] N. Zhang, Y. Dong, M. Jia, X. Bian, Y. Wang, M. Qiu, J. Xu, Y. Liu, L. Jiao, F. Cheng, *ACS Energy Lett.* **2018**, 3, 1366–1372.

[28] Z. Chen, C. Li, Q. Yang, D. Wang, X. Li, Z. Huang, G. Liang, A. Chen, C. Zhi, *Adv. Mater.* **2021**, 33, 2105426.

[29] S. Bai, X. Liu, K. Zhu, S. Wu, H. Zhou, *Nat. Energy* **2016**, 1, 16094.

[30] X. Liu, J.-Q. Huang, Q. Zhang, L. Mai, *Adv. Mater.* **2017**, 29, 1601759.

[31] A. Eftekhari, *Sustain. Energy Fuels* **2017**, 1, 14–29.

[32] K. Eom, J. T. Lee, M. Oschatz, F. Wu, S. Kaskel, G. Yushin, T. F. Fuller, *Nat. Commun.* **2017**, 8, 13888.

[33] M. Salama, Rosy, R. Attias, R. Yemini, Y. Gofer, D. Aurbach, M. Noked, *ACS Energy Lett.* **2019**, 4, 436–446.

[34] Y.-H. Wang, X.-T. Li, W.-P. Wang, H.-J. Yan, S. Xin, Y.-G. Guo, *Sci. China Chem.* **2020**, 63, 1402–1415.

[35] Z. Chen, Y. Zhao, F. Mo, Z. Huang, X. Li, D. Wang, G. Liang, Q. Yang, A. Chen, Q. Li, L. Ma, Y. Guo, C. Zhi, *Small Struct.* **2020**, 1, 2000005.

[36] Y. Boyjoo, H. Shi, Q. Tian, S. Liu, J. Liang, Z.-S. Wu, M. Jaroniec, J. Liu, *Energy Environ. Sci.* **2021**, 14, 540–575.

[37] S. Licht, Zinc sulfur battery, US Patent 6,207,324, **2001**.

[38] T. A. Bendikov, C. Yarnitzky, S. Licht, *J. Phys. Chem. B* **2002**, 106, 2989–2995.

[39] Y. Zhao, D. Wang, X. Li, Q. Yang, Y. Guo, F. Mo, Q. Li, C. Peng, H. Li, C. Zhi, *Adv. Mater.* **2020**, 32, 2003070.

[40] W. Li, K. Wang, K. Jiang, *Adv. Sci.* **2020**, 7, 2000761.

[41] Z. Chen, Q. Yang, F. Mo, N. Li, G. Liang, X. Li, Z. Huang, D. Wang, W. Huang, J. Fan, *Adv. Mater.* **2020**, 32, 2001469.

[42] Z. Chen, F. Mo, T. Wang, Q. Yang, Z. Huang, D. Wang, G. Liang, A. Chen, Q. Li, Y. Guo, X. Li, J. Fan, C. Zhi, *Energy Environ. Sci.* **2021**, 14, 2441–2450.

[43] G. I. Giles, K. M. Tasker, R. J. K. Johnson, C. Jacob, C. Peers, K. N. Green, *Chem. Commun.* **2001**, 2490–2491.

[44] J. R. He, W. Q. Lv, Y. F. Chen, K. C. Wen, C. Xu, W. L. Zhang, Y. R. Li, W. Qin, W. D. He, *ACS Nano* **2017**, 11, 8144–8152.

[45] Y. Yang, H. Yang, X. Wang, Y. Bai, C. Wu, *J. Energy Chem.* **2022**, 64, 144–165.

[46] Q. Pang, J. Meng, S. Gupta, X. Hong, C. Y. Kwok, J. Zhao, Y. Jin, L. Xu, O. Karahan, Z. Wang, S. Toll, L. Mai, L. F. Nazar, M. Balasubramanian, B. Narayanan, D. R. Sadoway, *Nature* **2022**, 608, 704–711.

[47] G. Chang, J. Liu, Y. Hao, C. Huang, Y. Yang, Y. Qian, X. Chen, Q. Tang, A. Hu, *Chem. Eng. J.* **2023**, 457, 141083.

[48] D. Liu, B. He, Y. Zhong, J. Chen, L. Yuan, Z. Li, Y. Huang, *Nano Energy* **2022**, 101, 107474.

[49] T. Zhou, H. Wan, M. Liu, Q. Wu, Z. Fan, Y. Zhu, *Mater. Today Energy* **2022**, 27, 101025.

[50] M. Cui, J. Fei, F. Mo, H. Lei, Y. Huang, *ACS Appl. Mater. Interfaces* **2021**, 13, 54981–54989.

[51] H. Zhang, Z. Shang, G. Luo, S. Jiao, R. Cao, Q. Chen, K. Lu, *ACS Nano* **2022**, 16, 7344–7351.

[52] A. Amiri, R. Sellers, M. Naraghi, A. A. Polycarpou, *ACS Nano* **2022**, 2, 1217–1228.

[53] W. Zhang, M. Wang, J. Ma, H. Zhang, L. Fu, B. Song, S. Lu, K. Lu, *Adv. Funct. Mater.* **2022**, 33, 2210899.

[54] C. Dai, X. Jin, H. Ma, L. Hu, G. Sun, H. Chen, Q. Yang, M. Xu, Q. Liu, Y. Xiao, X. Zhang, H. Yang, Q. Guo, Z. Zhang, L. Qu, *Adv. Energy Mater.* **2021**, 11, 2003982.

[55] C. Dai, L. Hu, L. Jin, H. Chen, X. Zhang, S. Zhang, L. Song, H. Ma, M. Xu, Y. Zhao, Z. Zhang, H. Cheng, L. Qu, *Adv. Mater.* **2021**, 33, 2105480.

[56] M. Yang, Z. Yan, J. Xiao, W. Xin, L. Zhang, H. Peng, Y. Geng, J. Li, Y. Wang, L. Liu, Z. Zhu, *Angew. Chem. Int. Ed.* **2022**, 61, e202212666.

[57] Z. Xu, Y. Zhang, W. Gou, M. Liu, Y. Sun, X. Han, W. Sun, C. Li, *Chem. Commun.* **2022**, 58, 8145–8148.

[58] L. Ma, Y. Ying, S. Chen, Z. Chen, H. Li, H. Huang, L. Zhao, C. Zhi, *Adv. Energy Mater.* **2022**, 12, 2201322.

[59] Z. Chen, H. Cui, Y. Hou, X. Wang, X. Jin, A. Chen, Q. Yang, D. Wang, Z. Huang, C. Zhi, *Chem* **2022**, 8, 2204–2216.

[60] C. Dai, L. Hu, H. Chen, X. Jin. Y. Han, Y. Wang, X. Li, X. Zhang, L. Song, M. Xu, H. Cheng, Y. Zhao, Z. Zhang, F. Liu, L. Qu, *Nat. Commun.* **2022**, 13, 1863.

[61] J. Wang, Y. Yang, Y. Wang, S. Dong, L. Cheng, Y. Li, Z. Wang, L. Trabzon, H. Wang, *ACS Nano*, **2022**, 16, 15770–15778.

[62] J. Wang, J. Du, J. Zhao, Y. Wang, Y. Tang, G. Cui, *J. Phys. Chem. Lett.* **2021**, 12, 10163–10168.

63] W. Li, Y. Ma, P. Li, X. Jing, K. Jiang, D. Wang, *Adv. Funct. Mater.* **2021**, 31, 2101237.

[64] W. Li, X. Jing, Y. Ma, M. Chen, M. Li, K. Jiang, D. Wang, *Chem. Eng. J.* **2021**, 420, 129920.

[65] N. N. Greenwood, A. Earnshaw, *Chemistry of the Elements*, Elsevier **2012**.

[66] S. Xin, Y. You, H.-Q. Li, W. Zhou, Y. Li, L. Xue, H.-P. Cong, *ACS Appl. Mater. Interfaces*, **2016**, 8, 33704–33711.

[67] A. Manthiram, S.-H. Chung, C. Zu, *Adv. Mater.* **2015**, 27, 1980–2006.

[68] C. Wu, W. H. Lai, X. Cai, S. L. Chou, H. K. Liu, Y. X. Wang, S. X. Dou, *Small*, **2021**, 17, 2006504.

[69] Y. Wang, E. Sahadeo, G. Rubloff, C.-F. Lin, S. B. Lee, *J. Mater. Sci.* **2019**, 54, 3671–3693.

[70] X. Fang, T. Zhai, U. K. Gautam, L. Li, L. Wu, Y. Bando, D. Golberg, *Prog. Mater Sci.* **2011**, 56, 175–287.

[71] J. Balach, J. Linnemann, T. Jaumann, L. Giebeler, *J. Mater. Chem. A* **2018**, 6, 23127–23168.

[72] X. Hong, J. Mei, L. Wen, Y. Tong, A. J. Vasileff, L. Wang, J. Liang, Z. Sun, S. X. Dou, *Adv. Mater.* **2019**, 31, 1802822.

[73] C.-P. Yang, S. Xin, Y.-X. Yin, H. Ye, J. Zhang, Y.-G. Guo, *Angew. Chem. Int. Ed.* **2013**, 52, 8363–8367.

[74] Y. Liu, J. Wang, Y. Xu, Y. Zhu, D. Bigio, C. Wang, *J. Mater. Chem. A* **2014**, 2, 12201–12207.

[75] S. Dong, D. Yu, J. Yang, L. Jiang, J. Wang, L. Cheng, Y. Zhou, H. Yue, H. Wang, L. Guo, *Adv. Mater.* **2020**, 32, 1908027.

[76] Z. Chen, Q. Yang, D. Wang, A. Chen, X. Li, Z. Huang, G. Liang, Y. Wang, C. Zhi, *ACS Nano* **2022**, 16, 5349–5357.

[77] R. Cooper, J. Culka, *J. Inorg. Nucl. Chem.* **1967**, 29, 1217–1224.

[78] M. Freni, D. Giusto, V. Valent, *J. Inorg. Nucl. Chem.* **1965**, 27, 755–756.

[79] M. F. Kotkata, S. A. Nouh, L. Farkas, M. M. Radwan, *J. Mater. Sci.* **1992**, 27, 1785–1794.

[80] R. C. Sharma, Y. A. Chang, *J. Phase Equilibria* **1996**, 17, 148–150.

[81] L. Vogel, P. Wonner, S. M. Huber, *Angew. Chem. Int. Ed.* **2019**, 58, 1880–1891.

[82] B. Krebs, F.-P. Ahlers, in *Adv. Inorg. Chem. Elsevier* **1990**, vol. 35, pp. 235–317.

[83] H.-J. Peng, J.-Q. Huang, Q. Zhang, *Chem. Soc. Rev.* **2017**, 46, 5237–5288.

[84] X. Huang, Y. Liu, C. Liu, J. Zhang, O. Noonan, C. Yu, *Chem. Sci.* **2018**, 9, 5178–5182.

[85] D. T. Li, R. C. Sharma, Y. A. Chang, *Bull. Alloy Phase Diagr.* **1989**, 10, 348–350.

[86] K. Xu, X. Liu, J. Liang, J. Cai, K. Zhang, Y. Lu, X. Wu, M. Zhu, Y. Liu, Y. Zhu, G. Wang, Y. Qian, *ACS Energy Lett.* **2018**, 3, 420–427.

[87] J.-U. Seo, G.-K. Seong, C.-M. Park, *Sci. Rep.* **2015**, 5, 7969.

[88] T. Chivers, R. S. Laitinen, K. J. Schmidt, J. Taavitsainen, *Inorg. Chem.* **1993**, 32, 337–340.

[89] H. Li, R. Meng, Y. Guo, B. Chen, Y. Jiao, C. Ye, Y. Long, A. Tadich, Q.-H. Yang, M. Jaroniec, S.-Z. Qiao, *Nat. Commun.* **2021**, 12, 5714.

[90] Z. Chen, X. Li, D. Wang, Q. Yang, L. Ma, Z. Huang, G. Liang, A. Chen, Y. Guo, B. Dong, X. Huang, C. Yang, C. Zhi, *Energy Environ. Sci.* **2021**, 14, 3492–3501.

[91] J. Ding, H. Zhang, W. Fan, C. Zhong, W. Hu, D. Mitlin, *Adv. Mater.* **2020**, 32, 1908007.

[92] H.-J. Kang, G. A. R. Bari, T.-G. Lee, T. T. Khan, J.-W. Park, H. J. Hwang, S. Y. Cho, Y.-S. Jun, *Nanomaterials* **2020**, 10, 2012.

[93] J. He, A. Manthiram, *Energy Storage Mater.* **2019**, 20, 55–70.

[94] B. Q. Li, L. Kong, C. X. Zhao, Q. Jin, X. Chen, H. J. Peng, J. L. Qin, J. X. Chen, H. Yuan, Q. Zhang, *InfoMat* **2019**, 1, 533–541.

[95] L. Peng, Z. Wei, C. Wan, J. Li, Z. Chen, D. Zhu, D. Baumann, H. Liu, C. S. Allen, X. Xu, *Nat. Catal.* **2020**, 3, 762–770.

[96] C. Deng, Z. Wang, S. Feng, S. Wang, J. Yu, *J. Mater. Chem. A*, **2020**, 8, 19704–19728.

[97] F. Liu, G. Sun, H. B. Wu, G. Chen, D. Xu, R. Mo, L. Shen, X. Li, S. Ma, R. Tao, *Nat. Commun.* **2020**, 11, 1–10.

[98] X. Liang, C. Hart, Q. Pang, A. Garsuch, T. Weiss, L. F. Nazar, *Nat. Commun.* **2015**, 6, 1–8.

[99] J. Zhang, X. Zhang, C. Xu, H. Yan, Y. Liu, J. Xu, H. Yu, L. Zhang, J. Shu, *Adv. Energy Mater.* **2022**, 12, 2103998.

[100] Y. Zhang, Q. Zhou, J. Zhu, Q. Yan, S. X. Dou, W. Sun, *Adv. Funct. Mater.* **2017**, 27, 1702317.

[101] S. Trasatti, *Pure Appl. Chem.* **1986**, 58, 955–966.

[102] L. Ma, J. Wu, Y. Li, Y. Lv, B. Li, Z. Jin, *Energy Storage Mater.* **2021**, 42, 723–752.

[103] X. F. Zhang, S. Q. Jiao, J. G. Tu, W. L. Song, X. Xiao, S. J. Li, M. Y. Wang, H. P. Lei, D. H. Tian, H. S. Chen, D. N. Fang, *Energy Environ. Sci.* **2019**, 12, 1918–1927.

[104] A. I. Alekperov, *Russ. Chem. Rev.* **1974**, 43, 235–250.

Chapter 11

# Sulfur-based Aqueous Batteries: Electrochemistry and Strategies[1]

Jiahao Liu,[a,b,*] Wanhai Zhou,[a,*] Ruizheng Zhao,[a] Zhoudong Yang,[a] Wei Li,[a] Dongliang Chao,[a,†] Shi-Zhang Qiao,[b] Dongyuan Zhao[a]

[a]Laboratory of Advanced Materials, Shanghai Key Laboratory of Molecular Catalysis and Innovative Materials, School of Chemistry and Materials, Fudan University, Shanghai, 200433, P. R. China
[b]School of Chemical Engineering and Advanced Materials, The University of Adelaide, Adelaide, SA 5005, Australia

While research interest in aqueous batteries has surged due to their intrinsic low cost and high safety, the practical application is plagued by restrictive capacity (less than 500 mAh g$^{-1}$) of electrode materials. Sulfur-based aqueous batteries (SABs) feature high theoretical capacity (1672 mAh g$^{-1}$), compatible potential, and affordable cost, arousing ever-increasing attention and intense efforts. Nonetheless, the underlying electrochemistry of SABs remains unclear, including complicated thermodynamic evolution and insufficient kinetic metrics. Consequently, multifarious irreversible reactions in various application systems imply the systematic complexity of SABs. Herein, rather than simply compiling recent progress, this chapter aims to construct a theory-to-application methodology. Theoretically, attention has been paid to a critical appraisal of the aqueous S-related electrochemistry, including fundamental properties evaluation, kinetic metrics with transient and steady-state analyses, and thermodynamic equilibrium and evolution. To put it into practice, current challenges and promising strategies are synergistically proposed. Practically, the above efforts are employed to evaluate and develop the device-scale applications, scilicet flow SABs, oxide SABs, and metal SABs. Last, chemical and engineering insights are rendered collectively for the future development of high-energy SABs.

---

\* These authors contributed equally.

† Corresponding author: chaod@fudan.edu.cn

[1] Adapted with permission from J. Liu, W. Zhou, R. Zhao, Z. Yang, W. Li, D. Chao, S.-Z. Qiao, D. Zhao, *J. Am. Chem. Soc.* **2021**, 143, 15475.

## 11.1 Introduction

A burgeoning battery industry has generated enormous achievements on smart devices, electric vehicles, and power grids over the past two decades. Thereinto, organic electrolytes underpin this industry due to their satisfactory performance, robust stability, and complete industrial supply chain. Nonetheless, expensive cost, rigorous operational conditions, and adverse properties (toxicity and flammability) underlie the insurmountable plights of organic batteries.[1] Therefore, researchers have turned attention back to a historically tried-and-true system, namely the aqueous electrolyte-based batteries (ABs). Equipped with low-cost, non-toxic and non-flammable merits,[2] ABs initiated a century-old era of commercialization with lead-acid batteries.[3] Furthermore, Jeff Dahn came up with the aqueous lithium-ion battery in 1994,[4] bringing in a new upsurge of research in ABs.

Admittedly, the existing intrinsic deficiencies in ABs have greatly hindered their further application. Specifically, the hydrogen evolution reaction (HER) and the oxygen evolution reaction (OER)[1c] extremely constrain the practical electrochemical stable window (ESW) of ABs, accompanying pH sensitivity, corrosion, and passivation problems. Although strategies have been implemented to widen the ESW, nevertheless, ABs still cannot compare against the high energy density of organic electrolyte-based Li-ion batteries. A pivotal limitation lies in the lack of aqueous-compatible electrodes with high capacity. As displayed in Figure 11.1, the majority of proposed aqueous electrodes possess restricted capacity and uncompetitive price. Thereinto, organics,[5] Prussian blue derivates,[6] and Mo-based materials[7] are short of specific capacities in comparison with electrodes for Li-ion batteries. Although manganese oxides[2,8] and vanadium oxides[9] are equipped with relatively high capacities, they also face challenges of elemental dissolution and structural instability due to their multi-valence states and complicated phases change in reactions. As a consequence, a historically proposed system, sulfur-based aqueous battery (SAB), has returned back to the view, as listed in Figure 11.1a.

S-based materials, including elementary sulfur and polysulfides (PSs), can fully fulfill the requirement of high specific capacity (1672 mAh g$^{-1}$) and low cost (*ca.* 52 Ah

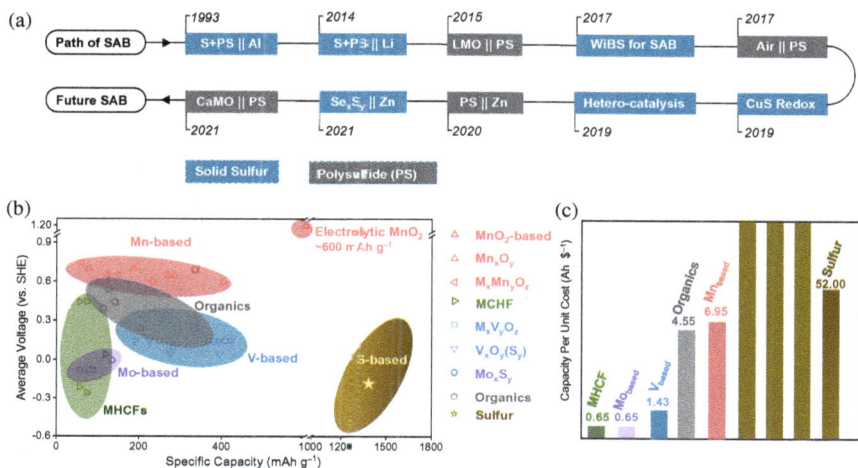

**Figure 11.1.** Status and significance of SABs. (a) Roadmap with major achievements. (b, c) Comparison of electrode materials applied in ABs on capacity (b) and cost (c).

per US dollar), as summarized in Figures 11.1b and 11.1c.[10] Nonetheless, subsequent progress has not been achieved until the past decade, ascribed to the technical limitations. From 2014,[12] articles on SABs have risen gradually, concentrating on redox flow SABs,[13] oxide SABs,[14] and metal SABs.[10–12,13b,15] Enlightened by these achievements, distinctive strengths and weaknesses of SABs are thus summarized and presented in Figure 11.2. Below is a brief description:

(a) The fundamental superiority of SABs is in the compatible potential in water,[13b,14b,16] serving as both cathode[15a,17] and anode[13c,14d] when matching with suitable counter electrodes;

(b) Sulfur exhibits top-level capacity *via* a two-electron transfer reaction;[10,18] some cases can even reach up to 3044 mAh $g_{(S)}^{-1}$ with extra redox contribution;[54]

(c) Furthermore, compared with the S-based organic battery (SOB), SAB features greater ionic conductivity and distinctive solubility of PSs due to the water-based solvents,[18] resulting in superior kinetics;

(d) Additionally, SAB shows excellent safety and non-toxicity properties.

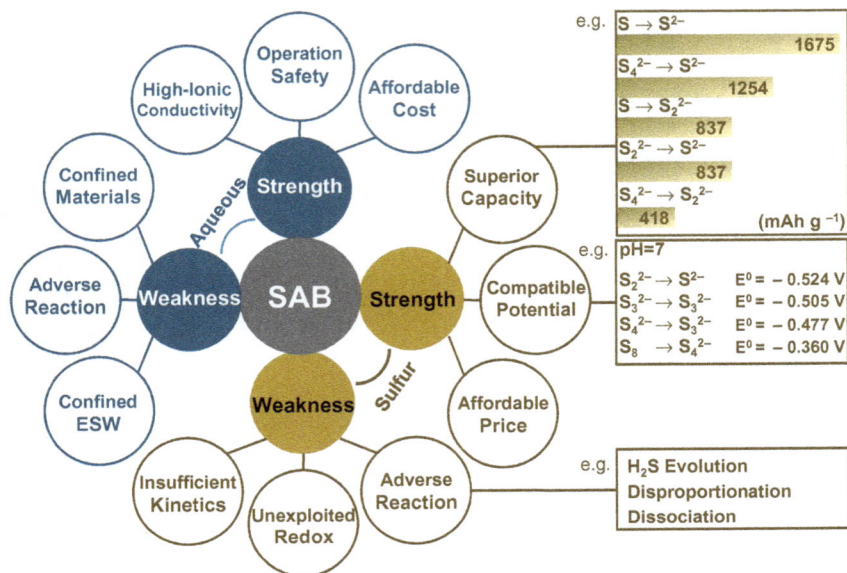

**Figure 11.2.** Strengths and weaknesses of SABs.

Unfortunately, SABs also have their own drawbacks. Except for the common HER/OER in ABs,[18] researchers have to cope with specific defects of SABs as summarized below:

(a) Complicated thermodynamic evolution in SABs would trigger various irreversible reactions, including $H_2S$ escape, disproportionation and dissociation, which degrade batteries doubtlessly;[14a]

(b) Diverse solid-liquid (S-L),[14d,15a] liquid-liquid (L-L),[13b,13d] and solid-solid (S-S) reactions[15d,15e] require a comprehensive evaluation on pros and cons;[13c]

(c) Moreover, the kinetics performance of SABs, which limits their power output, is still unsatisfactory.

The aforesaid strengths and weaknesses of SABs should be weighed prudently. The chapter reveals synergistic concerns of advanced SABs from their electrochemistry and challenges to evaluations and strategies. Critical analysis of recent advances in SABs is presented to establish a theory-to-application methodology at a device level, such as the redox flow SABs, oxide SABs, and metal SABs. Last, we provide a roadmap regarding the remaining issues and potential opportunities for the next-generation reliable SABs.

## 11.2 Electrochemistry of Sulfur-based Aqueous Batteries

### 11.2.1 Chemistry from Sulfur-based Organic Batteries to Sulfur-based Aqueous Batteries

#### 11.2.1.1 Chemistry of Sulfur-based Organic Batteries

To understand the electrochemistry of SABs on a full scale, a comparison is requisite from SOBs to SABs. For an exemplar Li-SOB, it commonly undergoes a two-stage S-L-S reduction from solid $S_8$ to the intermediates and to the final product $Li_2S$, as listed in Eqs. (11.1)–(11.4):[19]

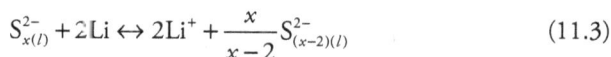

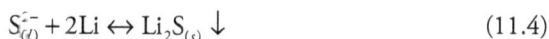

$$S_{8(s)} \leftrightarrow S_{8(l)} \tag{11.1}$$

$$S_{8(l)} + 2Li \leftrightarrow 2Li^+ + S_{8(l)}^{2-} \tag{11.2}$$

$$S_{x(l)}^{2-} + 2Li \leftrightarrow 2Li^+ + \frac{x}{x-2}S_{(x-2)(l)}^{2-} \tag{11.3}$$

$$S_{(l)}^{2-} + 2Li \leftrightarrow Li_2S_{(s)} \downarrow \tag{11.4}$$

This S-L-S reaction is determined by the solubility of PSs in organic solvents. According to Figure 11.3, the long-chain $Li_2S_n$ ($n \geq 4$) can be dissolved in most organic electrolytes. On the contrary, the short-chain $Li_2S_n$ ($n \leq 4$) are insoluble and would precipitate out at the end of discharge,[18] if without the usage of special solvents (considering some co-solvents[20] are able to dissolve PSs from $Li_2S_8$ to $Li_2S$).

Cañas et al.[19] have described this S-L-S redox in Li-SOBs following $S_{8(s)} \rightarrow S_{8(l)} \rightarrow S_{8(l)}^{2-} \rightarrow S_{6(l)}^{2-} \rightarrow S_{4(l)}^{2-} \rightarrow Li_2S_{2(s)} \rightarrow Li_2S_{(s)}$, as shown in Figure 11.3a. Obviously, SOBs show sluggish kinetics in view of two S-L/L-S conversions.[18],[21] Even worse, the soluble long-chain PSs can induce shuttle effect to trigger parasitic self-discharge and irreversible sediments. Consequently, the Coulombic efficiency, utilization, and lifespan of Li-SCBs are limited.

(a)

(b)

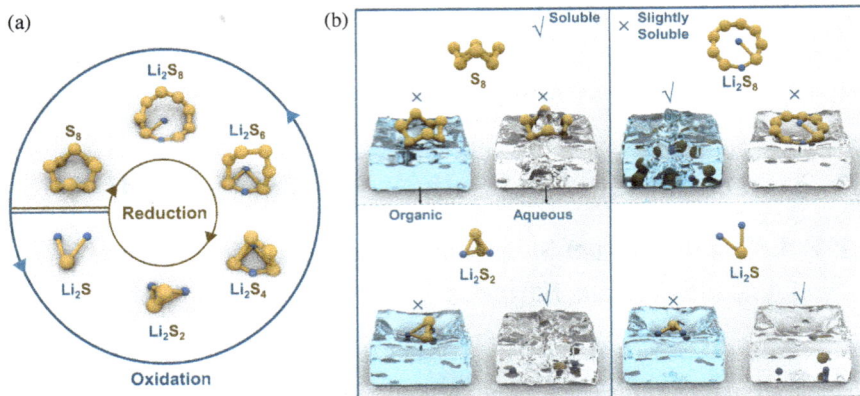

**Figure 11.3.** Comparison between SOBs and SABs. (a) Reduction and oxidation routes of Li-SOBs. (b) Solubility of S-based materials in Li-SOBs and Li-SABs.

### 11.2.1.2 *Chemistry of Sulfur-based Aqueous Batteries*

By contrast, the solubility of PS in an aqueous solvent is totally different, as shown in Figure 11.3b. Short-chain PSs (including $Li_2S_n$, $Na_2S_n$, and $K_2S_n$, n ≤ 4)[11,13a,14a,14e] present very high molar values of concentration in water,[10] while long-chain PSs are slightly dissolved. Therefore, the reaction route started from solid sulfur of SABs is a relatively simple one-step S-L redox. And the ionic conductivity in water is significantly greater (over 100 mS cm$^{-1}$, one or two orders of magnitude higher than that in organic solvents) to accelerate the mass transport.[18] Licht *et al.*[10–11,22] initially divided this one-step S-L redox into two phases: a full route (S-L redox) and a half route (L-L redox) in 1993. Very recently, researchers have also developed S-S redox. Therefore, three types of redox mechanisms can be sketched for SABs.

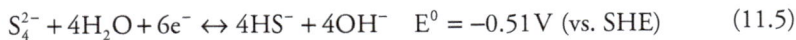

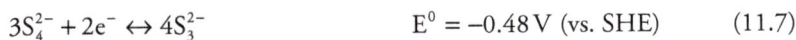

$$S_4^{2-} + 4H_2O + 6e^- \leftrightarrow 4HS^- + 4OH^- \quad E^0 = -0.51\,V \text{ (vs. SHE)} \quad (11.5)$$

$$S + H_2O + 2e^- \leftrightarrow HS^- + OH^- \quad E^0 = -0.51\,V \text{ (vs. SHE)} \quad (11.6)$$

$$3S_4^{2-} + 2e^- \leftrightarrow 4S_3^{2-} \quad E^0 = -0.48\,V \text{ (vs. SHE)} \quad (11.7)$$

$$S_3^{2-} + S \leftrightarrow S_4^{2-} \quad E^0 = -0.48\,V \text{ (vs. SHE)} \quad (11.8)$$

Eq. (11.5) describes the L-L redox of SAB with good kinetics but a relatively low capacity of 1254 mAh g$^{-1}$. Eq. (11.6) depicts the S-L redox of SAB, which is able to yield a high capacity$_{Sulfur}$ by completely utilizing solid sulfur, as explained in Eqs. (11.7) and (11.8).[10] Whilst there is a price, the S-L redox presents slower kinetics than that of L-L due to the extra solid-involved activation. Nevertheless, the kinetics of S-L redox in SAB still far surpasses the S-L-S redox in Li-SOBs.[14d,18] It is worth noting that the actual redox potential $E^0$ depends on the counter ions, especially in S-S redox, which determines the practical and available coupling system. In order to realize the full potential of SABs, it demands comprehensive electrochemical metrics on kinetics, thermodynamics, and a collection of exiting challenges in SABs.

## 11.2.2 *Electrochemistry of Sulfur-based Aqueous Batteries*

### 11.2.2.1 *Kinetic Metrics in Sulfur-based Aqueous Batteries*
Kinetic metrics are necessary to confirm the rate-determining step (RDS) and clarify the reaction mechanism for multiple-electron conversion reactions. Overall, a complete kinetic consideration includes mass transport and electrode process (chemical conversion and electrochemical step).[23]

The L-L redox, including $S_4^{2-} \leftrightarrow S_2^{2-}$, $S_4^{2-} \leftrightarrow S^{2-}$ or $S_2^{2-} \leftrightarrow S^{2-}$ as a relatively simple kinetic model, has been frequently employed in flow SABs. In the context of the static mode (flow rate = 0), the output power of flow SABs is highly limited due to the fact that the mass transport rate is lower than the redox rate of PSs.[24] To accelerate the mass transport in a dynamic mode by stirring (flow rate > 0), the relationship between output power and flow rate has to obey a logarithmic curve. It indicates that the RDS is synergistically shaped by electrode process and mass transport. Ultimately, if the mass transport rate exceeds the redox rate of the electrode process, the latter dominates the main RDS.[23a] As listed in Eq. (11.9), the intrinsic mass transport rate determined by diffusion coefficient $(D_i)$ is inversely proportional to the radii value of ions $(r_i)$ and the viscosity of the electrolyte media $(\eta)$, regardless of the external driving force. Hence the charge particles, solution concentration and solvent property can regulate the mass transport rate.

$$D_i = \frac{kT}{6\pi r_i \eta} \tag{11.9}$$

Lu *et al.* have initially revealed the kinetic feature in an actual L-L redox,[13a] as displayed in Figure 11.4a. Transient state cyclic voltammetry (CV) tests with variable scan rates

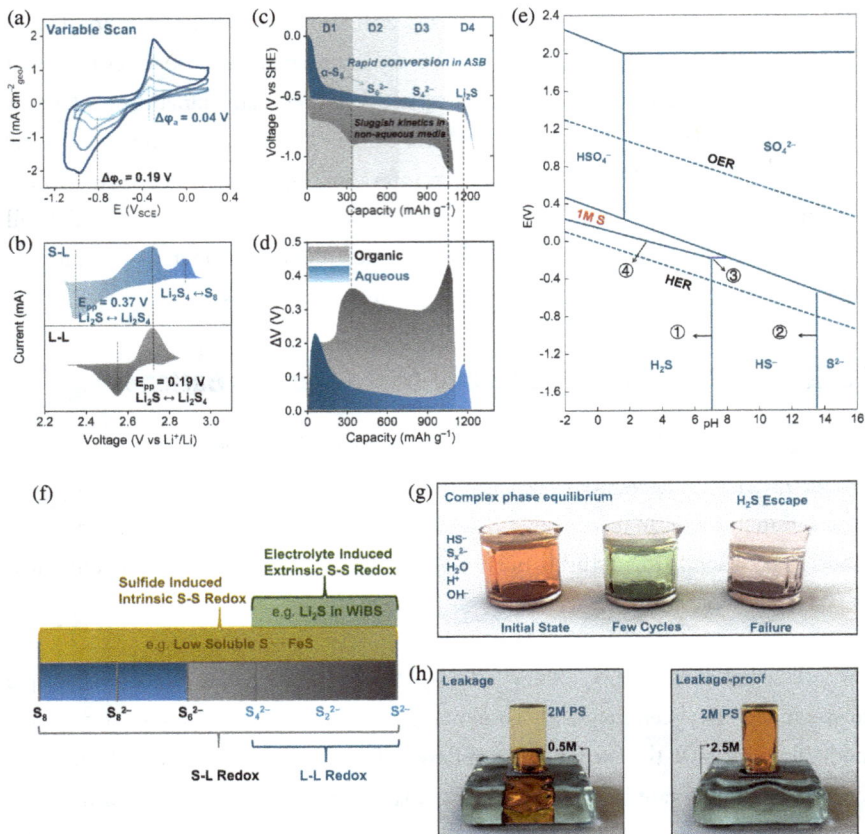

**Figure 11.4.** Electrochemical evaluation of SABs. (a) CV profiles of asymmetric kinetics in flow SABs. Reproduced with permission from Ref. [13a]. Copyright 2014 Elsevier. (b) Kinetic features in S-L redox and L-L redox. Reproduced with permission from Ref. [15a]. Copyright 2016 Royal Society of Chemistry. (c, d) Kinetic comparison of SOB and SAB electrodes. $\Delta V$ represents the voltage difference between reduction and oxidation. Reproduced with permission from Ref. [14d]. Copyright 2017 National Academy of Sciences. (e) Phase equilibrium for sulfur in aqueous solvent described by Pourbaix diagram. (f) Representative redox of S-L, L-L and S-S. (g) $H_2S$ evolution induced in cycling failure. (h) Chemical crossover and shuttle of soluble short-chain PSs.

highlight the asymmetric kinetics between reduction and oxidation stages of L-L type SAB. Thereinto, an obvious difference of potential change between anodic peak ($\Delta \varphi_a = 0.04$ V) and cathodic peak ($\Delta \varphi_c = 0.19$ V) exists, which is detrimental to the utilization of active materials and long-term stability of batteries. Additionally, the L-L redox-based SAB can also serve as a non-flow SAB, described in Figure 11.4b, where the gray curve represents L-L type and the blue curve is the S-L redox.[15a] The peak-to-peak potential $E_{pp}$ is 190 mV, much more than the transport RDS standard value (56.5/nmV). The calculated ratio of cathodic and anodic peak current $I_{pc}/I_{pa}$ in $S_4^{2-} \rightarrow S^{2-}$ is 0.85, much smaller than the transport RDS standard value of 1.[25] These demonstrate that the RDS for L-L redox is the electrode process, as is the S-L redox. To accurately evaluate the electrode process, detailed parameters of electrochemical polarization, concentration polarization, and Ohmic polarization are considered to describe the dominance at different stages of discharge/charge.

Exchange current density $i^0$ and rate constant $k^0$ are crucial to measuring the electrochemical polarization, which profiles the relationship between overpotential $\eta$ and net current density $I$, where variable $I$ can easily stimulate the variable $\eta$ at low $i^0$ and hardly impact on $\eta$ at high $i^0$. Generally, $i^0$ and $k^0$ can help to optimize battery design, especially for the interfacial properties between electrode and electrolyte. Nonetheless, the existing literature mainly focuses on macroscopic rate performance, with few concerns on the basic calculations of $i^0$ and $k^0$. Chiang et al.[26] precisely measured the $i^0$ and $k^0$ values of SOBs in various organic solvents. Accordingly, the $i^0$ values in the organic solvent are 0.01–1 mA cm$^{-2}$, determined by Eq. (11.10).[26] And its $\eta$-$I$ relationship shows that the $\eta$ of SOB is a small-polarization type (linear curve) rather than the high-polarization type (semi-logarithmic curve). Inspired by this work, the $i^0$ values of SABs are promising to exceed those in SOBs, in view of higher solvent conductivity in SABs. Furthermore, modification on electrodes and development of multicomponent aqueous solvents are considered to increase $i^0$ values significantly.

$$i^0 = \frac{RT}{zR_{ct}FA} \tag{11.10}$$

For an actual S-L redox-based SAB in Figure 11.4b, in view of the huge activation energy of $S(s) \leftrightarrow S_4^{2-}$, the subsequent chemical conversion is highly postponed. Therefore, the $E_{pp}$ values of $S^{2-} \rightarrow S_4^{2-}$ redox in the S-L and L-L curves are markedly different. The activation energy is partially associated with the inferior electronic conductivity of solid sulfur ($5 \times 10^{-28}$ S m$^{-1}$, $F_{ddd}$ space group),[27] whose bandgap is much higher than its congener element or solid solution (e.g., 3.442 eV for $S_8$ vs. 2.777 eV for $Se_2S_6$, calculated by Heyd-Scuseria-Ernzerhof hybrid functional).

The electrochemical polarization between the S-L-S redox SOB and the S-L redox SAB are well contrasted in Figures 11.4c and 11.4d.[14d] D1–D4 refer to different dominant PS species at corresponding discharge states. Three stages of $\Delta V$ (polarization between charge and discharge) can be found in Figure 11.4d, *i.e.*, the initial high $\Delta V$ of solid activation; the middle plateau $\Delta V$ of L-L redox; and the relatively high end $\Delta V$ of concentration polarization. Obviously, SABs exhibit better kinetics than SOBs. From a device-level perspective, Ohmic polarization can also be the RDS, which is mainly generated by the membrane with large impedance.[13b]

As a consequence, the metrics, involving RDSs from mass transport and electrode process, indicate that both the electrolyte conditions (concentration, component, and flow velocity), redox routes, and electrode properties are the dominating factors in designing power-oriented SABs.

### 11.2.2.2 *Thermodynamic Metrics in Sulfur-based Aqueous Batteries*

The Pourbaix diagram, as displayed in Figure 11.4e, reveals the complicated thermodynamic evolution, which is controlled by the pH condition and initial chemical activity of S-based materials. $S^{2-}$, $HS^-$, $H_2S_{(aq)}$, $SO_4^{2-}$ and $HSO_4^-$ are regarded as stable S-based species in water at an initial chemical activity of 1 M sulfur at 25°C. Thereinto, evolution ① refers to the $HS^-/H_2S$ at pH = 7; ② corresponds to the $HS^-/S^{2-}$ at pH = 13.90; ③ represents $S_{dissolved} \rightarrow HS^-$ at around pH > 7; and ④ indicates $S_{dissolved} \rightarrow H_2S$ in relatively acidic conditions. Accordingly, pH conditions dominate the evolution direction and map out the redox route at certain activity of S-based materials. A relatively high pH is suggested to avoid the $H_2S$ escape.

In addition, regulation of initial chemical activity (concentration) of PSs also plays an important role through affecting the chemical equilibrium.[28] Applicable concentration of PS is 2–25 M in aqueous solvent. A relatively high concentration of PSs leads to $S_x^{2-}/S^{2-}$ evolution and low concentration contribute to the $S_x^{2-}/HS^-$ evolution in electrochemical reduction.[15b]

Aside from the $H_2S$ escape, thermodynamically concomitant disproportionation and dissociation listed in Eqs. (11.11) and (11.12) also damage the lifespan of SABs. As a result, spontaneous thermodynamic evolution in SABs undoubtedly causes intrinsic loss of active materials. Fortunately, regulation of pH and chemical activity can facilitate an ideal redox route and reduce adverse reactions to some extent; however, exorbitant high pH value and concentration of PSs may induce other issues, such as restricted scope of materials selection and high viscosity of electrolyte.

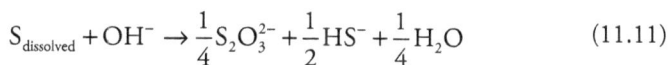

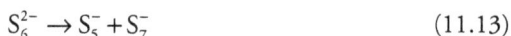

$$S_{dissolved} + OH^- \rightarrow \frac{1}{4}S_2O_3^{2-} + \frac{1}{2}HS^- + \frac{1}{4}H_2O \tag{11.11}$$

$$S_6^{2-} \rightarrow 2S_3^- \tag{11.12}$$

$$S_6^{2-} \rightarrow S_5^- + S_7^- \tag{11.13}$$

### 11.2.2.3 Redox Mechanisms in Sulfur-based Aqueous Batteries

As mentioned above, three types of redox mechanisms have been proposed in SABs, i.e., L-L type, S-L type, and S-S type redox, as depicted in Figure 11.4f.

#### 11.2.2.3.1 Liquid-liquid redox

L-L redox is driven by soluble short-chain PSs, covering $S_4^{2-} \leftrightarrow S^{2-}$, $S_4^{2-} \leftrightarrow S_2^{2-}$, $S_2^{2-} \leftrightarrow S^{2-}$, etc. Specifically, $S_4^{2-} \leftrightarrow S^{2-}$ performs well in many typical cases due to its considerable capacity of 1254 mAh g$^{-1}$.[22] Zhou et al.[15a] and Arumugam et al.[15b] successfully employed $S_4^{2-} \leftrightarrow S^{2-}$ as high-capacity cathode redox to match with metal anodes, thus outputting promising energy densities. In addition, $S_4^{2-} \leftrightarrow S_2^{2-}$ is mainly adopted in flow SABs as an anolyte, commonly coupled with haloid catholyte, such as Br-flow SABs and I-flow SABs.[13c,29] The shorter the redox chain is, the better the kinetics performs. However, the limited capacity ceiling of 418 mAh g$^{-1}$ for $S_4^{2-} \leftrightarrow S_2^{2-}$ dwindles the superiority of its SAB. Therefore, Lu et al.[13a] compromise between kinetics and capacity, adopting a nominal $S_2^{2-} \leftrightarrow S^{2-}$ redox to realize a capacity of 837 mAh g$^{-1}$ in their I-flow SABs. However, the kinetics gap between $S_2^{2-} \leftrightarrow S^{2-}$ and $I^- \leftrightarrow I^{3-}$ requests the usage of suitable catalysts. Extra attention needs to be paid here to control the voltage regime in L-L type SABs and avoid the formation of $S_8$.

#### 11.2.2.3.2 Solid-liquid redox

S-L redox starts from solid sulfur, which yields a substantial theoretical capacity of 1672 mAh g$^{-1}$. Licht et al.[1c] designed a hybrid cathode with 1:1 wt.% of elemental sulfur and $K_2S_2$ to form a solid solution. They successfully realized a measured capacity of 903 mAh g$^{-1}$, wherein the $K_2S_2$ provides the solubilizing interfaces for elemental sulfur. It is found that the dominant PSs are changed from $S_4^{2-}$ to shorter species, obeying the description of Eqs. (11.7) and (11.8), which can maximally consume the $S_8$. This mechanism was also promoted by Zhou *et al.* to enhance the capacity of $S_4^{2-} \leftrightarrow S^{2-}$ redox, which delivered 720 mAh g$^{-1}$ at a current of 0.4 mA cm$^{-2}$.[15a] The authors successfully achieved capacity as high as 1202 mAh g$^{-1}$ after mixing $Li_2S_4$ with solid sulfur at an optimized ratio of 2:8, demonstrating that almost all the $S_8$ species are utilized through Eqs. (11.6)–(11.8).

### 11.2.2.3.3 Solid-solid redox

As shown in Figure 11.4f, S-S redox can be artificially classified into two categories, *i.e.*, intrinsic and extrinsic. The intrinsic S-S redox is on the basis of a high affinity between sulfur and transition metals or low solubility of metal sulfides, such as the Fe, Ni, Cu, and Zn sulfides.[15d,15e,17] Consequently, intrinsic S-S redox is commonly employed in metal SABs. By contrast, extrinsic S-S redox frequently uses high-concentration electrolytes to strip free water from soluble PSs (LiPSs and NaPSs). The PSs thus present in solid states and serve in oxide/metal SABs.

Notably, the superiority of S-S redox is to stabilize the structure of active electrodes, avoid irreversible reactions ($H_2S$ escape) from the source, and construct robust redox pairs in water. Furthermore, the solid-induced polarization in S-S redox is acceptable (within 30–200 mV) in storing metal anions such as Li, Cu, Fe.[49,54–55]

As a consequence, for L-L, S-L, and S-S redox, the application scenarios are power-oriented SABs, capacity-oriented SABs, and performance-balanced SABs, respectively. Considering the kinetics, thermodynamics, and redox mechanisms of SABs, their drawbacks may overshadow the promise. Hence, it is time to establish an all-inclusive cognition of existing obstacles on SABs in order to put forward targeted strategies.

## 11.2.3 Challenges of Sulfur-based Aqueous Batteries

From electrode to battery, four issues are noteworthy in SABs:

(1) Narrow ESW. Taking the cathodic, anodic, and other polarizations into consideration, the ESW is only slightly larger than 1.23 V due to the HER/

OER.[1c,30] Worse still, this narrow ESW would not only limit the output voltage, but further exclude numerous high-capacity redox pairs reacting far below the HER or far above the OER potential. Therefore, Qiao *et al.*[1c] summarized positive strategies to widen ESW for general ABs, such as adjusting the pH value, introducing a solid electrolyte interface, employing suitable electrolyte additives,[31] modifying electrode structure, and utilizing competitive redox pairs. Usage of the abovementioned strategies for SABs needs further elaboration;

(2) Irreversible thermodynamic reactions. Tarascon *et al.*[46,47] have established the first assessment on irreversible $H_2S$ escape occurring at neutral and acidic solvents, as observed in Figure 11.4g. Further, Wan *et al.*[57] developed quantitative determination by using gas mass spectroscopy. Moreover, $H_2S$ can also corrode the stainless steel substrate in SABs. Besides, disproportionation deserves equal attention, which may cause capacity fading and degradation.

(3) Insufficient kinetics. The insufficient kinetics of SABs embodies huge activation energy in S-L conversion, asymmetrical polarization in L-L type SABs, and various polarizations for different S-S redox. Strategies should be tailored for diverse application scenarios.

(4) Device-level issues. As displayed in Figure 11.4h, diffusion of soluble PSs can cause leakage, shuttle, and chemical crossover under concentration gradient or electric field gradient, especially for the cases with Nafion membranes. Additionally, redox coupling of anode and cathode can make the most of the ESW range; hence, it is another technical issue that is associated with energy and power densities.

# 11.3 Strategies for Sulfur-based Aqueous Batteries

Considering that one strategy can target multiple issues, feasible strategies are discussed synoptically as below.

## 11.3.1 *pH Regulation*

It is a highly efficient method to implement pH regulation, due to its synergistic effect to handle the thermodynamics-induced problems and ESW.

For ESW, a neutral electrolyte is widely accepted to be equipped with the largest ESW compared with acid or alkali in the single electrolyte. However, in order to deliberately guide redox direction and shun the HER or OER potential, pH regulation is frequently adopted. An increase in pH can synergistically evade the HER potential and $H_2S$ escape. Specifically, Zhou et al. have successfully decreased the HER potential to 2.08 V (vs. Li/Li$^+$) and ensured the safe operation of PS redox from 2.3–2.8 V (vs. Li/Li$^+$) through adding 0.2 M LiOH.[15a] KOH[10,28] and NaOH[13b] are also employed to ensure the dominance of $S^{2-}$ rather than $H_2S$ at pH values above 12. Additionally, Arumugam et al.[15i] also verified that the alkaline (9 < pH < 14) electrolyte facilitates the degradation of thiosulfate to reduce irreversible reactions.

Decreasing pH value somewhile is crucial for counter electrodes such as air electrode. In this context, an acid-alkali separation battery is proposed to ensure an alkaline surrounding for the soluble PS side and an acidic environment for the counter side. For instance, Chiang et al.[13b] designed an Air||PS flow SAB, where the acid electrolyte contributed an energy density of 121 Wh L$^{-1}$. It is worth mentioning that the S-S redox, generally, is able to remain stable in neutral electrolytes without adverse evolution even if the counter electrode is not compatible with acid/alkali surroundings.

## 11.3.2 Redox Coupling

Redox coupling controls the voltages to inhibit HER/OER through limiting anodic and cathodic redox range of PSs synergistically within the ESW, as shown in Figure 11.5a.

Remick et al.[29] reported the first Br||PS flow SAB without strict voltage control; hence the unexpected $S_8$ gradually accumulated and resulted in battery failure. Learning from this case, Zhou et al.[15a] confined a voltage region (2.3–2.8 V vs. Li$^+$/Li) designedly to avoid the HER at low potential and refuse the precipitation of $S_8$ at high potential. Additionally, from a device-level perspective, redox coupling is feasible for designing an organic-aqueous separated battery through coupling the metal Li at organic side and the PSs at aqueous side with enlarged ESW. In order to control the PS reaction, it is suggested to confine the redox region within short chains (e.g., $S_4^{2-} \leftrightarrow S^{2-}$ rather than $S_4^{2-} \leftrightarrow HS^{2-}$) due to the kinetic consideration.[15c]

(a) Redox Coupling
$S_4^{2-}/S^8$ S-L Reaction
$S_4^{2-}/S^{2-}$ L-L Reaction
HER
Cell Potential (V)
Capacity (mAh g$^{-1}$)

(b) Energy Efficiency
Ohmic Resistance ($\Omega$ cm$^2$)
Negative Correlation
Combined Concentration
Energy Efficiency (%)
Combined Concentration (M)

(c) Li$_2$S$_4$ in WiBS    Li$_2$S$_4$ in Water
Li$_2$S$_4$
H$_2$O

(d) $\mu = \frac{1}{6\pi\eta r} + a_{solvent}$
Mobility ($10^{-5}$ cm$^2$ V$^{-1}$ s$^{-1}$)
Concentrated
Solvent Contribution
1/Viscosity (mPa$^{-1}$ s$^{-1}$)

(e) Mediator Redox
Zn(CH$_3$COO)$_2$  Zn(CH$_3$COO)$_2$+I$_2$
Equilibrium Potential
$\Delta E = 0.73$V
Potential (V vs Zn$^{2+}$/Zn)
Capacity (mAh g$^{-1}$)

(f) Power Density (mW cm$^{-2}$)
Current Density (mA cm$^{-2}$)
PS anion    I$_3^-$ anion

(g) Electrode modification
Ppy
S
Current (A g$^{-1}$)
S@Ppy-2$^{nd}$
S@Ppy-10$^{th}$
S@AC-2$^{nd}$
S@AC-10$^{th}$
Potential (V vs SHE)

(h) Catalysis
Hetero CoS/CoS$_2$
CoS
CoS$_2$
Current Density (mA cm$^{-2}$)
$E_{pp} = 0.1$V
Potential (V vs Ag/AgCl)

(i) S    SeS
Fermi level
Density of State

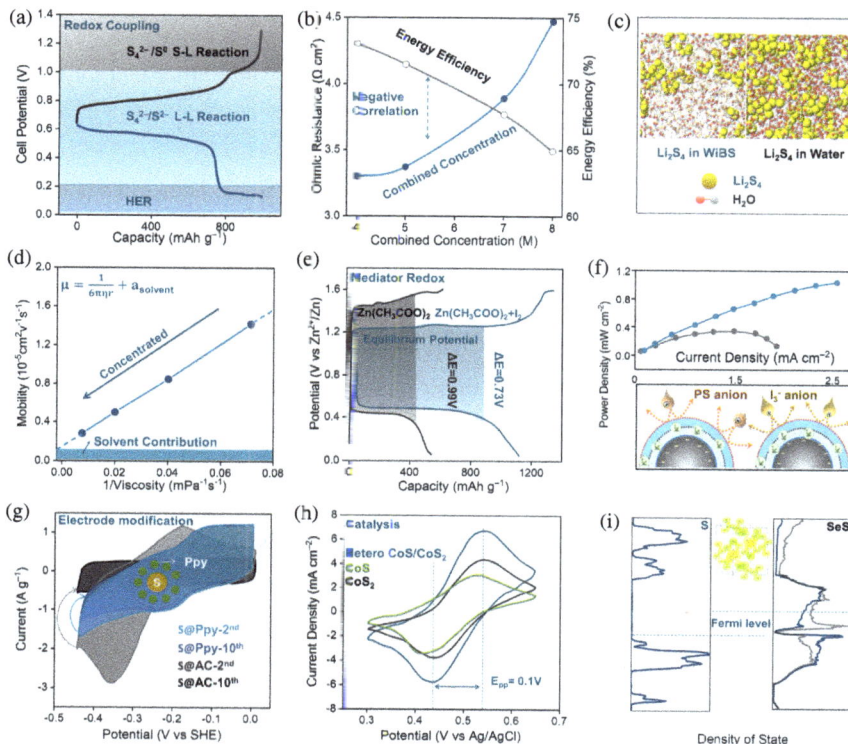

**Figure 11.5.** Summarized strategies for SABs. (a) Redox coupling to inhibit HER and confine redox regime. Reproduced with permission from Ref. [15b]. Copyright 2018 American Chemical Society. (b) Evaluation of activity of PSs. Reproduced with permission from Ref. [13a]. Copyright 2016 Elsevier. (c) Extrinsic S-S redox induced by WiBS strategy. Reproduced with permission from Ref. [14d]. Copyright 2017 National Academy of Sciences. (d) Evaluation of entire viscosity of concentrated electrolyte. Reproduced with permission from Ref. [26]. Copyright 2016 Electrochemical Society, Inc. (e) Polarization regulation by electrolyte additives. Reproduced with permission from Ref. [17]. Copyright 2020 John Wiley and Sons. (f) Membrane engineering of Nafion and hybrid modification. Reproduced with permission from Refs. [9h,32]. Copyright 2018 American Chemical Society and Copyright 2021 Springer Nature, respectively. (g) Electrode modification by organics and carbon. Reproduced with permission from Ref. [12]. Copyright 2013 Royal Society of Chemistry. (h) Electrochemical enhancement under single and heterojunction catalysts. Reproduced with permission from Refs. [2,13d]. Copyright 2019 Springer Nature. (i) Density of state result by solid-solution strategy for electrode modification.

### 11.3.3 *Polysulfide Activity Control*

A balanced consideration is essential to set the chemical activity (initial concentration) of PSs.

Li *et al.* supplied the advantages for high-concentrated PSs to output better energy density.[33] Positively, relatively high-concentrated PSs facilitate the dominant redox and avoid the adverse evolution.[15b] In addition, Tarascon *et al.* figured out that a slight excess of the concentration of PS can compensate for the loss of active sulfur caused by $H_2S$ escape.[14a–c] In contrast, as shown in Figure 11.5b, continuously increasing the concentration of PS would not only trigger higher Ohmic resistance but also hinder the mass transport due to enhanced viscosity. As a result, the kinetics of SABs would be seriously degraded. Moreover, the selectivity and isolation of the membrane can also be destroyed under relatively high PS concentration. Chiang *et al.* proposed that an empirical value to balance energy density and kinetics is to employ *ca.* 5 M PS in their flow SABs.[13b] Nevertheless, pre-experiments, such as the electrochemical impedance spectrum analysis and $i^0$ calculation, are suggested to determine the optimized concentration in corresponding types of SAB.

### 11.3.4 *Electrolyte Optimization*

Electrolyte engineering includes salt-oriented and additive-oriented modifications. For the former, water-in-salt (WiS), water-in-bisalt (WiBS),[34] and gelation electrolyte are widely practiced in ABs.

Wang *et al.* introduced the WiBS (21 M bistrifluoromethanesulfonimide lithium salt, *i.e.*, LiTFSI and 7 M lithium trifluoromethanesulfonate, *i.e.*, LiOTf) strategy into SAB to enlarge the ESW.[14d] Benefiting from that, HER potential observably dropped due to the formation of a solid electrolyte interface layer, and thus the ESW was extended to 1.7 V. Furthermore, WiBS confined the kinetics of free water molecules and solubility of PSs. In detail, molecular dynamics simulation reveals the underlying mechanism of WiBS, as shown in Figure 11.5c. With WiBS, an extrinsic S-S redox can be realized by anchoring the free water molecule with the salt molecule. As a result, the soluble PSs separated out in a solid state. Additionally, after further gelling by polyvinyl alcohol, the trace water-PS interaction can be inhibited, and the risk of $H_2S$

evolution thus has been eliminated. Later, Wan *et al.*[15g] and Zhi *et al.*[15h] implemented the WiBS strategy with $Al(OTf)_3$ − LiTFSI in Al-SABs and $Zn(TFSI)_2$ + LiTFSI in S||Zn SABs for the same purpose. Notably, concentrated LiTFSI is necessary to drive the WiBS scheme and shape the electrolyte with negligible free water. Nonetheless, as shown in Figure 11.5d, ionic migration has been limited in a concentrated electrolyte, and the expensive investment on LiTFSI is not affordable for commercialization. It is suggested that the future of WiBS lies in the exploration of cheap salts or construction of local high-concentrated electrolytes.

Among various additives, the redox mediator is noteworthy, which realizes infeasible redox under ordinary conditions. Trace $I_2$ (0.1–0.5 wt%) has been used in metal-air batteries as the redox mediator. The calculated equilibrium potential for S||Zn SABs is 1.04 V, as shown in Figure 11.5e. Without $I_2$ redox mediator, the charge polarization is about 0.4 V due to the high activation energy of ZnS. Interestingly, only trace $I_2$ can reduce the charge polarization to around 0.2 V. Therefore, exploiting more mediator redox is another effective route to introduce various materials into SABs.

## 11.3.5 *Membrane Engineering*

The membrane is designed to separate the same or different electrolyte phases (such as anolyte and catholyte) and prevent chemical crossover, especially in flow SABs. Generally, the membranes employed in SABs can be classified into four categories: conventional polymer porous membrane, fast-ion conductor ceramics, inorganic fiber membrane, and hybrid membrane.[35]

Polymer porous membranes are continually promoted from conventional sulfonated styrene & Teflon and sulfonated polyethylene[10–11,22] to advanced Nafion.[35] Generally, sulfonic acid functional groups are able to naturally repulse PSs and reduce the chemical crossover to some extent.[14b] Nonetheless, the Nafion membrane is unqualified to cope with it when it comes to obvious osmotic pressure, as mentioned in Figure 11.4h. Single-layer Nafion such as Nafion-NR-211 proved incapacitated under unbalanced osmotic pressure.[14b] Therefore, researchers tend to increase the thickness by combining Nafion-N117 with N115 together[13a,13c,13d,14c] to resist the osmotic pressure. The strategy performs well in short cycles with a low Ohmic polarization increase.

NASICON membranes thus are utilized to address osmotic pressure and even serve in acid-alkali separation batters for better isolation. Thereinto, Na-based NASICON membranes possess higher ionic conductivity than Li-based ones,[35] as shown in Figure 11.5f. Nevertheless, a mass of cases has manifested that NASICON is only valid at a large thickness (~1000 $\mu$m),[13b] with the costs of huge Ohmic polarization and unaffordable price.[13b,14a,15a,15b,35,36] Therefore, NASICON can only be transitional membranes. Intriguingly, membrane with high isolation is unnecessary if employing S-S redox or implementing WiBS strategy. In this case, an inorganic fiber membrane can be adequate for separation, reducing the cost and kinetic loss caused by Ohmic polarization.

The hybrid membrane is a matter of concern category recently, especially in flow SABs. As shown in Figure 11.5f, Lu *et al.* proposed an ingenious scheme to design a charge-reinforced ion-selective membrane, aiming at repulsing anionic groups by Coulombic force for the first time.[32] Inspired by this skilful work, future membranes should have special structures to repulse or confine PS anions. For instance, targeted anions cage based on free volume elements on polymer porous membrane,[37] specific functional groups to capture anions by porous aromatic frameworks,[38] and mesoporous metal in covalent organic framework (MOF-in-COF) are useful to limit the anions.[39]

## 11.3.6 *Electrode Engineering*

In the design of a performance-oriented SAB, optimization of electrodes with electronic conductivity and catalytic activity is necessary to overcome the huge polarization and adverse reactions.

Inorganic and organic skeleton, doping or compound layer have been proposed for the sulfur electrode.[25] Inorganic carbon is mainly utilized for high conductivity and low price. Thereinto, sulfur with Ketjenblack (S-KB),[14d,15d,15e] activated carbon[15d,15e] and MOF-derived carbon[15f,15g] are most popular to drive the electrode at a low polarization of 30–80 mV. However, the interaction force between non-polar carbon and polar PSs is weak,[26] leading to insufficient affinity. Therefore, as shown in Figure 11.5g, polar polymer electrodes such as sulfur@Polypyrrole (S@Ppy)[12] and *in situ* synthesized 3D organic framework[15h] are suggested to be introduced to provide strong interaction force and abundant reaction sites with lower $E_{pp}$ and better curve overlapping.

The catalytic host is another requisite to overcome the large overpotential of S-S type SABs and accelerate the L-L type SABs. Different from the catalysts in SOBs, SABs frequently adopt Co-based (CoS, $CoS_2$), Cu-based (CuS, $Cu_7S_4$), Ni-based (NiS, $NiS_2$), and Fe-based (FeS, $FeS_2$) catalysts.[40,41] Thereinto, Co-/Cu-based catalysts exhibited superior performances. For example, a CoS catalytic substrate (with electronic conductivity of $5.25 \times 10^{-6}$ S cm$^{-1}$) is able to reduce 95% impedance compared with the pure stainless steel substrate.[15a] Aside from a single catalyst, Qian *et al.* have intelligently devised the heterojunction of $CoS_2$/CoS based on the semiconductive theory as shown in Figure 11.5h.[13d] The contact interface was proved to shape two inverted space-charge regions and excite a built-in electric field. The uneven distribution of charge in the heterojunction thus enhanced charge transfer activity and established a highly efficient $CoS_2$/CoS catalyst. CV test illustrates a tiny $E_{pp}$ of 0.1 V. A relatively high peak current has been realized in Figure 11.5h, indicating the boosted redox activity. Additionally, this CV test exhibits no peak shift in the horizontal direction. The results demonstrate that all the kinetics enhancement can be realized by the chemical parallel conversion step, scilicet, and the catalysis effect. Aside from the Co-based catalyst, Cu-based catalysts also present satisfactory polarization reduction with lower prices. Lei *et al.* have successfully constructed a high-efficiency $Cu_7S_4$/carbon nanotube bifunctional catalyst to boost both PS and $I^-/I_3^-$ redox.[40] This well-designed work realizes the expectation to shorten the reaction rate between anolyte and catholyte.

It is worth noting that all these catalysts should not simultaneously boost the HER and/or OER evolution. Instead, some anti-catalysis electrodes can generate ultra-high HER overpotential. Related concepts are considered to be introduced into electrode modification in SABs.

Apart from these extrinsic strategies, for sulfur itself, solid solution can intrinsically enhance the electron conductivity, especially considering the high conductivity of selenium. The density of state in Figure 11.5i confirms that pure sulfur possesses a huge bandgap. Instead, the solid-solution SeS presents a conductor-like density of state.

Combining the above considerations together, an affordable and effective electrode should be a synergetic combination of active materials, catalytic host, and conductive framework. The aforesaid six strategies exhibit their own capacities to cope with single

**Figure 11.6.** A problem-to-strategy reference to SABs.

or combined problems in SABs. A problems-to-strategies diagram is illustrated in Figure 11.6 for a general overview.

## 11.4 Device-Scale Evaluation

With an integrated understanding of the electrochemistry, challenges, and strategies, the final step of the theory-to-application methodology is ready to be executed, *i.e.*, the device-scale SABs, including flow SABs, oxide SABs, and metal SABs. Table 11.1 summarizes the strategies and performances realized in some representative works. Furthermore, the potential distribution shown in Figure 11.7a proposes promising coupling modes for various SABs under different ESW regulation strategies (see illustrations in Figure 11.7b–d).

**Table 11.1.** Strategy and performance of representative SABs.

| Structure | Redox (L-L; S-L; S-S) | Voltage (V) | Electrode Modification | Membrane | Capacity$_{(S\text{-based})}$ (mAh g$^{-1}$) | Ref. |
|---|---|---|---|---|---|---|
| Naǀǀ[Na$_2$S$_2$] | Na$_2$S$_4$ ↔ Na$_2$S$_2$ (*L-L*) | 1.03 | CoS/Co$_2$S | N115 + N117 | Nominal 134 | [13d] |
| AirǀǀLi$_2$S$_4$ | Li$_2$S$_4$ ↔ Li$_2$S$_2$ (*L-L*) | ~1 | N/A | LIC-GC | N/A | [13b] |
| AirǀǀNa$_2$S$_2$ | Na$_2$S$_4$ ↔ Na$_2$S$_2$ (*L-L*) | 1.68 | CuS | NZSP | N/A | [42] |
| S@Ppy | S ↔ Na$_2$S (*S-L*) | 0.45$_{SHE}$ | Ppy | N/A | 473@0.1 A g$^{-1}$ | [12] |
| LMOǀǀK$_2$S | S$_2^2$ ↔ Li$_2$S (*S-L*) | 1.5 | Carbon | N-211 | 110$_{full}$@2 C | [14a] |
| LMOǀǀS | S ↔ Li$_2$S (*S-S*)$_{WBS}$ | 1.7 | Carbon | Glass fiber | 1327@0.2 C | [14d] |
| S + K$_2$S$_4$ǀǀAl | S ↔ KHS (*S-L*) | 1.28 | CoS | Porous polymer | 903@0.5 A cm$^{-2}$ | [10] |
| SǀǀAl | S ↔ AlS$_x$ (*S-S*) | 1.15 | Carbon | Glass fiber | 420@0.2 A g$^{-1}$ | [15g] |
| S + Li$_2$S$_4$ǀǀLi | S ↔ Li$_2$S (*S-L*) | 0.7 | CoS | LATP | 1202@0.2 mA cm$^{-2}$ | [15a] |
| SǀǀNa | Na$_2$S$_4$ ↔ Na$_2$S$_2$ (*L-L*) | ~1.0 | CuS | NZSP | Restrict at 418 | [15c] |
| SǀǀFe | S ↔ FeS$_2$ ↔ Fe$_3$S$_4$ ↔ FeS (*S-S*) | 0.8 | Carbon | Cellulose | ~1050@0.05 A g$^{-1}$ | [15e] |
| SǀǀZn | S ↔ CuS ↔ Cu$_2$S (*S-S*) | 1.15 | Carbon | Anion exchange | ~2000@0.05 A g$^{-1}$ | [15d] |
| Na$_2$S$_4$ǀǀZn | Na$_2$S$_4$ ↔ Na$_2$S (*L-L*) | 0.89 | CoS | NZSP | 822@0.5 mA cm$^{-2}$ | [15b] |
| SǀǀZn | S ↔ ZnS (*S-S*) | ~1.6 | CNT | Glass fiber | 435@2 A g$^{-1}$ | [17] |
| Li$_2$S$_6$ǀǀZn | Li$_2$S$_6$ ↔ Zn$_x$Li$_y$S$_{3-8}$ (*S-S*) | 1.93 | IL-PLSD | Non-wovens | 1148@0.3 A g$^{-1}$ | [15h] |

Ppy: polypyrrole; LMO: LiMn$_2$O$_4$; CNT: carbon nanotube; IL-PLSD: ionic liquid-poly(Li$_2$S$_6$-random-(1,3-diisopropenylbenzene)); LIC-GC: lithium-ion conducting glass ceramic; NZSP: Na$_3$Zr$_2$Si2PO$_{12}$; LATP: Li$_{1+x+y}$Al$_x$Ti$_{2-x}$P$_{3-y}$Si$_y$O$_{12}$.

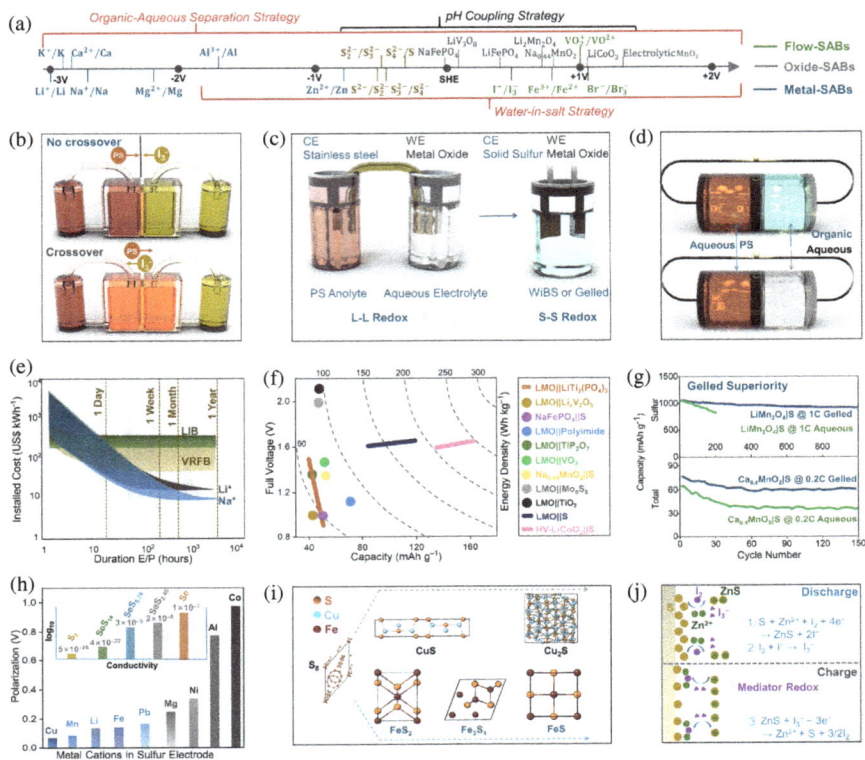

**Figure 11.7.** Summary of device-level evaluation in flow SABs, oxide SABs and metal SABs. (a) Potential distribution of redox in SABs. (b) Diagram of chemical crossover problems in redox flow SABs. (c) Evolution from Daniel-type to integrated oxide SABs by electrolyte engineering. (d) 3D model for organic-aqueous separation and all-aqueous metal SABs for metals at different depths of potential. (e) Cost metric of flow SABs employing Na⁺ and Li⁺ carrier ions. Reproduced with permission from Ref. [13b]. Copyright 2017 Elsevier. (f) Energy density profiles of different oxides in ABs. (g) Comparison of two types of non-free-water oxide SABs, concentrated and gelled. Reproduced with permission from Refs. [14d,43]. Copyright 2017 National Academy of Sciences and Copyright 2021 Springer Nature, respectively. (h) Polarization of sulfur electrode containing different metal ions and comparison of solid solution strategy. Reproduced with permission from Refs. [15d,27]. Copyright 2019 and 2021 John Wiley and Sons, respectively. (i) Representative S-S mechanisms in Cu-S battery and Fe-S battery. (j) Mediator redox mechanism to reduce huge activation energy.

## 11.4.1 Flow Sulfur-based Aqueous Batteries

As shown in Figure 11.7a, the S-based anodic redox in the yellow labels can be combined with cathodic redox in the green labels as flow SABs, which possess fast kinetics, decoupled power and energy output, and prolonged cycling life with acceptable price.

The primary problem that needs to be solved in flow SABs is the chemical crossover displayed in Figure 11.7b. A temporary plan is to use NASICON or multilayer Nafion membranes to cope with chemical crossover. Recently, the breakthrough on membranes has been achieved by designing ion cages to anchor, or ionic filtration to permeate, specific ions which can be introduced into SABs.[84–86] The selection of active materials is crucial to determine the specific capacity, volumetric energy density, and utilization of ESW. L-L redox such as $S_2^{2-} \leftrightarrow S^{2-}$ and $S_4^{2-} \leftrightarrow S_2^{2-}$ are employed in main flow SABs, matched with halogen-based or air catholytes. Thereinto, $Br^-/Br_3^-$ and $I^-/I_3^-$ redox are employed due to their higher solubility and faster kinetics. Furthermore, with an eye to the harmful bromine vapor of $Br_2$, $I^-/I_3^-$ redox is more attractive for its safer operation and a suitable redox potential at *ca.* 0.54 V. In practice, it is essential to eliminate the sediment of $I_2$ through the redox coupling strategy. In addition to halogen||PS flow SABs, Chiang *et al.* achieved a promising air-breathing air||PS flow SABs.[13b] An acid catholyte was successfully applied for its high energy density of 121 Wh L$^{-1}$. Taking the OER/ORR catalysts (IrO/Pt) into consideration, the installed cost (including energy and power cost, balance-of-plant cost, and additional cost) can also be controlled, as shown in Figure 11.7e. With the duration time lasting more than one month, the installation cost can be limited to under 10 US$ kWh$^{-1}$.

As for the performance, the energy efficiency $E_E$, Coulombic efficiency $E_C$ and voltage efficiency $E_V$ are pivotal indexes to evaluate the flow SABs, where $E_E = E_C \times E_V$. Obviously, kinetics promoting is the key to improve the three efficiencies concomitantly. The concentration of PSs, the external driving force (flow rate), and catalytic host underlie the kinetics-oriented flow SABs.

Flow SABs have already exhibited their potential as the substitute for exorbitant all-vanadium flow batteries. Further promotions on membrane and performance can be realized, for example, through a S-L compound fluid introduction to markedly increase the energy density for flow SABs.

## 11.4.2 Oxide Sulfur-based Aqueous Batteries

As summarized in Figures 11.7a and 11.7f, the oxide SABs commonly output a high voltage from 1.0 to 2.1 V by coupling the S-based anodic redox in yellow labels with the (Li/Na/Ca) oxide-based cathodic redox in gray labels. Thereinto, LiMn$_2$O$_4$ (LMO) and $LiCo_2O_4$ (LCO) have shown conspicuous superiority on capacity and energy density when matched with sulfur anodes.

In this context, battery construction is a crucial part of preliminary designs. Initially, as shown in Figure 11.7c, the researchers adopted Daniel cell[14a] to test the electrochemical parameters of LMO, $Na_{0.44}MnO_2$, $NaFePO_4$, and PS electrodes. However, the current density was seriously restricted at 1/100–1/50 C due to the huge impedance induced by the salt bridge. In order to boost the rate performance, Wang et al. have developed a new structure in Figure 11.7c by WiBS strategy.[14d] Based on the free-water competition effect, the sulfur anode operated 100 cycles with 1327 mAh $g^{-1}$ at 0.2 C. The related oxide SAB achieved a satisfactory energy density as high as 195 Wh $kg^{-1}$ or 454 Wh $L^{-1}$ in LCO‖S SABs. The results are close to and even surpass some non-aqueous SOBs. Nevertheless, the hidden trouble for this solid reaction is the trace dissolution of PSs, especially in the solid contact interface between PS and WiBS, which causes cumulative irreversible shuttle and disproportionation reaction. Therefore, a gelled quasi-solid-state battery in Figure 11.7g is used to isolate PSs from trace water, which also enlightens the design of multivalent $M_xMnO_2$ (M represents Ca, Mg, and Al)-based oxide SABs.[43]

A critical challenge for current oxide SABs is to exploit oxides with higher potential to fill the ESW widened by WiBS strategy. Further attention should lie on the affordable concentrated electrolyte or local-concentrated electrolyte.

### 11.4.3 *Metal Sulfur-based Aqueous Batteries*

The metal SABs refer to a combination of metal-based anodic redox (in blue labels) and S-based cathodic redox (in yellow labels), listed in Figure 11.7a. And two common structures of metal SABs are listed: an all-aqueous structure for Zn, Al, Fe, and Cu metals and an organic-aqueous separated battery (OASB) for active Li and Na metals,[44] as illustrated in Figure 11.7d.

Successful operation of an OASB requires the absolute inhibition of chemical crossover (including water molecules) and shuttle effect. Zhou et al. meticulously designed the first OASB with NASICON membrane to separate Li metal in an organic solvent from S-based materials in an aqueous solvent.[15a] This work paves the way to utilize the sensitive metal in SABs and fundamentally upgrade the energy density to 654 Wh kg$^{-1}$. Subsequently, Arumugam *et al.* applied a S||Na metal SAB based on the same OASB structure, achieved specific energy$_{(sulfur)}$ of *ca.* 600 Wh kg$^{-1}$.[15c] As for the all-aqueous batteries, the electrochemical polarization of sulfur electrodes containing different metal ions is profiled in Figure 11.7h. Thereinto, although Al possesses huge polarization, it is feasible to match with L-L/S-L due to its excellent operability. The first generation of PS||Al metal SAB yielded an energy density of 110 Wh kg$^{-1}$ with an output voltage of 1.3 V. The authors believed that the deposited Al(OH)$_3$ is a harmless layer. However, the second generation of S||Al SAB driven by WiBS strategy reveals the adverse effect of the Al-involved passivation layer. Affected by the surface passivation, Wan *et al.* observed a dramatic fading of capacity from 1410 mAh g$^{-1}$ at first discharge to 420 mAh g$^{-1}$.[15g] Detailed investigation on the high bandgap-passivated product was implemented and the result confirms that the passivation layer on the surface of Al can hinder the stripping-plating process of Al. Arguably, the passivation layer may be responsible for this abnormal fading due to the fact that the initial high capacity may synergistically be contributed from the electrolyte and Al stripping-plating. Therefore, more direct *in situ* characterizations are required to clarify the active mechanism of S||Al SABs in the future.

From Figures 11.7h and 11.7i, it is found that Cu and Fe can be selected as the anode for their unique redox of S $\leftrightarrow$ Cu$_2$S and S $\leftrightarrow$ FeS, respectively. Ji and co-workers have initiated a series of systematic works on S||Cu and S||Fe batteries.[15e] Acceptable electrode polarization of 0.05 V (Cu) and 0.16 V (Fe) were obtained. The full-cell polarizations are also limited in 0.35 V. Comparatively speaking, the higher polarization in S||Fe battery is related to the high stripping-planting potential for Fe/Fe$^{2+}$ between 0.04 V and –0.18 V. As shown in Figure 11.7i, the mechanism route of S||Fe SAB follows S$_8$ → FeS$_2$ → Fe$_3$S$_4$ → FeS, with a measured capacity of 1050 mAh g$^{-1}$. In comparison, the theoretical capacity of S $\leftrightarrow$ Cu$_2$S electrode reaches an overwhelming value of 3044 mAh g$^{-1}$ for its four-electron transfer process of S$_8$ → CuS → Cu$_2$S.[45] Interestingly, during the discharge, the incorporated Cu$^{2+}$ in hexagonal CuS serving as new redox center would facilitate the accommodation of new Cu$^{2+}$ to form monoclinic Cu$_2$S. In this case, metal Zn, instead of metal Cu, was suggested as the anode to fabricate the S/CuSO$_4$||ZnSO$_4$/Zn full battery to widen the output voltage. Similarly, S$_8$ → Cu$_7$S$_4$ → Cu$_2$S route worked smoothly with a rate capability of 497 mAh g$^{-1}$ at 7.5 A g$^{-1}$.[45]

Zn was recommended as the metal anode for its fair redox potential at −0.76 V.[36,46] However, the direct S||Zn SAB is of great challenge due to the passivated discharge product of ZnS.[17] Bendikov *et al.* only realized a primary S||Zn battery without separator in high pH of 15.[47] Another subtle strategy employed other metal ions as a carrier, such as Na+ ions, to prevent the shuttle of Zn ions.[15b] In the case of Na-based solid electrolyte, Na+ shuttles back and forth, hence the ZnS problem can be excluded. Recently, based on the $I_2$ mediator redox mechanism shown in Figure 11.7j, a more direct method has been proposed to refrain from the ZnS.[17] The generated overpotential in the system should activate the $I^-/I_3^-$ redox at around 1.3 V, which led to the oxidation of ZnS, resolving the kinetics problem of S ↔ ZnS redox. This mechanism can be described as a chemically assisted electrochemical process. Consequently, the S||Zn battery yields an energy density of 502 Wh kg$^{-1}$, calculated by the sulfur side. Apart from that, Wang *et al.* introduced a brand-new electrode in S/Se||Zn battery using the Se-S solid solution to reduce the intrinsic electrochemical polarization of the electrode.[27] The inset in Figure 11.7h gives the comparison result of electron conductivity from S → $SeS_{14}$ → $SeS_{5.76}$ → $SeS_{2.46}$ → Se. However, higher selenium content leads to relatively lower electrode capacity. As a result, $SeS_{5.76}$ is regarded as an optimal electrode to output satisfactory capacity and rate performance concurrently.

Metal SABs are believed to be the most promising application for S-based redox; however, obstacles herein need to be overcome. From the thermodynamics perspective, filling the ESW with low potential metal redox is critical. OASB strategy is promising to extend the scope of optional metal redox with ultra-low potential. From the kinetics perspective, the most pivotal consideration is to overcome the ionic conductivity difference between the organic and aqueous sides of OASB, *e.g.*, introducing ionic liquid into the organic solvent to regulate the ionic conductivity. Additionally, exploiting chemical mediators to reduce electrochemical polarization is another prospective method.

## 11.5 Summary and Outlook

### 11.5.1 *Summary*

This chapter establishes a theory-to-application methodology, taking the electrochemistry theory, current issues, optimized strategies, and applicable devices of SABs into account.

The special aqueous S-related electrochemistry, including thermodynamic evolution, kinetic metrics, and redox routes, are critically analyzed for the current SABs. They are actually interrelated. As a consequence, synergistic strategies are suggested to cope with narrow ESW, irreversible thermodynamic reactions, insufficient kinetics in L-L/S-L type SABs, and device-level issues. The following strategies should be contrived together, including pH regulation, redox coupling, activity adjustment, electrolyte optimization, membrane engineering, and electrode modification. Equipped with abovementioned methods, existing challenges in redox flow SABs, oxide SABs, and metal SABs can be handled properly.

## 11.5.2 *Future Perspective*

Despite the challenges existing in SABs, the outlook of SABs remains promising and exciting for their natural superiorities. In the future design of high-performance SABs, the following aspects could be given priority attention:

(1) Implementing direct electrochemical metrics through transient and steady state characterizations to confirm the accurate RDS in SABs. Additionally, the thermodynamic metrics on activity, pH, depth of discharge/charge still require further characterizations;

(2) Constructing chemically assisted electrochemical theory. The mechanistic understanding is desired in realizing the "impossible" electrochemical reactions (with huge polarization) or thermodynamic reactions (from $M_xS_y$ to $M^{y+}$ and $S^{n-}$) by medium chemical conversion steps, such as the $I^-/I_3^-$ mediator redox;

(3) Clarifying the chemical structure-function relationship, especially for the porous structural control on electrodes and membranes to confine the PS anions. In addition, electrostatic Coulombic force and covalent bond are applicable forces to build the ionic cage or ionic filtration;

(4) Exploiting bandgap-adjustable heterojunction for active materials and catalysts based on semiconductive chemistry, in order to improve the conductivity and affinity. Moreover, rate-tunable catalytic hosts to balance the rates of anode and cathode are pivotal for extending the lifespan of batteries;

(5) Expanding the S-S redox intrinsically and extrinsically. Intrinsically, highly cohesive and low-soluble sulfides deserve broader attention. Extrinsically, local-concentrated electrolytes can be designed as a transitional strategy to anchor

sulfides. In the long run, affordable single or multiple aqueous solvents are promising for the next generation of scalable SABs.

## Acknowledgment

This work was supported by National Key R&D Program of China (2018YFE0201701 and 2018YFA0209401), and the National Natural Science Foundation of China (22088101, 21733003 and 21975050). The authors also thank the financial supports from Fudan University (No. JIH2203010 and No. IDH2203008/003).

## References

[1] a) J. Xie, Z. Liang, Y.-C. Lu, *Nat. Mater.* **2020**, 19, 1006; b) C. Wessells, R. A. Huggins, Y. Cui, *J. Power Sources* **2011**, 196, 2884; c) D. Chao, W. Zhou, F. Xie, C. Ye, H. Li, M. Jaroniec, S.-Z. Qiao, *Sci. Adv.* **2020**, 6, eaba4098.

[2] D. Chao, W. Zhou, C. Ye, Q. Zhang, Y. Chen, L. Gu, K. Davey, S. Z. Qiao, *Angew. Chem. Int. Ed.* **2019**, 58, 7823.

[3] H. Bode, *Lead-Acid Batteries*, **1978**.

[4] W. Li, J. R. Dahn, D. S. Wainwright, *Science* **1994**, 264, 1115.

[5] a) D. Kundu, P. Oberholzer, C. Glaros, A. Bouzid, E. Tervoort, A. Pasquarello, M. Niederberger, *Chem. Mater.* **2018**, 30, 3874; b) G. Dawut, Y. Lu, L. Miao, J. Chen, *Inorg. Chem. Front.* **2018**, 5, 1391; c) Z. Guo, Z. Ma, X. Dong, J. Huang, Y. Wang, Y. Xia, *Angew. Chem. Int. Ed.* **2018**, 57, 11737; d) B. Häupler, C. Rössel, A. M. Schwenke, J. Winsberg, D. Schmidt, A. Wild, U. S. Schubert, *NPG Asia Mater.* **2016**, 8, e283.

[6] a) R. Trócoli, F. La Mantia, *ChemSusChem.* **2015**, 8, 481; b) T. Gupta, A. Kim, S. Phadke, S. Biswas, T. Luong, B. J. Hertzberg, M. Chamoun, K. Evans-Lutterodt, D. A. Steingart, *J. Power Sources* **2016**, 305, 22; c) L.-P. Wang, P.-F. Wang, T.-S. Wang, Y.-X. Yin, Y.-G. Guo, C.-R. Wang, *J. Power Sources* **2017**, 355, 18; d) K. Lu, B. Song, J. Zhang, H. Ma, *J. Power Sources* **2016**, 321, 257; e) Z. Hou, X. Zhang, X. Li, Y. Zhu, J. Liang, Y. Qian, *J. Mater. Chem. A* **2017**, 5, 730; f) L. Zhang, L. Chen, X. Zhou, Z. Liu, *Adv. Energy Mater.* **2015**, 5, 1400930.

[7] a) Y. Cheng, L. Luo, L. Zhong, J. Chen, B. Li, W. Wang, S. X. Mao, C. Wang, V. L. Sprenkle, G. Li, *ACS Appl. Mater. Interfaces* **2016**, 8, 13673; b) J. Zhao, Y. Li, X. Peng, S. Dong, J. Ma, G. Cui, L. Chen, *Electrochem. Commun.* **2016**, 69, 6; c) W. Liu, J. Hao, C. Xu, J. Mou, L. Dong, F. Jiang, Z. Kang, J. Wu, B. Jiang, F. Kang, *Chem. Commun.* **2017**, 53, 6872; d) W. Xu, C. Sun, K. Zhao, X. Cheng, S. Rawal, Y. Xu, Y. Wang, *Energy Storage Mater.* **2019**, 16, 527.

[8] a) B. Zhang, Y. Liu, X. Wu, Y. Yang, Z. Chang, Z. Wen, Y. Wu, *Chem. Commun.* **2014**, 50, 1209; b) M. H. Alfaruqi, J. Gim, S. Kim, J. Song, D. T. Pham, J. Jo, Z. Xiu, V. Mathew, J. Kim, *Electrochem. Commun.* **2015**, 60, 121; c) H. Pan, Y. Shao, P. Yan, Y. Cheng, K. S. Han, Z. Nie, C. Wang, J. Yang, X. Li, P. Bhattacharya, *Nat. Energy* **2016**, 1, 1; d) S. Islam, M. H. Alfaruqi, V. Mathew, J. Song, S. Kim, S. Kim, J. Jo, J. P. Baboo, D. T. Pham, D. Y. Putro, *J. Mater. Chem. A* **2017**, 5, 23299; e) C. Zhu, G. Fang, J. Zhou, J. Guo, Z. Wang, C. Wang, J. Li, Y. Tang, S. Liang, *J. Mater. Chem. A* **2018**, 6, 9677; f) M. Song, H. Tan, D. Chao, H. J. Fan, *Adv. Funct. Mater.* **2018**, 28, 1802564.

[9] a) D. Kundu, B. D. Adams, V. Duffort, S. H. Vajargah, L. F. Nazar, *Nat. Energy* **2016**, 1, 1; b) M. H. Alfaruqi, V. Mathew, J. Song, S. Kim, S. Islam, D. T. Pham, J. Jo, S. Kim, J. P. Baboo, Z. Xiu, *Chem. Mater.* **2017**, 29, 1684; c) P. He, M. Yan, G. Zhang, R. Sun, L. Chen, Q. An, L. Mai, *Adv. Energy Mater.* **2017**, 7, 1601920; d) P. He, G. Zhang, X. Liao, M. Yan, X. Xu, Q. An, J. Liu, L. Mai, *Adv. Energy Mater.* **2018**, 8, 1702463; e) W. Li, K. Wang, S. Cheng, K. Jiang, *Energy Storage Mater.* **2018**, 15, 14; f) Z. Peng, Q. Wei, S. Tan, P. He, W. Luo, Q. An, L. Mai, *Chem. Commun.* **2018**, 54, 4041; g) F. Wan, L. Zhang, X. Dai, X. Wang, Z. Niu, J. Chen, *Nat. Commun.* **2018**, 9, 1; h) F. Wang, E. Hu, W. Sun, T. Gao, X. Ji, X. Fan, F. Han, X.-Q. Yang, K. Xu, C. Wang, *Energy Environ. Sci.* **2018**, 11, 3168; i) G. Yang, T. Wei, C. Wang, *ACS Appl. Mater. Interfaces* **2018**, 10, 35079; j) J. Lai, H. Zhu, X. Zhu, H. Koritala, Y. Wang, *ACS Appl. Energy Mater.* **2019**, 2, 1988; k) D. Chao, C. Zhu, M. Song, P. Liang, X. Zhang, N. H. Tiep, H. Zhao, J. Wang, R. Wang, H. Zhang, *Adv. Mater.* **2018**, 30, 1803181; l) D. Chao, H. J. Fan, *Chem.* **2019**, 5, 1359.

[10] D. Peramunage, S. Licht, *Science* **1993**, 261, 1029.

[11] S. Licht, D. Peramunage, *J. Electrochem. Soc.* **1993**, 140, L4.

[12] J. Shao, X. Li, L. Zhang, Q. Qu, H. Zheng, *Nanoscale* **2013**, 5, 1460.

[13] a) Z. Li, G. Weng, Q. Zou, G. Cong, Y.-C. Lu, *Nano Energy* **2016**, 30, 283; b) Z. Li, M. S. Pan, L. Su, P.-C. Tsai, A. F. Badel, J. M. Valle, S. L. Eiler, K. Xiang, F. R. Brushett, Y.-M. Chiang, *Joule* **2017**, 1, 306; c) L. Su, A. F. Badel, C. Cao, J. J. Hinricher, F. R. Brushett, *Ind. Eng. Chem. Res.* **2017**, 56, 9783; d) D. Ma, B. Hu, W. Wu, X. Liu, J. Zai, C. Shu, T. T. Tsega, L. Chen, X. Qian, T. L. Liu, *Nat. Commun.* **2019**, 10, 1.

[14] a) R. Demir-Cakan, M. Morcrette, J.-B. Leriche, J.-M. Tarascon, *J. Mater. Chem. A* **2014**, 2, 9025; b) R. Demir-Cakan, M. Morcrette, J.-M. Tarascon, *J. Mater. Chem. A* **2015**, 3, 2869; c) B. Tekin, S. Sevinc, M. Morcrette, R. Demir-Cakan, *Energy Technol.* **2017**, 5, 2182; d) C. Yang, L. Suo, O. Borodin, F. Wang, W. Sun, T. Gao, X. Fan, S. Hou, Z. Ma, K. Amine, K. Xu, C. Wang, *Proc. Natl. Acad. Sci.* **2017**, 114, 6197; e) S. Sevinc, B. Tekin, A. Ata, M. Morcrette, H. Perrot, O. Sel, R. Demir-Cakan, *J. Power Sources* **2019**, 412, 55.

[15] a) N. Li, Z. Weng, Y. Wang, F. Li, H.-M. Cheng, H. Zhou, *Energy Environ. Sci.* **2014**, 7, 3307; b) M. M. Gross, A. Manthiram, *ACS Appl. Mater. Interfaces* **2018**, 10, 10612; c) M. M. Gross, A. Manthiram, *Energy Storage Mater.* **2019**, 19, 346; d) X. Wu, A. Markir, L. Ma, Y. Xu, H. Jiang, D. P. Leonard, W. Shin, T. Wu, J. Lu, X. Ji, *Angew. Chem.* **2019**, 131, 12770; e) X. Wu, A. Markir, Y. Xu, E. C. Hu, K. T. Dai, C. Zhang, W. Shin, D. P. Leonard, K. I. Kim, X. Ji, *Adv. Energy Mater.* **2019**, 9, 1902422; f) C. Dai, X. Jin, H. Ma, L. Hu, G. Sun, H. Chen, Q. Yang, M. Xu, Q. Liu, Y. Xiao, X. Zhang, H. Yang, Q. Guo, Z. Zhang, L. Qu, *Adv. Energy Mater.* **2020**, 11, 2003982; g) Z. Hu, Y. Guo, H. Jin, H. Ji, L.-J. Wan, *Chem. Commun.* **2020**, 56, 2023; h) Y. Zhao, D. Wang, X. Li, Q. Yang, Y. Guo, F. Mo, Q. Li, C. Peng, H. Li, C. Zhi, *Adv. Mater.* **2020**, 32, 2003070; i) M. M. Gross, A. Manthiram, *ACS Appl. Energy Mater.* **2019**, 2, 3445.

[16] D. Zheng, G. Wang, D. Liu, J. Si, T. Ding, D. Qu, X. Yang, D. Qu, *Adv. Mater. Technol.* **2018**, 3, 1700233.

[17] W. Li, K. Wang, K. Jiang, *Adv. Sci.* **2020**, 7, 2000761.

[18] S. Yun, S. H. Park, J. S. Yeon, J. Park, M. Jana, J. Suk, H. S. Park, *Adv. Funct. Mater.* **2018**, 28, 1707593.

[19] N. A. Cañas, S. Wolf, N. Wagner, K. A. Friedrich, *J. Power Sources* **2013**, 226, 313.

[20] Q. Cheng, W. Xu, S. Qin, S. Das, T. Jin, A. Li, A. C. Li, B. Qie, P. Yao, H. Zhai, *Angew. Chem. Int. Ed.* **2019**, 58, 5557.

[21] D. Liu, C. Zhang, G. Zhou, W. Lv, G. Ling, L. Zhi, Q. H. Yang, *Adv. Sci.* **2018**, 5, 1700270.

[22] S. Licht, *J. Electrochem. Soc.* **1987**, 134, 2137.

[23] a) Z. Quanxing, *Introduction of Kinetics of Electrode Process*, Science Press Beijing **2002**; b) M. Wild, L. O'Neill, T. Zhang, R. Purkayastha, G. Minton, M. Marinescu, G. J. Offer, *Energy Environ. Sci.* **2015**, 8, 3477.

[24] D. Reed, E. Thomsen, B. Li, W. Wang, Z. Nie, B. Koeppel, J. Kizewski, V. Sprenkle, *J. Electrochem. Soc.* **2015**, 163, A5211.

[25] H. Peng, Y. Zhang, Y. Chen, J. Zhang, H. Jiang, X. Chen, Z. Zhang, Y. Zeng, B. Sa, Q. Wei, *Mater. Today Energy* **2020**, 18, 100519.

[26] F. Y. Fan, M. S. Pan, K. C. Lau, R. S. Assary, W. H. Woodford, L. A. Curtiss, W. C. Carter, Y.-M. Chiang, *J. Electrochem. Soc.* **2016**, 163, A3111.

[27] W. Li, Y. Ma, P. Li, X. Jing, K. Jiang, D. Wang, *Adv. Funct. Mater.* **2021**, 31, 2101237.

[28] S. Licht, J. Davis, *J. Phys. Chem. B* **1997**, 101, 2540.

[29] R. J. Remick, P. G. P. Ang, US Patent 4485154, **1984**.

[30] Z. Liu, Y. Huang, Y. Huang, Q. Yang, X. Li, Z. Huang, C. Zhi, *Chem. Soc. Rev.* **2020**, 49, 180.

[31] a) M. Yu, Y. Lu, H. Zheng, X. Lu, *Chem. Eur. J.* **2018**, 24, 3639; b) L. Suo, O. Borodin, T. Gao, M. Olguin, J. Ho, X. Fan, C. Luo, C. Wang, K. Xu, *Science* **2015**, 350, 938; c) C. Yang, J. Chen, X. Ji, T. P. Pollard, X. Lü, C.-J. Sun, S. Hou, Q. Liu, C. Liu, T. Qing, *Nature* **2019**, 569, 245.

[32] Z. Li, Y.-C. Lu, *Nat. Energy* **2021**, 6, 517.

[33] S. Zhang, W. Guo, F. Yang, P. Zheng, R. Qiao, Z. Li, *Batteries Supercaps* **2019**, 2, 627.

[34] D. Chao, S.-Z. Qiao, *Joule* **2020**, 4, 1846.

[35] E. Allcorn, G. Nagasubramanian, H. D. Pratt III, E. Spoerke, D. Ingersoll, *J. Power Sources* **2018**, 378, 353.

[36] J. Hao, L. Yuan, C. Ye, D. Chao, K. Davey, Z. Guo, S. Qiao, *Angew. Chem.* **2021**, 60, 7366.

[37] M. J. Baran, M. E. Carrington, S. Sahu, A. Baskin, J. Song, M. A. Baird, K. S. Han, K. T. Mueller, S. J. Teat, S. M. Meckler, *Nature* **2021**, 592, 225.

[38] A. A. Uliana, N. T. Bui, J. Kamcev, M. K. Taylor, J. J. Urban, J. R. Long, *Science* **2021**, 372, 296.

[39] H. Fan, M. Peng, I. Strauss, A. Mundstock, H. Meng, J. Caro, *Nat. Commun.* **2021**, 12, 1.

[40] Y. Qin, X. Li, W. Liu, X. Lei, *Mater. Today Energy* **2021**, 21, 100746.

[41] M. S. Faber, M. A. Lukowski, Q. Ding, N. S. Kaiser, S. Jin, *J. Phys. Chem. C* **2014**, 118, 21347.

[42] M. M. Gross, A. Manthiram, *ACS Appl. Energy Mater.* **2018**, 1, 7230.

[43] X. Tang, D. Zhou, B. Zhang, S. Wang, P. Li, H. Liu, X. Guo, P. Jaumaux, X. Gao, Y. Fu, *Nat. Commun.* **2021**, 12, 1.

[44] a) C. Ye, D. Chao, J. Shan, H. Li, K. Davey, S.-Z. Qiao, *Matter* **2020**, 2, 323; b) C. Ye, Y. Jiao, D. Chao, T. Ling, J. Shan, B. Zhang, Q. Gu, K. Davey, H. Wang, S. Z. Qiao, *Adv. Mater.* **2020**, 32, 1907557; c) B. Chen, D. Chao, E. Liu, M. Jaroniec, N. Zhao, S.-Z. Qiao, *Energy Environ. Sci.* **2020**, 13, 1096; d) B. W. Zhang, T. Sheng, Y. X. Wang, S. Chou, K. Davey, S. X. Dou, S. Z. Qiao, *Angew. Chem. Int. Ed.* **2019**, 58, 1484; e) C. Ye, Y. Jiao, H. Jin, A. D. Slattery, K. Davey, H. Wang, S. Z. Qiao, *Angew. Chem. Int. Ed.* **2018**, 57, 16703.

[45] Y. Wang, D. Chao, Z. Wang, J. Ni, L. Li, *ACS Nano* **2021**, 15, 5420.

[46] W. Zhou, D. Zhu, J. He, J. Li, H. Chen, Y. Chen, D. Chao, *Energy Environ. Sci.* **2020**, 13, 4157.

[47] T. A. Bendikov, C. Yarnitzky, S. Licht, *J. Phys. Chem. B* **2002**, 106, 2989.

https://doi.org/10.1142/9789811278327_0012

Chapter 12

# Design Strategies for High-Performance Aqueous Zn/Organic Batteries[1]

Zhiwei Tie, Zhiqiang Niu*

*Key Laboratory of Advanced Energy Materials
Chemistry (Ministry of Education), College of Chemistry,
Nankai University, Tianjin, 300071, P. R. China*

Organic electroactive compounds are attractive to serve as the cathode materials of aqueous zinc-ion batteries (ZIBs) because of their resource renewability, environmentally friendliness and structural diversity. Up to now, various organic electrode materials have been developed and different redox mechanisms are observed in aqueous Zn/organic battery systems. In this chapter, we present recent developments in the energy storage mechanisms and design of the organic electrode materials of aqueous ZIBs, including carbonyl compounds, imine compounds, conductive polymers, nitronyl nitroxides, organosulfur polymers and triphenylamine derivatives. Furthermore, we highlight the design strategies to improve their electrochemical performance in the aspects of specific capacity, output voltage, cycle life and rate capability. Finally, we discuss the challenges and future perspectives of aqueous Zn/organic batteries.

## 12.1 Introduction

Rechargeable aqueous zinc-ion batteries (ZIBs) are attracting tremendous attention because the Zn anode possesses intrinsic merits such as high specific capacity ($820$ mAh g$^{-1}$ and $5855$ mAh cm$^{-3}$), low redox potential ($-0.76$ V vs. standard

---

* Corresponding author: zqniu@nankai.edu.cn
[1] Adapted with permission from Z. Tie, Z. Niu, *Angew. Chem. Int. Ed.* **2020**, 59, 21293.

hydrogen electrode), dramatic stability in water and low cost.[1] Recently, the development of aqueous ZIBs mainly focuses on design of cathode materials,[2, 3] optimization of electrolytes,[4] and modification of Zn anode.[5, 6] Among them, the design of cathode materials plays a critical role in enhancing the electrochemical performance of aqueous ZIBs. Various cathode materials of aqueous ZIBs have been developed, such as transition metal oxides or sulfides (based on manganese,[7–10] vanadium[11–16] and molybdenum[17]), Prussian blue analogues,[18] V-based NASICONs[19] and organic compounds.[20] Organic electrode materials are only composed of sustainable elements (such as C, H, O, N and S) and can be obtained from artificial synthesis or abundant biomass resources.[21] Moreover, their structures can also be designed flexibly, which contributes to the controllable tailoring of their physical properties (e.g., solubility and electrical conductivity) and electrochemical performance (e.g., specific capacity and output voltage).[22] Furthermore, the redox chemistry in organic electrode materials of ZIBs is only accompanied by the rearrangement of chemical bonds instead of the insertion/extraction of large hydrated $Zn^{2+}$, avoiding large structural changes that often exist in inorganic compounds.[23] Therefore, organic materials are considered as promising cathode materials of aqueous ZIBs.[24, 25]

Aqueous Zn/organic battery was first reported based on conductive polymer polyaniline (PANI), where Zn served as reference electrode (Figure 12.1).[26] After that, some strategies such as electrolyte optimization,[27] carbon materials hybridization,[28] and copolymerization[29] were further developed to improve the electrochemical performance of Zn/PANI batteries. However, they still suffered from low energy density due to their limited capacities. It was not until 2009 that a p-type compound nitronyl nitroxide with high voltage was designed as the cathode of aqueous ZIBs.[30] Unfortunately, it still displayed limited capacity and suffered from poor cycle life at high-voltage state. As a result, in 2016, a $\pi$-conjugated organosulfur polymer with stable electrochemical performance was developed.[31] To further enhance the capacity of organic materials, carbonyl compounds were designed to reversibly store $Zn^{2+}$ in 2018.[32] However, the discharge products of small carbonyl molecules generally suffer from high solubility in aqueous electrolytes, inevitably leading to poor cycle stability. Therefore, these small molecules were further polymerized into corresponding polymers with high molecular weight, which have low solubility in aqueous

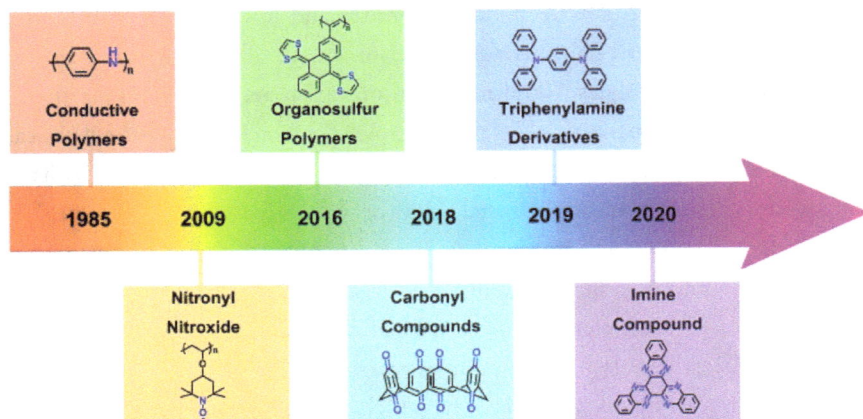

**Figure 12.1.**   The development of organic electrode materials of aqueous ZIBs.

electrolytes.[33, 34] Up to now, a variety of organic materials have been designed as the cathodes of ZIBs.

The energy storage mechanisms of Zn/organic batteries depend on the molecular structures of organic compounds and the selection of electrolytes. Apart from common $Zn^{2+}$, $H^+$ could also serve as the charge carrier in aqueous ZIBs, which exhibits high specific capacity, excellent rate performance and long cycle life due to its small ionic radius and low relative atomic mass.[35] To realize $H^+$ uptake/removal, an imine compound was developed in aqueous ZIBs.[36] In addition to the design of molecular structures, other strategies such as addition of carbon materials and optimization of electrolyte have been also developed to improve their electrochemical performance. Although significant effort has been made in the design of aqueous Zn/organic batteries, a comprehensive review focusing on aqueous Zn/organic batteries is still absent.

In this chapter, we will summarize the redox mechanisms of aqueous Zn/organic batteries and recent progress in the design of various organic cathode materials, including carbonyl compounds, imine compounds, conductive polymers, nitronyl nitroxides, organosulfur polymers and triphenylamine derivatives. Subsequently, we discuss the strategies to enhance the electrochemical performance of aqueous Zn/organic batteries. Finally, we present the challenges and future perspectives of Zn/organic batteries.

## 12.2 Motivation of Designing Organic Cathode Materials

As discussed above, both inorganic and organic compounds can serve as the active cathode materials of aqueous ZIBs. Inorganic cathode materials feature transition metal oxides or sulfides (e.g., manganese, vanadium and molybdenum), Prussian blue analogues and V-based NASICONs, which are primarily produced from mineral deposits instead of renewable resources (Figure 12.2a). Furthermore, the contents of Mn, V and Mo elements in the earth's crust are low. In the long term, ZIBs based on inorganic cathode materials could suffer from limited resources for their large-scale application.

With the increased concerns about natural resources and environmental issues, it is highly desired to develop renewable electrode materials of aqueous ZIBs. Compared with inorganic counterparts, organic cathode materials are mainly made up of C, H, O, N and S elements, which are abundant on the earth. Importantly, they can be obtained by artificial synthesis or from abundant biomass resources. In artificial synthesis, most reactants can be obtained from the natural resources directly. For example, p-aminobenzoic acid is an essential precursor for the synthesis of calix[4] quinone (C4Q), which can be prepared from wheat bran.[32] Furthermore, some redox-active organic molecules even exist in biomass resources. For instance, riboflavin that can act as the electrode material of aqueous ZIBs is widely distributed in our food, such as eggs, milk and fruit.[37] Therefore, resource availability will promote the utilization of organic compounds in aqueous ZIBs in the future.

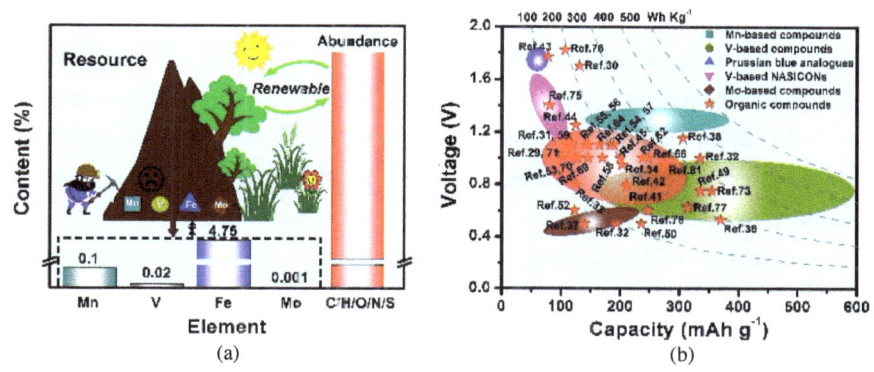

**Figure 12.2.** (a) The contents of redox-active elements on the earth's crust used in aqueous ZIBs. (b) A comparison of organic electrode materials with typical inorganic cathodes of aqueous ZIBs in mild electrolytes.

Apart from renewability, organic electrode materials possess the tunability of molecular structures, which can be achieved by grafting desired redox-active moieties onto different carbon skeletons. It endows organic electrodes with the ability to adjust their electrochemical behaviors. The voltages of organic electrode materials can be effectively adjusted by the introduction of electron-withdrawing groups (such as –CN, –F, –Cl and –Br) in organic molecular skeletons to reduce their lowest unoccupied molecular orbital (LUMO) energy levels. In addition to voltage, capacity is also an important parameter to evaluate the electrochemical performance of aqueous ZIBs. As is known, the theoretical capacity of an organic compound mainly depends on the number of active sites per weight. Therefore, theoretical capacity can be enhanced by increasing the number of active sites and reducing relative molecular mass. Furthermore, covalently interconnected carbon chains can be constructed as carbon skeletons to fabricate the redox-active polymers, whose low solubility in aqueous electrolyte would contribute to the long cycle stability of aqueous Zn/organic batteries. These are different from the conventional inorganic electrode materials of aqueous ZIBs.

The electrochemical performance of aqueous ZIBs mainly depends on their cathode materials. Figure 12.2b displays the comparison of the electrochemical performance between organic and inorganic cathode materials of aqueous ZIBs in mild electrolytes. It is clear that the average discharge voltage of most organic electrode materials is usually about 1 V. However, in some cases, their average discharge voltage can reach to about 1.7 V, which is generally higher than most inorganic compounds and comparable with Prussian blue analogues. In addition, although the average specific capacity of organic compounds for aqueous ZIBs is lower than V-based and Mn-based substances, it is higher than Mo-based compounds and even more than twice that of Prussian blue analogues and V-based NASICONs. Impressively, it is noted that the energy densities of C4Q and poly(aniline-co-azure C) can be more than 300 Wh kg$^{-1}$ by the mass of active substance, which are comparable to that of some Mn- and V-based inorganic electrode materials.[32, 38] However, the research on organic electrode materials of aqueous ZIBs are far behind inorganic compounds. Therefore, there is a broad development space to design redox-active organic molecules to meet the requirement of aqueous Zn/organic batteries with high performance in the future.

## 12.3 Energy Storage Mechanism

In general, organic electrode materials of aqueous ZIBs can be categorized into n-type, p-type and bipolar-type according to the charge-state change of the redox-active groups

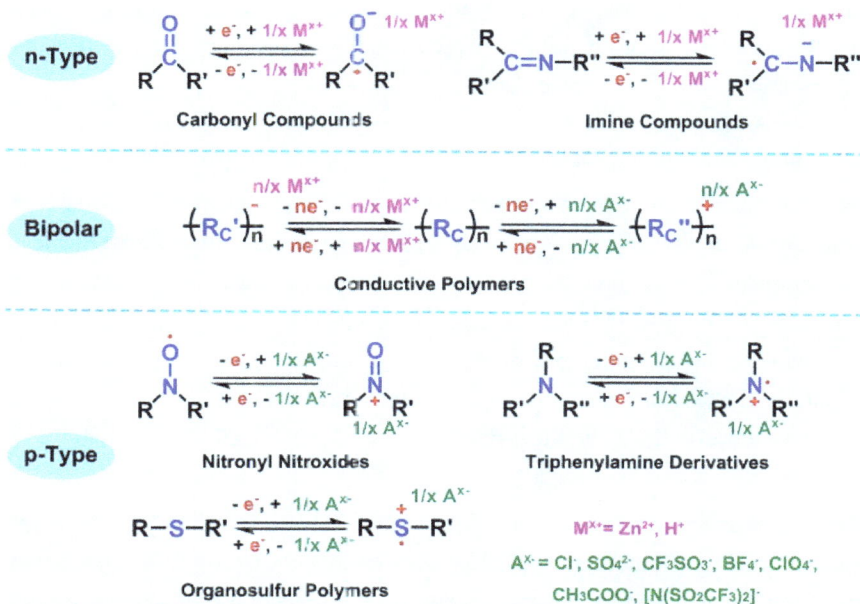

**Figure 12.3.** The energy storage mechanisms of different types of organic electrode materials in aqueous ZIBs.

during the charge/discharge process, as shown in Figure 12.3.[39, 40] During the discharge process, n-type organic compounds (e.g., carbonyl and imine compounds) often experience a reduction process first, where they will be reduced and converted into anions. Subsequently, these anions will combine with the cations (such as $Zn^{2+}$ and $H^+$) of the electrolyte. In this process, the selection of cations depends on the molecular structures of n-type organic molecules. For example, organic compounds with carbonyl groups are often reduced and combine with $Zn^{2+}$ by the coordination reactions during the discharge process.[32, 41] However, sulfur heterocyclic quinone dibenzo[b, i] thianthrene-5,7,12,14-tetraone could reversibly store both $Zn^{2+}$ and $H^+$.[42] Different from carbonyl compounds, an imine compound, diquinoxalino [2, 3-a : 2', 3'-c] phenazine (HATN), can achieve $H^+$ uptake/removal by coordination/incoordination reaction in mild electrolytes.[36]

Different from n-type counterparts, p-type organic electrode materials will first undergo an oxidation process. During oxidation, they will lose electrons and change into cations.

Thereafter, the cations will react with anions (e.g., $Cl^-$, $ClO_4^-$, $SO_4^{2-}$, $CF_3SO_3^-$ and $BF_4^-$) in the electrolyte to maintain charge neutrality. Generally, nitronyl nitroxide, organosulfur polymers and triphenylamine derivatives belong to p-type organic electrode materials. Among them, nitronyl nitroxide is the promising candidate because of its high voltage and fast kinetics.[43] During the charge process, nitronyl nitroxides lose electrons and turn into oxoammonium cations of $-N^+=O$, which will combine with the anions in the electrolyte simultaneously. Furthermore, such process is reversible in the subsequent discharge process. In the case of organosulfur polymers,[31] the conversion between the $-S-$ and $-S^+-$ moieties will take place during the charge/discharge process, accompanying the uptake/removal of anions in aqueous electrolytes. In triphenylamine derivatives, their tertiary nitrogens will convert into cations and react with anions in the charge process, displaying the p-type energy storage mechanism.[44]

Bipolar-type organic electroactive materials have the features of both n-type and p-type cases. Conductive polymers are typical bipolar-type organic electrode materials. They can be not only reduced from half-oxidation state to reduction state in the discharge processes, but also oxidized from half-oxidation state to oxidation state in the charge processes. For example, the as-prepared half-oxidation PANI possesses doped $=NH^+-$ and undoped $=N-$ groups. In the initial discharge process, the redox-active moieties of $=NH^+-$ and $=N-$ would obtain electrons to be reduced to $-NH-$ and $-N^--$, respectively. At the same time, the doped anions (such as $Cl^-$) will be removed from $=NH^+-$ groups and cations (such as $Zn^{2+}$) coordinate with $-N^--$ groups. During the following charge process, the $-NH-$ groups can be reversibly oxidized to $=NH^+-$ groups, which will interact with anions in the electrolyte. Simultaneously, the cations will also dissociate from $-N^--$ groups, accompanied by the formation of $=N-$ groups in PANI.[45] Similarly, poly(benzoquinonyl sulfide) with both $C=O$ and $-S-$ moieties displayed bipolar-type redox chemistry in aqueous ZIBs. Non-hydrated zinc ions were found to be the cation species associated with the carbonyl-related redox reaction, while the counter anions ($CF_3SO_3^-$) participated in the reaction between the $-S-$ and $-S^+-$ moieties at a higher voltage.[46]

The electrochemical performance of Zn/organic batteries depends on their energy storage mechanisms. In general, the voltages of p-type organic electrode materials are higher than the cases of n-type and bipolar-type organic compounds.[47] Moreover, p-type organic redox-active materials often exhibit faster reaction kinetics since there are nearly no bond rearrangements during their charge and discharge processes.[48] However, p-type organic cathodes usually exhibit low specific capacities because most parts of their molecular structures are redox-inactive.[30] Compared with p-type cathode materials, n-type counterparts generally possess higher specific capacity since more active sites per weight exist in them. However, both p-type and n-type cathode materials suffer from inferior electronic conductivity, limiting their rate performance. Different from p-type and n-type compounds, bipolar-type conductive polymers show high electronic conductivity, which would be beneficial for electron transfer and thus results in excellent rate performance. However, they often deliver limited specific capacity because of low available doping level.

## 12.4 Design of Organic Cathode Materials

Recently, various organic cathode materials, including carbonyl compounds, imine compounds, conductive polymers, nitronyl nitroxide, organosulfur polymers and triphenylamine derivatives, have been developed. Owing to the variety of their molecular structures, they display remarkably different behaviors in the aspects of capacity, voltage, rate, cycle number, capacity retention and preparation complexity, as shown in Figure 12.4. In this section, these organic cathode materials will be discussed in detail.

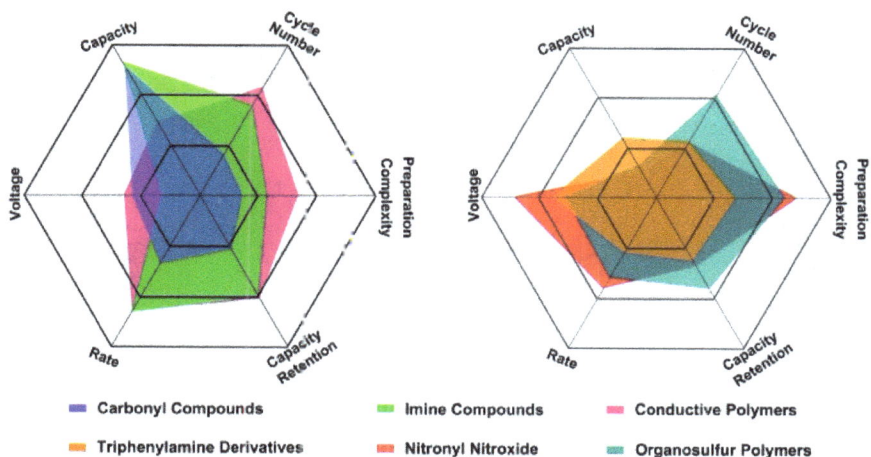

**Figure 12.4.** An overview of properties of carbonyl compounds, imine compounds, conductive polymers, triphenylamine derivatives, nitronyl nitroxide and organosulfur polymers.

## 12.4.1 *Carbonyl Compounds*

Carbonyl compounds are the typical n-type electrode materials and are widely used in ZIBs because they often show high specific capacity (Figure 12.5). Some carbonyl compounds are almost insoluble in aqueous electrolytes; however, their discharge products display high dissolution in water, which inevitably leads to inferior cycle stability. In addition, the poor electronic conductivity also limits their rate capability. Compared with monovalent cations (e.g., Li$^+$, Na$^+$ and K$^+$), divalent zinc ions prefer to combine with a couple of redox sites by coordination reaction, which will determine the redox chemistry and design strategies of n-type organic electrode materials. The reported carbonyl cathode compounds of aqueous ZIBs include quinones,[34] ketones[49] and imides.[50] In their molecules, carbonyl groups are considered as the highly redox-active sites, which are favorable for Zn$^{2+}$ uptake. Chen *et al.* developed a series of organic quinones as the cathode materials of aqueous ZIBs for the first time.[32] Owing to the effect of steric hindrance, the zinc ions have strong interactions with two carbonyls at the top and four carbonyls at the bottom in the C4Q molecule. As a result, C4Q

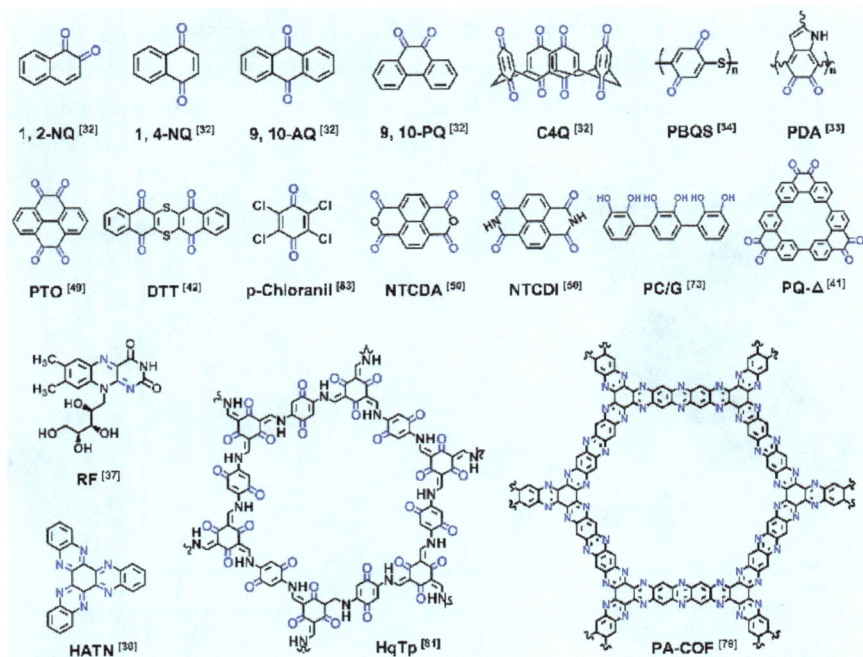

**Figure 12.5.** Molecular structures of n-type organic electrode materials used in aqueous ZIBs.

exhibited a high specific capacity of 335 mAh g$^{-1}$, corresponding to uptake of three Zn$^{2+}$ and utilization of six carbonyls. However, its discharge products faced serious dissolution in aqueous electrolytes, leading to poor cycle life. Therefore, a cation-selective membrane had to be employed to enhance its cycle performance. Different from quinones, pyrene-4,5,9,10-tetraone (PTO), a ketone compound, could achieve a capacity retention of 70% over 1,000 cycles at 3 A g$^{-1}$, and cation-selective membrane was not even utilized.[49] Compared with quinones and ketones, imides compounds have imino moieties around the carbonyl groups. The nitrogen atoms with lone pair of electrons could improve the redox activity of carbonyl groups, which will be beneficial for Zn$^{2+}$ uptake. Moreover, the π-conjugated aromatic structure could strengthen π–π intermolecular interactions, further suppressing their dissolution in aqueous electrolytes. As a result, 1,4,5,8-naphthalene diimide displayed a highly reversible specific capacity of 240 mAh g$^{-1}$ at 0.1 A g$^{-1}$ and retained 73.7% of its initial capacity after 2,000 cycles at 1 A g$^{-1}$.[50] Furthermore, the dissolution of carbonyl cathode compounds and their discharge products could be limited by the polymerization of small carbonyl molecules, improving their cycling stability.[33] However, the introduction of covalently interconnected carbon skeletons will increase molecular mass and thus lead to reduced capacity and poor rate performance. Therefore, both cycle life and rate capability have to be further considered in the design of carbonyl cathode compounds.

## 12.4.2 Imine Compounds

As the cathode materials of aqueous ZIBs, the imine compounds with C=N groups also exhibit high redox activity, which can be reduced and combined with the cations (e.g., Zn$^{2+}$ and H$^+$) of electrolytes during discharge processes. Generally, imine compounds possess high specific capacity, fast reaction kinetics and long cycle life. However, they often have a low average discharge voltage, which limits their practical application. We developed a π-conjugated aromatic compound, HATN, to serve as the cathode material of ZIBs.[36] The Zn/HATN batteries not only deliver an initial discharge capacity of 405 mAh g$^{-1}$ at a current density of 100 mA g$^{-1}$, but also display excellent rate performance and long cycle life with a capacity retention of 93.3% after 5,000 cycles at 5 A g$^{-1}$. Such excellent performance is ascribed to the fact that the Zn/HATN battery displays H$^+$ uptake/removal rather than conventional Zn$^{2+}$ insertion/extraction since the π-conjugated aromatic compounds with imine groups had the ability to chelate protons by coordination reactions. Therefore, if their voltages could be enhanced effectively by introducing electron-withdrawing groups, the imine compounds would be promising cathode materials of aqueous ZIBs.

### 12.4.3 Conductive Polymers

Conductive polymers usually show high electronic conductivity due to their long π-electron conjugated system, which will contribute to their fast redox kinetics.[51] Among various conductive polymers, PANI and polypyrrole have been used to serve as the electrode materials of aqueous ZIBs (Figure 12.6).[52–56] They often display bipolar-type redox chemistry in aqueous ZIBs. In the case of Zn/PANI batteries, $Zn^{2+}$ could serve as the cation species associated with the imine-related redox reaction, while the counter anions will participate in the reaction between the =NH+– and –NH– moieties during charge/discharge processes. Although they often deliver limited specific capacity (<200 mAh g$^{-1}$) due to their low available doping level, they usually display

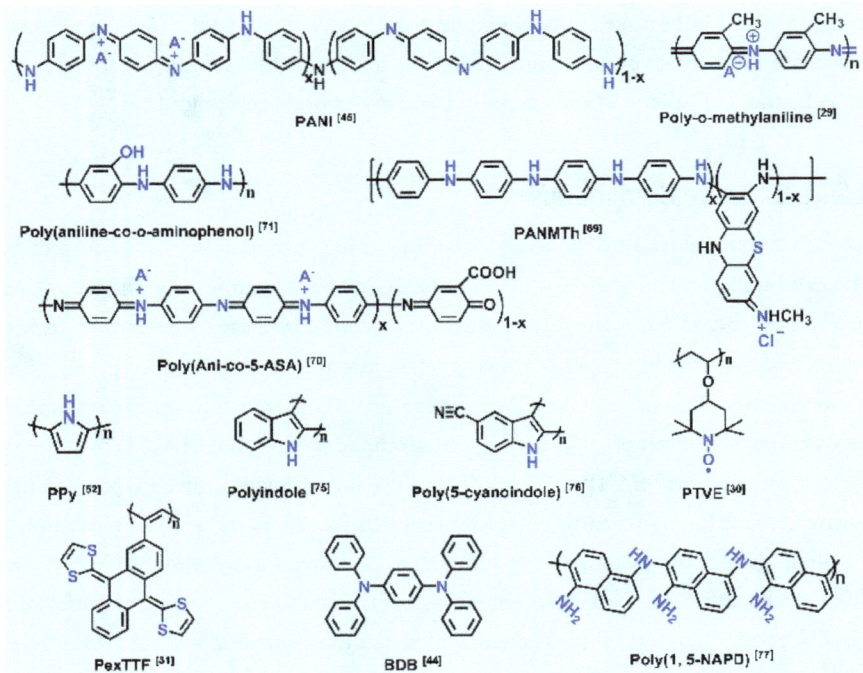

**Figure 12.6.** Molecular structures of bipolar and p-type organic electrode materials used in aqueous ZIBs.

more stable electrochemical performance in comparison with most other previously reported cathode materials. As a result, they were usually utilized to act as the electrodes of flexible ZIBs with various configurations.[57–62] The electrochemical performance of conductive polymers could be further improved by the optimization of electrolyte,[27, 63] hybridization with inorganic materials,[28, 64–67] morphology regulation[68] and copolymerization.[38, 69–71] For instance, the conductive polymers could be intercalated into the interlayers of metal oxides to effectively eliminate the structural changes of them during discharge/charge processes, resulting in excellent rate capability and long cycling life.[67]

### 12.4.4 *Nitronyl Nitroxides*

Nitronyl nitroxides are common $p$-type electrode materials of aqueous ZIBs. Owing to their low electron energy level and small electron rearrangement during redox processes, they often exhibit high discharge voltages and fast redox kinetics.[48] More importantly, they can display flat charge/discharge plateaus, which will be beneficial to provide a stable voltage output. For example, poly(2,2,6,6-tetramethylpiperidinyloxy-4-yl vinyl ether) (PTVE) has a discharge voltage plateau at 1.73 V, which is higher than other organic and most inorganic cathode materials.[30] Moreover, the discharge voltage plateau of nitronyl nitroxide also depends on the electrolytes in the ZIB systems.[43] The discharge voltages in $ZnSO_4$ electrolyte would be higher in comparison with the cases of $Zn(CF_3SO_3)_2$ and $Zn(ClO_4)_2$ electrolytes. However, at such high voltage, oxygen evolution usually occurs and some ions can be decomposed in the case of aqueous electrolytes. Therefore Zn/PTVE battery showed poor cycle life in a wide potential window. Aqueous electrolytes with high stable potential window have to be designed to match these high-voltage organic cathodes of aqueous ZIBs. In addition to high discharge voltage, nitronyl nitroxide could also exhibit excellent rate performance since there are almost no bond rearrangements during redox processes. For instance, the Zn/PTVE battery exhibited a specific capacity of 131 mAh $g^{-1}$ even at a high current density of 60 C (1 C = 136 mA $g^{-1}$).[30] However, their molecular structures are complicated and the amount of active groups in a molecule is limited. As a result, their specific capacities are lower and their synthesis process is more complicated compared with carbonyl compounds, imine compounds and conductive polymers (Figure 12.4).

### 12.4.5 *Organosulfur Polymers*

Organosulfur polymers could also lose electrons during the charge process, accompanying the coordination reaction with anions in the electrolytes. Thus, they are also p-type cathode materials of ZIBs with high voltage. For example, poly(acetylene)-based 9,10-di(1,3-dithiol-2-ylidene)-9,10-dihydroanthracene (PexTTF) was designed to act as the cathode material of aqueous ZIBs, which displayed an average discharge voltage of 1.1 V.[31] During oxidation, its 1,3-dithiol-2-ylidene ring converted into a stable thiophene ring, which will contribute to the long cycle stability of aqueous ZIBs. As a result, the Zn/PexTTF battery exhibited a specific capacity of 111 mAh $g^{-1}$ at 120 C (1 C = 133 mA $g^{-1}$) and maintained a capacity of 105 mAh $g^{-1}$ over 1,000 cycles.

### 12.4.6 *Triphenylamine Derivatives*

Triphenylamine derivatives are another typical p-type compounds. They will first undergo oxidation and then bind anions from the electrolytes to maintain charge neutrality in the charge process. Recently, 1,4-bis(diphenylamino)benzene (BDB) was used as the cathode material of aqueous ZIBs.[44] BDB possesses two tertiary nitrogens, which could be oxidized successively in two steps accompanied by the insertion of anions during the charging process. Furthermore, such oxidation reaction is highly reversible in discharging. The Zn/BDB batteries exhibited a specific capacity of 125 mAh $g^{-1}$ at 26 mA $g^{-1}$ with an average discharge voltage of 1.25 V. However, the discharge products of BDB suffer from serious dissolution in aqueous electrolytes, which seriously degraded the cycle stability of Zn/BDB batteries. Compared with other organic compounds, triphenylamine derivatives display lower electrochemical properties (e.g., capacity, rate, cycle number and capacity retention), as shown in Figure 12.4.

## 12.5 Design Strategies for High Performance

In general, the electrochemical performance of aqueous ZIBs mainly focuses on energy density, power density and cycle performance. Among them, energy density is decided by specific capacity and voltage; power density depends on the charge/discharge capability of organic electrode materials at different rates. In this section, we will highlight the strategies to improve specific capacity, output voltage, cycle life and rate performance of aqueous Zn/organic batteries.

## 12.5.1 Strategies for High Specific Capacity

The theoretical capacity ($Q_{theo}$) of organic electrode materials can be calculated as[72]:

$$Q_{theo} = \frac{nF}{3.6M_{theo}} \qquad (12.1)$$

where $n$ is the number of transferred electrons in the reaction process, $F$ is the Faraday constant, and $M_{theo}$ represents molecular mass. Clearly, organic molecules with high specific capacity can be designed by increasing the number of redox-active groups and reducing their relative molecular mass. Generally, carbonyl groups are considered as the redox-active sites for $Zn^{2+}$ uptake in quinone compounds. Therefore, the theoretical capacities of quinone compounds could be enhanced by increasing the number of carbonyl groups per molecular weight. For example, 9,10-plastoquinone (PQ) with a molecular mass of 208 g mol$^{-1}$ could deliver a two-electron reaction, leading to a theoretical capacity of 258 mAh g$^{-1}$. After introducing two carbonyl groups, the resultant PTO can transfer four electrons per molecule, corresponding to an enhanced theoretical capacity of 409 mAh g$^{-1}$.[32, 49] In addition, after removing some inactive carbon skeleton from 9,10-PQ, the 1,2-naphthoquinone (NQ) would be achieved, which can deliver a higher theoretical capacity of 339 mAh g$^{-1}$.

The above two strategies mainly focus on optimizing the theoretical capacities of organic materials. In fact, their practical specific capacities will also depend on many factors, such as morphology, electronic conductivity and electrolytes (Figure 12.7). For example, the specific capacity of "sponge-like" PANI was higher than the "pebble-like" case because of the better penetration of electrolyte in the porous structure than compact construction.[68] Furthermore, after incorporating PANI into conductive carbon architectures, the practical capacities of PANI could also be increased. The PANI-graphene oxide (GO)/carbon nanotube (CNT) composite could display a high specific capacity of 247 mAh g$^{-1}$ at a current density of 100 mA g$^{-1}$.[66] Such increased practical capacity depends not only on the high electronic conductivity supplied by the interconnected CNT network, but also the improvement of doping level by GO. In addition, the concentration of electrolytes plays an important role in their ionic conductivity. At different electrolyte concentrations, the specific capacities of Zn/organic batteries would be varied.[73]

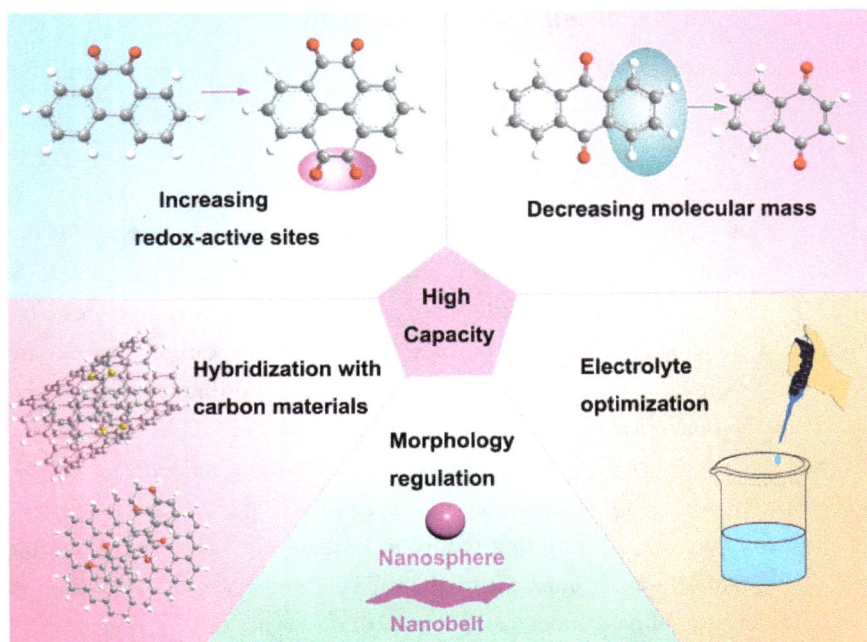

**Figure 12.7.** Design strategies for high specific capacity.

## 12.5.2 *Strategies for High Output Voltage*

Voltage is also an important aspect for the application of organic electrode materials in aqueous ZIBs. According to the electrochemical reaction process, the voltage ($E$) can be calculated through[72]:

$$E = \frac{\Delta G}{nF} \tag{12.2}$$

where $\Delta G$ is the change of Gibbs free energy, $n$ is the number of transferred electrons, and $F$ is the Faraday constant. Therefore, the voltage of an organic electrode material can be determined by the $\Delta G/n$ of the reaction process. In the case of the Zn/PTVE battery system, density functional theory calculations indicated that the $\Delta G/n$ of $[2(PTVE^+)SO_4^{2-}]$ was larger than that of $(PTVE^+CF_3SO_3^-)$ and $(PTVE^+ClO_4^-)$.[43] As a result, the voltages of Zn/PTVE batteries in $ZnSO_4$ electrolyte were superior in comparison with the cases in $Zn(CF_3SO_3)_2$ and $Zn(ClO_4)_2$ electrolytes.

[12] D. L. Chao, C. Y. Zhu, M. Song, P. Liang, X. Zhang, N. H. Tiep, H. F. Zhao, J. Wang, R. M. Wang, H. Zhang, H. J. Fan, *Adv. Mater.* **2018**, 30, 1803181.

[13] Y. Zhang, F. Wan, S. Huang, S. Wang, Z. Q. Niu, J. Chen, *Nat. Commun.* **2020**, 11, 2199.

[14] F. Wan, X. Y. Wang, S. S. Bi, Z. Q. Niu, J. Chen, *Sci. China Chem.* **2019**, 62, 609.

[15] P. He, G. B. Zhang, X. B. Liao, M. Y. Yan, X. Xu, Q. Y. An, J. Liu, L. Q. Mai, *Adv. Energy Mater.* **2018**, 8, 6.

[16] F. Wan, Z. Q. Niu, *Angew. Chem. Int. Ed.* **2019**, 58, 16358.

[17] H. F. Liang, Z. Cao, F. W. Ming, W. L. Zhang, D. H. Anjum, Y. Cui, L. Cavallo, H. N. Alshareef, *Nano Lett.* **2019**, 19, 3199.

[18] Q. Yang, F. N. Mo, Z. X. Liu, L. T. Ma, X. L. Li, D. L. Fang, S. M. Chen, S. J. Zhang, C. Y. Zhi, *Adv. Mater.* **2019**, 31, 9.

[19] P. Hu, T. Zhu, X. P. Wang, X. F. Zhou, X. J. Wei, X. H. Yao, W. Luo, C. W. Shi, K. A. Owusu, L. Zhou, L. Q. Mai, *Nano Energy.* **2019**, 58, 492.

[20] J. Cui, Z. W. Guo, J. Yi, X. Y. Liu, K. Wu, P. C. Liang, Q. Li, Y. Y. Liu, Y. G. Wang, Y. Y. Xia, J. J. Zhang, *ChemSusChem.* **2020**, 13, 2160.

[21] M. Zhang, Y. Zhang, W. W. Huang, Q. C. Zhang, *Batteries & Supercaps* **2020**, 3, 476.

[22] Y. L. Liang, Z. L. Tao, J. Chen, *Adv. Energy Mater.* **2012**, 2, 742.

[23] Y. Lu, Q. Zhang, L. Li, Z. Q. Niu, J. Chen, *Chem.* **2018**, 4, 2786.

[24] J. H. Huang, X. L. Dong, Z. W. Guo, Y. G. Wang, *Angew. Chem. Int. Ed.* **2020**, 59, 18322.

[25] Y. L. Liang, Y. Yao, *Joule* **2018**, 2, 1690.

[26] A. G. Macdiarmid, J.-C. Chiang, M. Halpern, W.-S. Huang, S.-L. Mu, L. D. Nanaxakkara, S. W. Wu, S. I. Yaniger, *Mol. Cryst. Liq. Cryst.* **1985**, 121, 173.

[27] M. Sima, T. Visan, M. Buda, *J. Power Sources* **1995**, 56, 133.

[28] K. Ghanbari, M. F. Mousavi, M. Sharnsipur, H. Karami, *J. Power Sources* **2007**, 170, 513.

[29] S. L. Mu, B. D. Qian, *Synth. Met.* **1989**, 32, 129.

[30] K. Koshika, N. Sano, K. Oyaizu, H. Nishide, Macromol. *Chem. Phys.* **2009**, 210, 1989.

[31] B. Haupler, C. Rossel, A. M. Schwenke, J. Winsberg, D. Schmidt, A. Wild, U. S. Schubert, *NPG Asia Mater.* **2016**, 8, e283.

[32] Q. Zhao, W. W. Huang, Z. Q. Luo, L. J. Liu, Y. Lu, Y. X. Li, L. Li, J. Y. Hu, H. Ma, J. Chen, *Sci. Adv.* **2018**, 4, eaao1761.

[33] X. J. Yue, H. D. Liu, P. Liu, *Chem. Commun.* 2019, 55, 1647.

[34] G. Dawut, Y. Lu, L. C. Miao, J. Chen, *Inorg. Chem. Front.* **2018**, 5, 1391.

[35] X. Y. Wu, J. J. Hong, W. Shin, L. Ma, T. C. Liu, X. X. Bi, Y. F. Yuan, Y. T. Qi, T. W. Surta, W. X. Huang, J. Neuefeind, T. P. Wu, P. A. Greaney, J. Lu, X. L. Ji, *Nat. Energy* **2019**, 4, 123.

[36] Z. W. Tie, L. J. Liu, S. Z. Deng, D. B. Zhao, Z. Q. Niu, *Angew. Chem. Int. Ed.* **2020**, 59, 4920.

[37] L. W. Cheng, Y. H. Liang, Q. N. Zhu, D. D. Yu, M. X. Chen, J. F. Liang, H. Wang, *Chem-Asian. J.* **2020**, 15, 1290.

[38] P. Li, Z. S. Fang, Y. Zhang, C. S. Mo, X. H. Hu, J. H. Jian, S. Y. Wang, D. S. Yu, *J. Mater. Chem. A* **2019**, 7, 17292.

[39] S. Muench, A. Wild, C. Friebe, B. Haupler, T. Janoschka, U. S. Schubert, *Chem. Rev.* **2016**, 116, 9438.

[40] P. Poizot, J. Gaubicher, S. Renault, L. Dubois, Y. L. Liang, Y. Yao, *Chem. Rev.* **2020**, 120, 6490.

[41] K. W. Nam, H. Kim, Y. Beldjoudi, T. W. Kwon, D. J. Kim, J. F. Stoddart, *J. Am. Chem. Soc.* **2020**, 142, 2541.

[42] Y. R. Wang, C. X. Wang, Z. G. Ni, Y. M. Gu, B. L. Wang, Z. W. Guo, Z. Wang, D. Bin, J. Ma, Y. G. Wang, *Adv. Mater.* **2020**, 32, 2000338.

[43] Y. W. Luo, F. P. Zheng, L. J. Liu, K. X. Lei, X. S. Hou, G. Xu, H. Meng, J. F. Shi, F. J. Li, *ChemSusChem.* **2020**, 13, 2239.

[44] H. Glatz, E. Lizundia, F. Pacifico, D. Kundu, *ACS Appl. Energy. Mater.* **2019**, 2, 1288.

[45] F. Wan, L. L. Zhang, X. Y. Wang, S. S. Bi, Z. Q. Niu, J. Chen, *Adv. Funct. Mater.* **2018**, 28, 1804975.

[46] Y. Zhang, Y. L. Liang, H. Dong, X. J. Wang, Y. Yao, *J. Electrochem. Soc.* **2020**, 167, 5.

[47] Y. Lu, J. Chen, *Nat. Rev. Chem.* **2020**, 4, 127.

[48] C. Friebe, U. S. Schubert, *Top. Curr. Chem.* **2017**, 375, 35.

[49] Z. W. Guo, Y. Y. Ma, X. L. Dong, J. H. Huang, Y. G. Wang, Y. Y. Xia, *Angew. Chem. Int. Ed.* **2018**, 57, 11737.

[50] X. S. Wang, L. Chen, F. Lu, J. Y. Liu, X. C. Chen, G. J. Shao, *ChemElectroChem.* **2019**, 6, 3644.

[51] S. Wang, S. Huang, M. J. Yao, Y. Zhang, Z. Q. Niu, *Angew. Chem. Int. Ed.* **2020**, 59, 11800.

[52] J. Q. Wang, J. Liu, M. M. Hu, J. Zeng, Y. B. Mu, Y. Guo, J. Yu, X. Ma, Y. J. Qiu, Y. Huang, *J. Mater. Chem. A* **2018**, 6, 11113.

[53] S. Huang, F. Wan, S. S. Bi, J. C. Zhu, Z. Q. Niu, J. Chen, *Angew. Chem. Int. Ed.* **2019**, 58, 4313.

[54] H. Y. Shi, Y. J. Ye, K. Liu, Y. Song, X. Q. Sun, *Angew. Chem. Int. Ed.* **2018**, 57, 16359.

[55] M. S. Rahmanifar, M. F. Mousavi, M. Shamsipur, *J. Power Sources* **2002**, 110, 229.

[56] S. Q. Li, G. L. Zhang, G. L. Jng, J. Q. Kan, *Synth. Met.* **2008**, 158, 242.

[57] H. H. Yi, Y. Ma, S. Zhang, B. Na, R. Zeng, Y. Zhang, C. Lin, *ACS Sustain. Chem. Eng.* **2019**, 7, 18894.

[58] J. C. Zhu, M. J. Yao, S. Huang, J. L. Tian, Z. Q. Niu, *Angew. Chem. Int. Ed.* **2020**, 59, 16480.

[59] H. Yu, G. Liu, M. Wang, R. Ren, G. Shim, J. Y. Kim, M. X. Tran, D. Byun, J. K. Lee, *ACS Appl. Mater. Interfaces.* **2020**, 12, 5820.

[60] H. M. Cao, F. Wan, L. L. Zhang, X. Dai, S. Huang, L. L. Liu, Z. Q. Niu, *J. Mater. Chem. A* **2019**, 7, 11734.

[61] Y. Ma, X. L. Xie, R. H. Lv, B. Na, J. B. Ouyang, H. S. Liu, *ACS Sustain. Chem. Eng.* **2018**, 6, 8697.

[62] Y. Liu, L. Y. Xie, W. Zhang, Z. W. Dai, W. Wei, S. J. Luo, X. Chen, W. Chen, F. Rao, L. Wang, Y. Huang, *ACS Appl. Mater. Interfaces.* **2019**, 11, 30943.

[63] B. Z. Jugovic, T. L. Trisovic, J. Stevanovic, M. Maksimovic, B. N. Grgur, *J. Power Sources* **2006**, 160, 1447.

[64] C. Kim, B. Y. Ahn, T. S. Wei, Y. Jo, S. Jeong, Y. Choi, I. D. Kim, J. A. Lewis, *ACS Nano* **2018**, 12, 11838.

[65] J. H. Huang, Z. Wang, M. Y. Hou, X. L. Dong, Y. Liu, Y. G. Wang, Y. Y. Xia, *Nat. Commun.* **2018**, 9, 2906.

[66] W. C. Du, J. F. Xiao, H. B. Geng, Y. Yang, Y. F. Zhang, E. H. Ang, M. H. Ye, C. C. Li, *J. Power Sources* **2020**, 450, 227716.

[67] D. Bin, W. C. Huo, Y. B. Yuan, J. H. Huang, Y. Liu, Y. X. Zhang, F. Dong, Y. G. Wang, Y. Y. Xia, *Chem* **2020**, 6, 968.

[68] Z. Mandic, M. K. Rokovic, T. Pokupcic, *Electrochim. Acta* **2009**, 54, 2941.

[69] C. X. Chen, X. Z. Hong, A. K. Chen, T. T. Xu, L. Lu, S. L. Lin, Y. H. Gao, *Electrochim. Acta* **2016**, 190, 240.

[70] S. L. Mu, Q. F. Shi, *Synth. Met.* **2016**, 221, 8.

[71] S. L. Mu, *Synth. Met.* **2004**, 143, 269.

[72] Y. Lu, Y. Y. Lu, Z. Q. Niu, J. Chen, *Adv. Energy Mater.* **2018**, 8, 1702469.

[73] S. Q. Zhang, W. T. Zhao, H. Li, Q. Xu, *ChemSusChem.* **2020**, 13, 188.

[74] Y. L. Liang, P. Zhang, S. Q. Yang, Z. L. Tao, J. Chen, *Adv. Energy Mater.* **2013**, 3, 600.

[75] Z. J. Cai, C. W. Hou, *J. Power Sources* **2011**, 196, 10731.

[76] Z. J. Cai, J. Guo, H. Z. Yang, Y. Xu, *J. Power Sources* **2015**, 279, 114.

[77] Y. Zhao, Y. N. Wang, Z. M. Zhao, J. W. Zhao, T. Xin, N. Wang, J. Z. Liu, *Energy Storage Mater.* **2020**, 28, 64.

[78] W. X. Wang, V. S. Kale, Z. Cao, S. Kandambeth, W. L. Zhang, J. Ming, P. T. Parvatkar, E. Abou-Hamad, O. Shekhah, L. Cavallo, M. Eddaoudi, H. N. Alshareef, *ACS Energy Lett.* **2020**, 5, 2256.

[79] Z. Q. Luo, L. J. Liu, J. X. Ning, K. X. Lei, Y. Lu, F. J. Li, J. Chen, *Angew. Chem. Int. Ed.* **2018**, 57, 9443.

[80]  S. Gu, S. F. Wu, L. J. Cao, M. C. Li, N. Qin, J. Zhu, Z. Q. Wang, Y. Z. Li, Z. Q. Li, J. J. Chen, Z. G. Lu, *J. Am. Chem. Soc.* **2019**, 141, 9623.

[81]  M. A. Khayum, M. Ghosh, V. Vijayakumar, A. Halder, M. Nurhuda, S. Kumar, M. Addicoat, S. Kurungot, R. Banerjee, *Chem. Sci.* **2019**, 10, 8889.

[82]  R. K. Nagarale, G. S. Gohil, V. K. Shahi, *Adv. Colloid Interface Sci.* **2006**, 119, 97.

[83]  D. Kundu, P. Oberholzer, C. Glaros, A. Bouzid, E. Tervoort, A. Pasquarello, M. Niederberger, *Chem. Mater.* **2018**, 30, 3874.

[84]  C. L. Wang, Y. Xu, Y. G. Fang, M. Zhou, L. Y. Liang, S. Singh, H. P. Zhao, A. Schober, Y. Lei, *J. Am. Chem. Soc.* **2015**, 137, 3124.

[85]  C. M. Das, L. X. Kang, Q. L. Ouyang, K.-T. Yong, *InfoMat* **2020**, 2, 698.

[86]  Z. D. Lei, Q. S. Yang, Y. Xu, S. Y. Guo, W. W. Sun, H. Liu, L. P. Lv, Y. Zhang, Y. Wang, *Nat. Commun.* **2018**, 9, 576.

[87]  X. L. Fan, F. Wang, X. X. Ji, R. X. Wang, T. Gao, S. Y. Hou, J. Chen, T. Deng, X. G. Li, L. Chen, C. Luo, L. N. Wang, C. S. Wang, *Angew. Chem. Int. Ed.* **2018**, 57, 7146.

[88]  L. L. Liu, Z. Q. Niu, J. Chen, *Chem. Soc. Rev.* **2016**, 45, 4340.

# Index

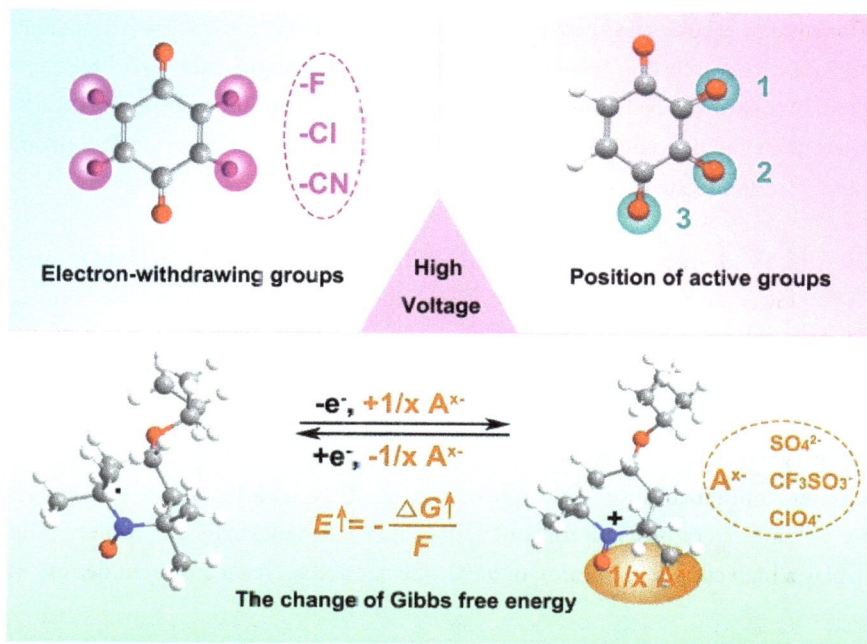

**Figure 12.8.** Design strategies for output voltage.

The voltages of organic electrode materials of ZIBs can also be tuned by engineering molecular structures (Figure 12.8). As is known, low energy level of LUMO indicates high electron affinity, contributing to high voltage.[74] The introduction of electron-withdrawing groups (such as –CN, –F, –Cl and –Br) can reduce the LUMO energy level of organic electrode materials of ZIBs, which is beneficial for improving their voltages.[23] For example, poly(5-cyanoindole) can reach a higher discharge plateau around 2.0 V in ZIBs by introducing electron-withdrawing –CN group to polyindole.[75,76] Additionally, the position of active groups could also affect the voltage of Zn/organic batteries. The voltages of quinone compounds with two carbonyl groups in ortho-position (1,2-NQ and 9,10-PQ) were higher than those in para-position (1,4-NQ and 9,10-AQ).[32]

## 12.5.3 Strategies for Long Cycle Life

Although some organic compounds generally exhibit low solubility in water, their discharge products often dissolve in aqueous electrolytes of ZIBs easily. The solubility of organic molecules and their discharged products in aqueous electrolytes mainly determine their cycle life. Therefore, the main strategy to improve the cycle life of Zn/organic batteries is to avoid this dissolution issue. Besides, the volume changes of organic electrode materials could also have influence on the cycling stability of Zn/organic batteries to some extent.

Compared with the small organic compounds, their corresponding polymers often display lower solubility in water because of their covalently connected structures. Therefore, the electroactive small molecules can be linked or grafted on a polymer skeleton during the polymerization process to achieve the covalently interconnected structure, where the redox-active sites are still available (Figure 12.9). This is to say that polymer molecules can retain highly redox activity and lower their solubility in water simultaneously, which endows aqueous ZIBs with long cycle stability. As the polymer electrode materials of ZIBs, poly(1,5-naphthalenediamine) could display a high capacity retention of 91% after 10,000 cycles at a current density of

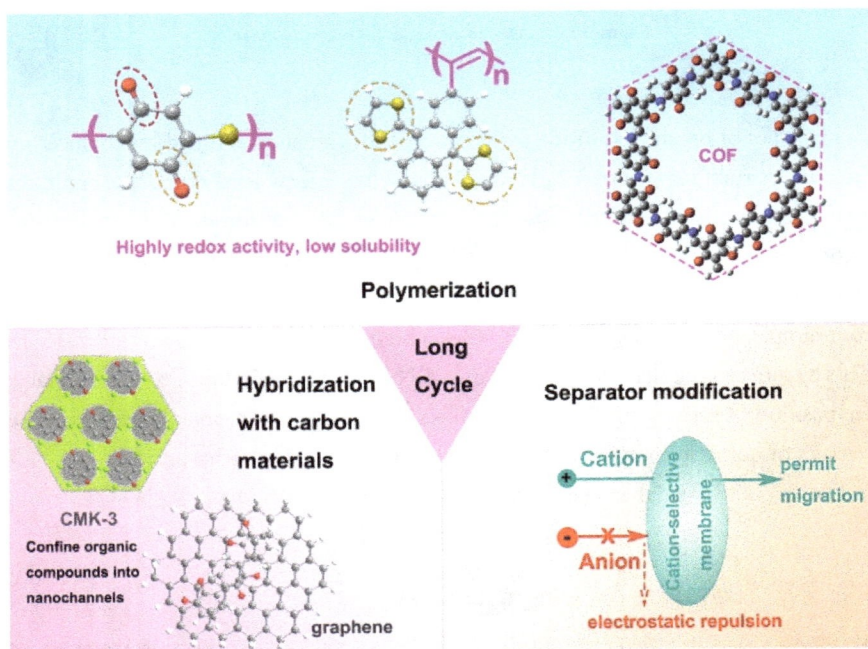

**Figure 12.9.** Design strategies for long cycle life.

10 A g$^{-1}$.[77] In addition to chain-like polymers, small organic compounds can also be assembled into covalent organic frameworks (COFs) with redox-active groups to act as the electrode materials of ZIBs.[78] COFs often possess high porosity and excellent chemical stability, which will be beneficial to fast ion diffusion and long cycle stability in ZIBs, respectively.[79,80] Recently, a COF, 2,5-diaminohydroquinone dihydrochloride was designed to serve as the cathodes of ZIBs, which exhibited a high capacity retention of 95% over 1,000 cycles at a current density of 3750 mA g$^{-1}$.[81]

After the discharge products are dissolved in electrolytes, they will migrate to Zn anode in a working ZIBs and induce the irreversible formation of by-products on the anode. Separator modification can be used to prevent the ions of discharge products from moving through the separator to avoid the formation of by-products on Zn anodes. The cation-selective membrane has been utilized in aqueous Zn/organic battery systems.[32] Owing to the negatively charged groups (e.g. $SO_3^{2-}$) on the membrane, it can prevent anions from going through the separator by electrostatic repulsion, while permitting cation migration.[82] In the case of Zn/C4Q battery, after using cation-selective membranes, the battery displayed remarkably enhanced cycling stability with a high capacity retention of 93% after 100 cycles, which is much higher than the case of conventional filter paper-based ZIBs.[32] Although cation-selective membrane can effectively enhance the cycle stability of Zn/organic batteries, its cost is generally much higher than common separators, such as filter and glass fiber.

In some cases, the volume changes of organic electrode materials caused by phase transformation during charge/discharge process could also lead to poor cycling stability of Zn/organic battery. An organic compound, tetrachloro-1,4-benzoquinone (p-chloranil), suffered phase evolution accompanied by large volume change during the charge/discharge process, leading to serious capacity decay. Interestingly, after confining them into the nanochannels of mesoporous carbon CMK-3, the phase evolution process could be tamed, and thus its cycle stability could also be improved.[83]

In addition, the cycle life of Zn/organic batteries could also be restricted by the Zn anode to some extent. In the Zn/organic batteries, Zn anodes usually suffer from dendrite, hydrogen evolution and passivation, which will degrade their cycle life. In general, the Zn dendrites display a needle-like morphology, whose tips could function as charge centers and induce a tip effect in subsequent reaction process. More seriously, the detrimental Zn dendrites could puncture the separator, ultimately resulting in a short circuit of batteries. Apart from Zn dendrites, the corrosion of Zn anodes will also occur in aqueous electrolytes, which leads to hydrogen evolution. The hydrogen evolution inevitably increases the inner pressure of Zn/organic battery devices. Accompanying this side reaction, a large amount of $OH^-$ will yield simultaneously, which triggers irreversible formation of by-products (e.g., $ZnO$, $Zn(OH)_2$ and $Zn_4(OH)_6SO_4 \cdot nH_2O$) on the surface of Zn anodes, resulting in the passivation of the Zn anodes after a long cycle. Therefore, the modification of Zn anodes could also be considered. For example, functional interfacial layers or 3D-structured Zn anodes could be constructed to induce uniform zinc nucleation on the Zn anodes during the charging process, restraining dendrite formation. In addition, the design of solid-state electrolytes and the introduction of functional additives into electrolytes would be promising strategies to restrict the hydrogen evolution and the passivation of Zn anodes, enhancing the cycle stability of Zn/organic batteries. Apart from these strategies, high-performance organic anode materials could also be taken into consideration in the future design of Zn/organic batteries.

## 12.5.4 *Strategies for High Rate Capability*

The rate performance of Zn/organic batteries depends on the fast transfer of electrons and ions in the ZIB systems. However, the electronic conductivity of organic electroactive materials is often inferior. As a result, their rate performance is usually unsatisfactory. There are two main approaches to improve the conductivity of organic cathodes: introduction of conductive moieties into the carbon skeleton and combination of organic molecules with carbon architectures (Figure 12.10). Generally, the conductive moieties of organic compounds are π-conjugated units. The addition of π-conjugated units can extend the π-electron conjugated systems in the molecules of redox-active compounds, leading to the enhancement of π-electron delocalization.[84] It will promote the migration of the electron and further improve the conductivity of the organic compounds. Nanocarbon-based materials often possess excellent conductivity and large surface area. They are promising hosts of organic cathode materials.[85–87]

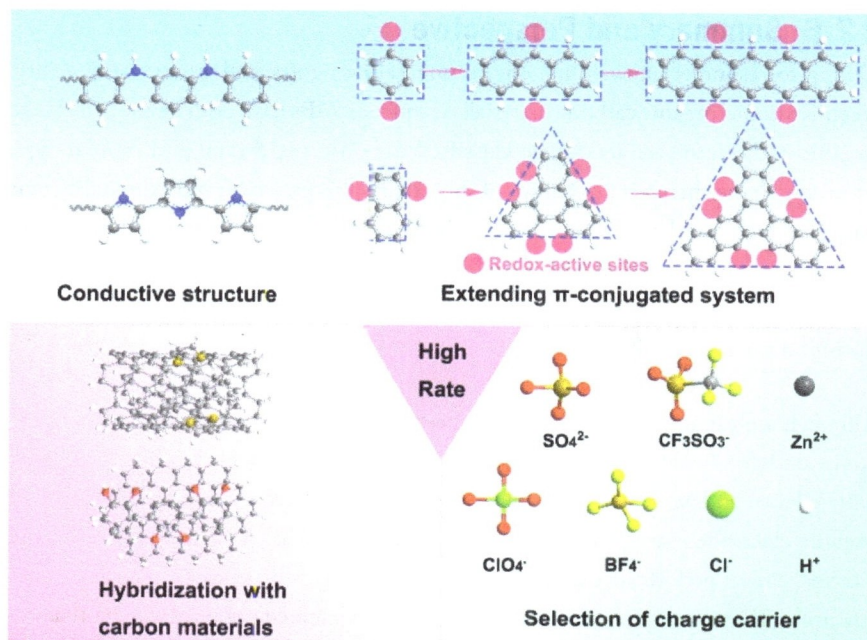

**Figure 12.10.** Design strategies for high rate capability.

Furthermore, owing to their unique structures, nanocarbon materials can form conductive porous networks in nanocarbon-based cathodes.[88] The electrons can be transported in the networks easily even as the organic compounds are incorporated into the composite electrodes. After adding multi-walled CNTs into organosulfur polymer PexTTF electrodes, PexTTF could display a specific capacity of 111 mAh g$^{-1}$ even at a high current density of 120 C (1 C = 133 mA g$^{-1}$).[31] In some cases, H$^+$-based aqueous Zn/organic batteries also exhibited excellent rate performance. For instance, Zn/HATN batteries with H$^+$ coordination/incoordination reaction could deliver a high capacity of 123 mAh g$^{-1}$ even at a high current density of 20 A g$^{-1}$, maintaining 33.2% of that at 0.1 A g$^{-1}$.[36] However, the insertion of H$^+$ into organic cathodes would increase the local pH of the electrolyte near the electrodes. It could deteriorate some organic electrodes materials, degrading the electrochemical performance of ZIBs. In addition, in some cases, the energy densities of H$^+$-based aqueous Zn/organic batteries are not superior to the Zn$^{2+}$-based aqueous Zn/organic batteries. Therefore, the design of organic electrode materials with high energy densities and rate performance has to be considered in the development of H$^+$-based aqueous Zn/organic batteries.

## 12.6 Summary and Perspective

Owing to resource renewability, environmental benignity and structural diversity, recently various organic cathode materials of aqueous ZIBs have been developed. These organic cathode materials can be classified into n-type, p-type and bipolar-type according to the charge-state changes of their active groups, which will exhibit different energy storage mechanisms and electrochemical performance. Furthermore, some strategies were developed to further improve their electrochemical performance, such as regulation of molecular structures, addition of conductive carbon materials and optimization of electrolytes.

Although a great amount of progress has been achieved in the design of active organic compounds for the electrodes of aqueous ZIBs, much work remains to be done. Compared with V-based and Mn-based inorganic compounds, the electrochemical performance of organic electrode materials is still lower in aqueous ZIBs. As discussed above, the electrochemical performance of organic electrode materials depends on their molecular structures. Therefore, the rational molecular design of organic compounds with more active groups and electron-withdrawing groups has to be considered to increase their specific capacity and voltages, respectively. Furthermore, the rational molecular design can be first predicted by theoretical calculations. In addition, like inorganic electrode materials, particle size and morphology also play an important role in the practical capacity, rate performance, cycle life, and even phase behaviors of the organic compounds. However, in most cases, the microstructures of the organic electrode materials were not considered. There is no doubt that the polymerization of small molecule compounds is essential for suppressing the dissolution of discharging products. However, these covalently interconnected carbon skeletons will inevitably increase molecular weight and decrease electron conductivity, further leading to reduced capacity and inferior rate performance. It is desired to optimize the polymer structure and further improve their polymerization degree, enhancing their capacity and cycle life. In addition, organic electrode materials can be introduced into carbon materials (e.g., carbon black, CNTs and graphene) to enhance the electronic conductivity of electrodes and limit the dissolution of discharged products in aqueous electrolytes. However, the microstructures in these carbon materials are usually disordered and random. Compared with the disordered microstructures, ordered and controllable microstructures with the optimized meso- and micro-pores will display higher electronic conductivity and larger specific area of carbon materials. Therefore, carbon materials with controllable microstructures could be considered to design the organic electrodes of ZIBs.

Recently, the common electrolytes of aqueous Zn/organic batteries are $ZnSO_4$ and $Zn(CF_3SO_3)_2$ solutions. However, $ZnSO_4$ electrolyte often leads to the formation of $Zn_4(OH)_6SO_4 \cdot nH_2O$ on the surface of the Zn anode, which will fade the electrochemical performance of batteries. Although this process can be restricted in $Zn(CF_3SO_3)_2$ electrolyte, its cost is very high. Furthermore, $ZnSO_4$ and $Zn(CF_3SO_3)_2$ solutions cannot serve as the electrolytes of high-voltage ZIBs. To match the high-voltage cathodes, "water-in-salt" (e.g., 19 M $Li-N(SO_3CF_3)_2$ + 1 M $Zn(CF_3SO_3)_2$) electrolytes were developed. However, the low ionic conductivity of "water-in-salt" electrolytes will retard the reaction kinetics of Zn/organic batteries. Therefore, high-performance solid-state or quasi-solid-state polymer electrolytes could be considered.

## Acknowledgements

This work was supported by the National Key R&D Program of China (2019YFA0705600), and National Natural Science Foundation of China (51822205).

## References

[1] D. L. Chao, W. H. Zhou, F. X. Xie, C. Ye, H. Li, M. Jaroniec, S. Z. Qiao, *Sci. Adv.* **2020**, 6, 19.

[2] M. Song, H. Tan, D. L. Chao, H. J. Fan, *Adv. Funct. Mater.* **2018**, 28, 27.

[3] B. Y. Tang, L. T. Shan, S. Q. Liang, J. Zhou, *Energy Environ. Sci.* **2019**, 12, 3288.

[4] F. Wan, J. C. Zhu, S. Huang, Z. Q. Niu, *Batteries & Supercaps* **2020**, 3, 323.

[5] Z. M. Zhao, J. W. Zhao, Z. L. Hu, J. D. Li, J. J. Li, Y. J. Zhang, C. Wang, G. L. Cui, *Energy Environ. Sci.* **2019**, 12, 1938.

[6] X. S. Xie, S. Q. Liang, J. W. Gao, S. Guo, J. B. Guo, C. Wang, G. Y. Xu, X. W. Wu, G. Chen, J. Zhou, *Energy Environ. Sci.* **2020**, 13, 503.

[7] D. L. Chao, W. H. Zhou, C. Ye, Q. H. Zhang, Y. G. Chen, L. Gu, K. Davey, S. Z. Qiao, *Angew. Chem. Int. Ed.* **2019**, 58, 7823.

[8] W. Sun, F. Wang, S. Y. Hou, C. Y. Yang, X. L. Fan, Z. H. Ma, T. Gao, F. D. Han, R. Z. Hu, M. Zhu, C. S. Wang, *J. Am. Chem. Soc.* **2017**, 139, 9775.

[9] F. N. Mo, G. J. Liang, Q. Q. Meng, Z. X. Liu, H. F. Li, J. Fan, C. Y. Zhi, *Energy Environ. Sci.* **2019**, 12, 706.

[10] Y. L. Zhao, Y. H. Zhu, X. B. Zhang, *InfoMat* **2020**, 2, 237.

[11] F. Wan, L. L. Zhang, X. Dai, X. Y. Wang, Z. Q. Niu, J. Chen, *Nat. Commun.* **2018**, 9, 1656.